MODERN ASPECTS OF ELECTROCHEMISTRY, No. 49:

MODELING AND DIAGNOSTICS OF POLYMER ELECTROLYTE FUEL CELLS

For other titles published in this series, go to
http://www.springer.com/series/6251

Modern Aspects of Electrochemistry

Series Editors:
Ralph E. White
Department of Chemical Engineering
University of South Carolina
Columbia, SC 29208

Constantinos G. Vayenas
Department of Chemical Engineering
University of Patras
Patras 265 00
Greece

Managing Editor:
Maria E. Gamboa-Aldeco
1107 Raymer Lane
Superior, CO 80027

MODERN ASPECTS OF ELECTROCHEMISTRY, No. 49:

MODELING AND DIAGNOSTICS OF POLYMER ELECTROLYTE FUEL CELLS

Edited by

Ugur Pasaogullari
University of Connecticut, USA

Chao-Yang Wang
The Pennsylvania State University, USA

Springer

Editors
Ugur Pasaogullari
University of Connecticut
Center for Clean Energy Engineering
44 Weaver Road U-5233
Storrs CT 06269-5233
USA
ugurpasa@engr.uconn.edu

Chao-Yang Wang
Department of Mechanical and Nuclear Engineering
Penn State University
16802 University Park
Pennsylvania
301C Reber Bldg.
USA
cxw31@psu.edu

Series Editors
Ralph E. White
Department of Chemical Engineering
University of South Carolina
Columbia, SC 29208

Constantinos G. Vayenas
Department of Chemical Engineering
University of Patras
Patras 265 00
Greece

Managing Editor
Maria E. Gamboa-Aldeco
1107 Raymer Lane
Superior, CO 80027

ISSN 0076-9924
ISBN 978-0-387-98067-6 e-ISBN 978-0-387-98068-3
DOI 10.1007/978-0-387-98068-3
Springer New York Dordrecht Heidelberg London

Library of Congress Control Number: 2010930615

Printed on acid-free paper

Springer is part of Springer Science+Business Media (www.springer.com)

Preface

Polymer electrolyte fuel cells (PEFCs) or proton exchange membrane fuel cells (PEMFCs) have been suggested as alternatives to replace many existing energy conversion technologies, including internal combustion engines and batteries. The most significant advances in PEFC technology achieved in the last decade have occurred in areas related to automotive applications, namely cold-start capabilities, enhanced durability and better understanding of water management and mass transport losses.

This volume of Modern Aspects of Electrochemistry is intended to provide an overview of advancements in experimental diagnostics and modeling of polymer electrolyte fuel cells. Chapters by Huang and Reifsnider and Gu et al. provide an in-depth review of the durability issues in PEFCs as well as recent developments in understanding and mitigation of degradation in the polymer membrane and electrocatalyst.

Enabling cold start, the startup of PEFC stacks from subzero temperatures, is a very important capability achieved only within the last few years. Tajiri and Wang provide a tutorial overview of the requirements for cold start, and provide a summary of experimental diagnostics and cold-start modeling studies.

Chapters 4–6 address specific diagnostic methods in PEFCs. Martin et al. provide a detailed review of methods for distributed diagnostics of species, temperature, and current in PEFCs in Chapter 4. In Chapter 5, Hussey and Jacobson describe the operational principles of neutron radiography for in-situ visualization of liquid water distribution, and also outline issues related to temporal and spatial resolution. Tsushima and Hirai describe both magnetic resonance imaging (MRI) technique for visualization of water in PEFCs and tunable diode laser absorption spectroscopy (TDLAS) for measurement of water vapor concentration in Chapter 6.

Diffusion media (DM) are prone to flooding with liquid water. Although the DM is an essential component of PEFCs that enable distribution of species and collection of current and heat, little was known about capillary transport in DMs until recently. In Chapters 7 Gostick et al. provide a description of liquid water transport in porous DM due to capillarity and describe experimental techniques used to characterize DM properties.

The final two chapters discuss modeling of PEFCs. Mukherjee and Wang provide an in-depth review of meso-scale modeling of two-phase transport, while Zhou et al. summarize both the simulation of electrochemical reactions on electrocatalysts and the transport of protons through the polymer electrolyte using atomistic simulation tools such as molecular dynamics and Monte Carlo techniques.

Each chapter in the volume is self-contained; therefore they do not need to be read in a certain order.

Special thanks are due to 23 authors who contributed to this volume.

U. Pasaogullari
University of Connecticut, Storrs, CT, USA

C.-Y. Wang
The Pennsylvania State University, University Park, PA, USA

Contents

Chapter 1

DURABILITY OF PEM FUEL CELL MEMBRANES

Xinyu Huang and Ken Reifsnider

Chapter 2

MODELING OF MEMBRANE-ELECTRODE-ASSEMBLY DEGRADATION IN PROTON-EXCHANGE-MEMBRANE FUEL CELLS – LOCAL H_2 STARVATION AND START–STOP INDUCED CARBON-SUPPORT CORROSION

Wenbin Gu, Paul T. Yu, Robert N. Carter, Rohit Makharia, and Hubert A. Gasteiger

Chapter 3

COLD START OF POLYMER ELECTROLYTE FUEL CELLS

Kazuya Tajiri and Chao-Yang Wang

Chapter 4

SPECIES, TEMPERATURE, AND CURRENT DISTRIBUTION MAPPING IN POLYMER ELECTROLYTE MEMBRANE FUEL CELLS

Jonathan J. Martin, Jinfeng Wu, Xiao Zi Yuan, and Haijiang Wang

Chapter 5

HIGH-RESOLUTION NEUTRON RADIOGRAPHY ANALYSIS OF PROTON EXCHANGE MEMBRANE FUEL CELLS

D.S. Hussey and D.L. Jacobson

Chapter 6

MAGNETIC RESONANCE IMAGING AND TUNABLE DIODE LASER ABSORPTION SPECTROSCOPY FOR IN-SITU WATER DIAGNOSTICS IN POLYMER ELECTROLYTE MEMBRANE FUEL CELLS

Shohji Tsushima and Shuichiro Hirai

Chapter 7

CHARACTERIZATION OF THE CAPILLARY PROPERTIES OF GAS DIFFUSION MEDIA

Jeffrey T. Gostick, Marios A. Ioannidis, Michael W. Fowler, and Mark D. Pritzker

Chapter 8

MESOSCOPIC MODELING OF TWO-PHASE TRANSPORT IN POLYMER ELECTROLYTE FUEL CELLS

Partha P. Mukherjee and Chao-Yang Wang

Chapter 9

ATOMISTIC MODELING IN STUDY OF POLYMER ELECTROLYTE FUEL CELLS – A REVIEW

Xiangyang Zhou, Juanjuan Zhou, and Yijin Yin

List of Contributors, MAE 49

Robert N. Carter
Electrochemical Energy Research Laboratory; General Motors Research and Development, 10 Carriage Street, Honeoye Falls, NY 14472, USA

Michael W. Fowler
Department of Chemical Engineering, University of Waterloo, Waterloo, ON, Canada; E-mail: mfowler@uwaterloo.ca

Hubert A. Gasteiger
Department of Chemistry, Technische Universität München, Lichtenbergstrasse 4, D-85747, Garching, Germany

Jeffrey T. Gostick
Department of Chemical Engineering, McGill University, Montreal, QC, Canada; E-mail: jeff@gostick.ca

Wenbin Gu
Electrochemical Energy Research Laboratory, General Motors Research and Development, 10 Carriage Street, Honeoye Falls, NY 14472, USA; E-mail: wenbin.gu@gm.com

Shuichiro Hirai
Research Center for Carbon Recycling and Energy, Tokyo Institute of Technology, 2-12-1, O-okayama, Meguro-ku, Tokyo 152-8552, Japan

Xinyu Huang
Department of Mechanical Engineering, University of South Carolina, Columbia, SC, USA; E-mail: huangxin@cec.sc.edu

D.S. Hussey
National Institute of Standards and Technology, 100 Bureau Dr., MS 8461, Gaithersburg, MD 20899-8461, USA; E-mail: daniel.hussey@nist.gov

Marios A. Ioannidis
Department of Chemical Engineering, University of Waterloo, Waterloo, ON, Canada

D.L. Jacobson
National Institute of Standards and Technology, 100 Bureau Dr., MS 8461, Gaithersburg, MD 20899-8461, USA

Rohit Makharia
Electrochemical Energy Research Laboratory, General Motors Research and Development, 10 Carriage Street, Honeoye Falls, NY 14472, USA

Jonathan J. Martin
Institute for Fuel Cell Innovation, National Research Council Canada, Vancouver, BC, Canada V6T 1W5

Partha P. Mukherjee
Oak Ridge National Laboratory, One Bethel Valley Road, P.O. Box 2008, MS 6164, Oak Ridge, TN 37831-6164, USA; E-mail: mukherjeepp@ornl.gov

Mark D. Pritzker
Department of Chemical Engineering, University of Waterloo, Waterloo, ON, Canada; E-mail: pritzker@uwaterloo.ca

Ken Reifsnider
Department of Mechanical Engineering, University of South Carolina, Columbia, SC, USA; E-mail: reifsnid@cec.sc.edu

Kazuya Tajiri
Argonne National Laboratory, 9700 S. Cass Avenue, Argonne, IL 60439, USA

Shohji Tsushima
Research Center for Carbon Recycling and Energy, Tokyo Institute of Technology, 2-12-1, O-okayama, Meguro-ku, Tokyo 152-8552, Japan

Chao-Yang Wang
Electrochemical Engine Center (ECEC), and Department of Mechanical and Nuclear Engineering, The Pennsylvania State University, University Park, PA 16802, USA

Haijiang Wang
Institute for Fuel Cell Innovation, National Research Council Canada, Vancouver, BC, Canada V6T 1W5

Jinfeng Wu
Institute for Fuel Cell Innovation, National Research Council Canada, Vancouver, BC, Canada V6T 1W5

Yijin Yin
Department of Mechanical and Aerospace Engineering, University of Miami, P.O. Box 248294, Coral Gables, FL 33157, USA

Paul T. Yu
Electrochemical Energy Research Laboratory, General Motors Research and Development, 10 Carriage Street, Honeoye Falls, NY 14472, USA

Xiao Zi Yuan
Institute for Fuel Cell Innovation, National Research Council Canada, Vancouver, BC, Canada V6T 1W5

Xiangyang Zhou
Department of Mechanical and Aerospace Engineering, University of Miami, P.O. Box 248294, Coral Gables, FL 33157, USA

Juanjuan Zhou
Department of Mechanical and Aerospace Engineering, University of Miami, P.O. Box 248294, Coral Gables, FL 33157, USA

Modern Aspects of Electrochemistry

Topics in Number 47 include:

- Corrosiveness of supercritical water oxidation (SCWO) media and technical issues that need to be resolved for practical use of SCWO technology
- Optical techniques – including ellipsometry, Raman spectroscopy, potential modulation reflectance, and photo-electrochemical technique – and an understanding of what really happens during the passivation process
- Experimental characterization and theoretical simulation of galvanic coupling phenomena in localized corrosion
- Fabrication of hierarchical three-dimensional porous structures for energy storage and conversion by an electrochemical deposition process, an area in which electrochemistry and materials science are intertwined

Topics in Number 48 include:

- Effects of codeposition of hydrogen on the structure of electrodeposited copper
- New classes of electrode materials introduced by spontaneous deposition of Ru and Os on Au(111) and Pt(111) single crystal surfaces
- Recent developments on metal, metal oxide, and conductive polymer electrodeposition for energy device applications
- Surface morphology of activated electrodes on their electrochemical properties
- Electrochemical micromachining and surface microstructuring based on porous-type anodization of patterned films
- In-depth review of the latest developments in electroless deposition

1

Durability of PEM Fuel Cell Membranes

Xinyu Huang and Ken Reifsnider

Department of Mechanical Engineering, University of South Carolina, Columbia, SC, USA

1. SUMMARY

Durability is still a critical limiting factor for the commercialization of polymer electrolyte membrane (PEM) fuel cells, a leading energy conversion technology for powering future hydrogen fueled automobiles, backup power systems (e.g., for base transceiver station of cellular networks), portable electronic devices, etc. Ionic conducting polymer (ionomer) electrolyte membranes are the critical enabling materials for the PEM fuel cells. They are also widely used as the central functional elements in hydrogen generation (e.g., electrolyzers), membrane cell for chlor-alkali production, etc. A perfluorosulfonic acid (PFSA) polymer with the trade name Nafion® developed by DuPont™ is the most widely used PEM in chlor-alkali cells and PEM fuel cells. Similar PFSA membranes have been developed by Dow Chemical, Asahi Glass, and lately Solvay Solexis. Frequently, such membranes serve the dual function of reactant separation and selective ionic conduction between two otherwise separate compartments. For some applications, the compromise of the "separation" function via the degradation and mechanical failure of the electrolyte membrane can be the life-limiting factor; this is particularly the case for PEM in hydrogen/oxygen fuel cells.

U. Pasaogullari, C.-Y. Wang (eds.), *Modeling and Diagnostics of Polymer Electrolyte Fuel Cells*, Modern Aspects of Electrochemistry 49,
DOI 10.1007/978-0-387-98068-3_1,

Despite decades of research efforts, there is still a lack of fundamental understanding of the underlying degradation processes of PEM that involve mechanical, physical, chemical, and electrochemical phenomena. As a result, reliable accelerated testing and life prediction methodologies remain elusive. Investigations on the failure mechanisms of PEM frequently break down the degradation processes into two categories: mechanical degradation and chemical degradation. This approach has yielded improved understanding of these processes separately. On the other hand, the mechanical and chemical degradation appear to be highly intertwined. For example, evidences have shown that chemical degradation processes can result in premature mechanical failure (rupture) of the membrane, mechanical degradation processes (e.g., tensile stress and stress cycling) can accelerate chemical degradation, etc. It has also been observed that the localized chemical degradation process is particularly harmful to the mechanical strength and toughness of the membrane electrode assemblies (MEAs). Much work needs to be done in elucidating these degradation and localization phenomena and their consequences. Reliable accelerated testing and life prediction approaches must capture the correct or relevant degradation and failure modes, and the degree of degradation must be quantified and tracked. Methodologies that can be checked and validated experimentally are the most desirable. The authors will discuss the current state-of-the-art understanding of PEM fuel cell membrane degradation and failure mechanisms, particularly the interaction and coupling of various degradation processes, and will suggest alternative concepts for the implementation of accelerated testing and life prediction methods for PEM fuel cell devices.

2. REVIEW OF PEM FUEL CELL DEGRADATION PHENOMENA AND MECHANISMS

Two major barriers to the commercialization of PEM fuel cells are high cost and poor durability. The US Department of Energy has established the durability target of electrolyte membranes for automotive fuel cells at 5,000 h and for stationary fuel cells at 40,000 h with additional cost constraints and operation requirements. In commercial applications, the integrity of fuel cell membranes must be maintained throughout the usable life of the PEM fuel cell stacks because of the potential safety hazard and the associated liability issues. PEM fuel cells for automotive applications drive a significant portion of current fuel cell research. The durability requirements for automotive fuel cells are arguably much more challenging than those for other applications due to some unique operation characteristics of automotive fuel cells, such as power output transients due to acceleration and deceleration, startup–shutdown cycles, freeze–thaw cycles, etc. The better understandings of the degradation of PEM fuel cells for automotive applications will almost certainly benefit the development of durable PEM fuel cells for other applications.

While the focus of this chapter is on the degradation of PEM fuel cell *membranes*, we would like to start with a brief review of the degradation of PEM fuel cells because the membrane degradation is closely related to the degradation of other PEM fuel cell components, such as electrocatalysts, gas diffusion media, etc. Degradation behavior of PEM fuel cells frequently manifests itself as a gradual, unrecoverable performance loss followed by a sudden failure of one or multiple cells in a stack, as shown schematically in Fig. 1. This is indicative of multiple parallel degradation and failure mechanisms occurring inside PEM fuel cell stack. For a particular fuel cell design and an intended application, it is critical to identify and focus on the life-limiting degradation and failure mode as the basis for durability projection and life prediction.

Borup et al.[1] published a comprehensive review on various aspects of durability and degradation issues of PEM fuel cells. Table 1 summarizes commonly observed degradation symptoms and

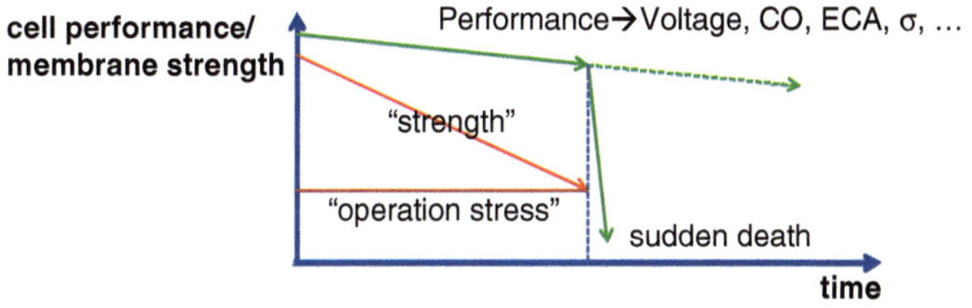

Figure 1. Competing degradation mechanisms in PEM fuel cells. (CO: hydrogen crossover rate; ECA: electrochemical area; σ: membrane ionic conductivity.)

Table 1
MEA Degradation and Failure Mechanisms.

Symptoms \ Mechanisms		Carbon corrosion (air-air start)	Gas impurities (e.g., CO, H_2S, SO_2 ...)	Contaminants (e.g. some transition metal cations, anions)	Catalyst instability (pt sintering, dissolution, re-crystallization)	GDL loss of wet-proof (flooding)	Seal failure (gross leaking)	Membrane failure (pinholing, and tear)
Gradual performance degradation	Activation loss (cathode ORR & anode HOR)	*	*	*	*			
	Ohmic loss (ionic and electronic resistance)	*		*				
	Mass transport loss	*				*		
Sudden failure	OCV drop (H_2, O_2 X-over)						* “Sudden death”	*
Recoverable ?		No	Somewhat	Somewhat	No	No	Somewhat	No
Mitigation approach		Mitigation via. potential control	air bleed, CO tolerant catalysts, …	avoid potential contaminants in conduits and in new alloy catalysts	new alloy catalysts, avoid OCV & potential-cycling	Surface modifica-tion	Alternative material, increase compression	avoid high T, dry & RH-cycling
Cross-cutting effects			oxygen into anode → H_2O_2	membrane de-polymerization via. Fenton reaction	Pt catalyzed membrane de-polymerization	flooding → fuel or oxygen starvation	oxygen into anode → H_2O_2	local heating via. exothermal reactions

the related degradation mechanisms for MEAs in PEM fuel cells. The gradual performance loss is typically associated with the reduction of catalyst activity by platinum catalyst sintering, dissolution, poisoning of electrocatalysts by various impurities (gaseous and ionic) and carbon corrosion, etc. Ionic contaminants can enter the electrolyte membrane and exchange with the proton associated with acid group in the ionomer membrane, reducing its effective proton conductivity. The gradual loss of the hydrophobicity of gas diffusion layer/media (GDL/GDM) can result in increasing gas transport losses and decreasing limiting current. Severe carbon corrosion can occur during the startup after long-term idling; it not only affects catalyst activity, but also affects mass transport properties of the electrodes and the electrical contact resistance between the electrode and the gas diffusion media. Several decay mechanisms that can lead to the gradual performance loss have been extensively studied. These include carbon corrosion, electrocatalyst instability, effects of certain contaminates (CO, H_2S) on fuel cell performance, etc. The performance decay due to some of the aforementioned degradation mechanisms can be recoverable. The absorbed impure gas on the catalyst surface can sometime be removed (cleaned up) by raising/cycling the potential of the platinum catalysts and/or exposing the contaminated electrode to oxygen.[2] Some ionic contaminants can be flushed out. Effective mitigation method to reduce carbon corrosion by potential control has been reported.[3]

In contrast to the gradual performance loss, mechanical breach of membrane typically results in the sudden failure of cell/stack due to massive gas crossover in between the anode and the cathode compartments through holes and tears. The loss of gas sealing around edge seal can also lead to catastrophic stack failure in a similar fashion. The ability of the membrane and edge seal to resist mechanical breach can be a limiting factor in the durability of PEM fuel cell stacks. There exist crosscutting effects that couple one degradation mechanism to another. Pozio et al.[4] reported that Iron contaminants from the end plates can accelerate the rate of membrane decomposition. Recently, it was reported that Pt band formed inside the membrane can also accelerate membrane degradation.[5] The source of the platinum in the platinum band is the dissolved platinum from the electrode (particularly cathode) during high potential hold and potential cycling; the platinum ions then diffuse into the membrane and recrystallize by interacting with the crossover hydrogen.[6–9] In addition to potential cycling, certain contaminants (e.g., chlorides,[10] sulfates[11]) can also accelerate Pt dissolution, and thus expedite the formation of Pt band inside the membrane. Local fuel starvation[12] can lead to carbon corrosion and local high potential which accelerates the loss of platinum in the catalyst layer. Hence, it is suspected that membrane degradation can be related to the local flooding. These crosscutting effects greatly complicate the PEM fuel cell membrane degradation processes as a function of operating history.

Compared to some of the extensively studied PEM fuel cell degradation mechanisms, such as catalyst instability and carbon corrosion, the mechanisms that result in the "sudden death" (Fig. 1 and Table 1) of the fuel cell due to membrane fracture or pinholing are less understood. Even less understood are the crosscutting effects that interrelate membrane degradation mechanisms. The lack of fundamental understanding of membrane degradation mechanisms have resulted in many *technical barriers*. These include the lack of guidelines for synthesizing new membrane materials, the lack of membrane durability benchmark for the evaluation of new membranes, the lack of effective mitigation approaches for PEM membrane degradation, and the lack of accelerated testing methodologies. Substantial advancement of our understanding of PEM fuel cell degradation mechanisms is necessary for the engineering of commercially viable PEM fuel cells. In the following sections, we will review the membrane degradation mechanisms, in particular, the crosscutting effects that relate membrane degradation to other forms of fuel cell degradation. Assuming that membrane failure is the life-limiting mode, a conceptual life prediction method using accelerated membrane degradation test data will be presented. The mitigation approaches for membrane degradation will be briefly discussed at the end of this chapter.

3. MEMBRANE DEGRADATION

Membrane degradation has been loosely categorized into "chemical degradation" and "mechanical degradation. "There is a level of ambiguity as to whether the classification is based on the cause or the effect. The authors believe that it is advantageous to rationalize membrane degradation as one complex process that has chemical and mechanical driving forces, and these driving forces acting together result in the change of the chemical, physical, and mechanical properties of the membrane. The chemical driving forces include all factors that promote the radical attack inside the membrane, and the mechanical driving forces include mainly hygrothermal stress and stress cycling. The results of the membrane degradation process include chemical effects, such as the decomposition of the polymer chain; physical effects, such as the change of crystallinity, and entanglement density; and mechanical effects, such as the loss of the strength. Frequently the single largest impact of chemical degradation on the membrane is the loss of membrane mechanical toughness, i.e., producing a mechanical degradation effect. Catastrophic cell failure due to membrane breach is believed to be the result of a combined chemical and mechanical driving forces acting together. In the following sections, we will discuss the mechanical characterization of fuel cell membrane, the membrane degradation driven by chemical factors, the membrane degradation driven by mechanical factors, and the interaction between the membrane degradation driven by chemical factors and by mechanical factors.

3.1. Stress in Membrane and MEAs

Mechanical stresses exist in the MEA as a combination of *built-in stresses*, *assembly stresses*, and *operation stresses*. The cell/stack fabrication and assembly process result in the built-in and the assembly stresses. During operation, stresses can be induced by the change of water content in the membrane. There are possibly other sources of stresses due to various "off design" conditions.

The assembly stress is largely due to stack clamping. In the assembled cell/stack, the MEA is constrained by the sealing material on the edge and is in contact with gas diffusion media which is supported by the bipolar plates with gas flow channels. To lower the electrical contact resistance between the cell and the bipolar plate, a nominal clamping pressure from 0.34 to 1.38 MPa (50–200 psi) is typically applied to the active area of the MEA. The clamping pressure also helps maintain the gas-tightness of the edge seals. This clamping pressure is transferred to the MEA through a fibrous gas diffusion media (carbon paper or carbon cloth) and a solid bipolar plate with flow channels.

The local clamping pressure on the MEA can be much higher than the nominal clamping pressure. The ribs of flow channels on the flow plate reduce the effective contact area with the MEA. In a cross-flow arrangement, the contact area is only 25% of the electrochemical active area of the MEA. Figure 2 shows a "checker board" pattern formed on a pressure sensitive film inserted into a single cell assembly. The pattern shows the non-uniformity of the clamping pressure distribution on the active area of the MEA.

The coarse texture of the fibrous gas diffusion media can further amplify the contact stress exerted on the MEA. Figure 3 shows the relative size of a carbon fiber with respect to the typical thickness of the electrode and the electrolyte membrane. It can be seen that the diameter of the carbon fiber in the gas diffusion media is comparable to the thickness of the electrode. The rigid carbon fiber pressed onto the porous electrode layer can produce in-prints which can later become a stress-concentration and defect-initiation sites at the electrode–electrolyte interface. A microporous layer, if used, tends to smooth out the surface of the GDM and reduces fiber in-print. Thicker electrode layer also offers protection against fiber in-prints.

Figure 2. A "checker board" pattern of clamping pressure distribution on the active area of cell measured by Pressurex® pressure sensitive film; the flow channels in bipolar plates are arranged in a cross-flow pattern; red indicates higher pressure.

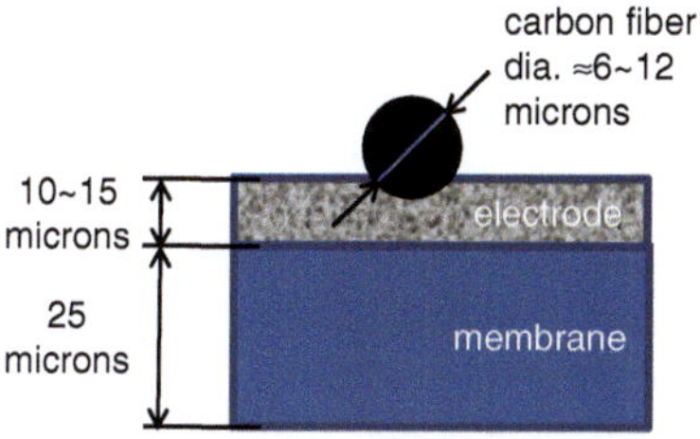

Figure 3. The relative size (diameter) of a carbon fiber in the GDM in comparison with the thickness of electrode layer and the thickness of membrane.

Local discontinuities due to the buildup of geometric tolerances of the components that come into direct contact with membrane or MEA can induce local stress concentrations. As shown in Fig. 4, membrane wrinkling can occur during cell/stack assembly or during operation. Wrinkles on the MEA can disrupt the clamping pressure distribution and the local electrical contact conditions. Near the edge seal, MEA can be subjected to a shearing force if the gas diffusion media and the edge seal do not line up properly. This can be mitigated by using sub-gaskets and/or alternative edge seal configurations.

The built-in and operation stresses are the consequences of the large swelling and shrinkage of the ionomer membrane when it uptakes and loses water. This is frequently referred to as "dimensional instability" in the literature. Water in the PFSA membrane is an essential ingredient of its proton conduction behavior. Water affects the morphology[13,14] of the ionic clusters (at nanoscale) which

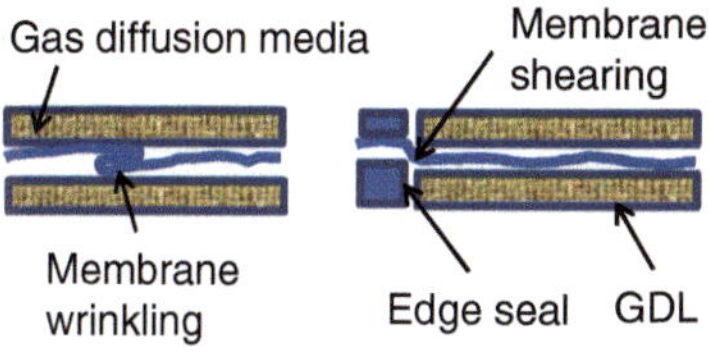

Figure 4. Membrane wrinkling and shearing under the gas diffusion layer in the assembled cell/stack.

is believed to be the major proton transport pathway. Higher water content in the membrane is desirable for obtaining higher proton conductivity. Macroscopic dimensional changes can be regarded as an undesirable "side effect," which cannot be completely eliminated. As membrane gains or losses water, its proton conductivity can change by orders of magnitudes; at the same time its geometric size can swell or shrink by about 5–15%. In industrial production, MEA is usually integrated together with the edge seal, the gas diffusion medium, or even the rigid bipolar plate. Depending on the fabrication procedure and the environmental condition, when the integrated part is fabricated, the MEA corresponding to certain stress-free size is "frozen" into the integrated cell package; when the MEA in the cell experience an environment with a different relative humidity (RH), the water content corresponding to a different stress-free size is acquired; due to the constrained configuration, the actual size of the MEA cannot change much; and hence, the built-in stress is generated.

During cell/stack operation, water content in the membrane is affected by the local intensive variables, such as local temperature, water vapor concentration in the gas phase, gas temperature and velocity in the channel, and the properties of the electrode and gas diffusion media. The power fluctuation can result in temperature variation inside the cell/stack, which will subsequently change the local membrane water content. As the water content in the membrane tends to be non-uniform and unsteady, this results in operation stresses. When the membrane uptakes water from a dry state, it tends to expand as there is no space for it to extend in plane and it can wrinkle up as schematically shown in Fig. 4; when the membrane dries out, the wrinkled part may not flatten out, and this "ratcheting" effect can cause the pile up of wrinkles at regions where membrane can find space to fold. The operation stress is typically cyclic in nature due to startup–shutdown cycles, freeze–thaw cycles, and power output cycles.

At the microscopic scale, mechanical stress can also develop due to the property-mismatch of the electrode and the membrane. The porous electrode is typically much weaker than the membrane and has lower hygrothermal expansion coefficient than the membrane; hence it can develop mud-cracks and delaminates from the membrane after RH cycling. This will perturb the local electrochemical processes and results in non-uniform decay of membranes.

Based on the authors' experience, membrane fracture frequently occurs during events when membrane is dried out; the in-plane hygrothermal stress is a critical driving force in membrane mechanical failure. The in-plane hygrothermal stress can be lowered by reducing the swelling/shrinkage ratio of the membrane, i.e., the "dimensional instability" of the membrane.

3.2. Mechanical Characterization of Membranes

At the beginning of life (BOL), the mechanical strength and toughness of fuel cell membrane are typically adequate to sustain various mechanical stresses encountered during fuel cell operation without failure. After prolonged service, the membrane gradually weakens; it may fracture at locations where the membrane material tends to decay relatively faster and/or high mechanical stresses tend to develop. This is schematically illustrated in Fig. 1. *Good BOL mechanical strength/toughness does not necessarily imply good membrane endurance; rather what affects the usable life of a fuel cell membrane is how well it can preserve its mechanical strength/toughness during the fuel cell operation history.* The mechanical stress–strain curve obtained for fuel cell membranes or MEAs can be very informative regarding the decay trend of membrane strength/toughness, which can serve as an indicator of the state of membrane degradation.

Figure 5 shows the stress–strain responses of an MEA with Nafion (NR-111) membrane tested at 25°C and with four RH levels.

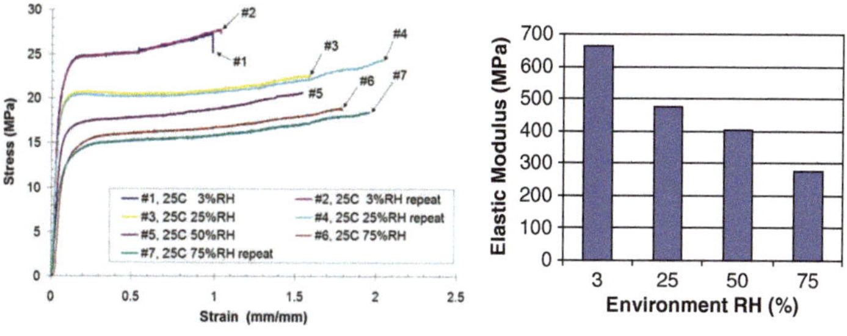

Figure 5. Left: stress–strain behavior of MEA based on NR111 membrane tested at 25°C and with different RH levels; right: calculated elastic modulus as a function of RH [reprinted with permission from reference 34].[34]

Replicate tests were conducted at 3, 25 and 75% RH and good repeatability was observed. The elastic modulus as function of RH calculated from these stress–strain curves is also shown. The shape of the stress–strain curve can be approximated by two linear segments. It is clear that RH affects the elastic modulus and the yield stress of these MEAs with Nafion-type membrane. Note that the elastic modulus more than doubled when the MEA was dried from 75 to 3% RH. However, the yield strain and the slope of the second linear segment are affected to a lesser degree; it is notable that the 3% RH condition exhibited the lowest strain-to-failure. Despite some variations, the MEAs tested at all four RH levels were found to be fairly ductile, with strain-to-failure exceeding 100%. The yield stress varies from approximately 12 MPa to 17.5 MPa and the strain-to-failure varies from 86.4 to 152.7%. This is indicative of the initial non-uniformity of the MEA and the presence of initial random defects in the as-fabricated membrane or MEAs.

Because the elastic modulus and the yield strength of the ionomer membranes are greatly affected by the water contents and the environment temperature, for generating mechanical property data for modeling purposes, it is essential to test membrane and MEA in an environment with precisely controlled temperature and RH. The authors have developed and utilized a unique mechanical testing facility for testing fuel cell membranes. The schematic of the system and pictures of several system components are shown in Fig. 6. It consists of an MTS Tytron® horizontal load frame, a home-built environmental chamber with temperature and RH control, and an optical strain measurement instrument (Messphysik ME460 video-extensometer).

The authors have found that for the evaluation of the strength/toughness of the membrane (as a metric of the state of membrane degradation), it is not necessary to test the membrane or MEAs in the tightly controlled environment. As long as the membrane or the MEA samples are tested in the same ambient environment under the same strain rate, one can obtain repeatable and reliable results that reflect the change of mechanical strength/toughness of the membrane. The stress–strain curves of an MEA with Nafion N-112 membrane tested at ambient environment is shown in Fig. 7. A total of four strips cut from one MEA were tested. The stress–strain data in the elastic range are fairly

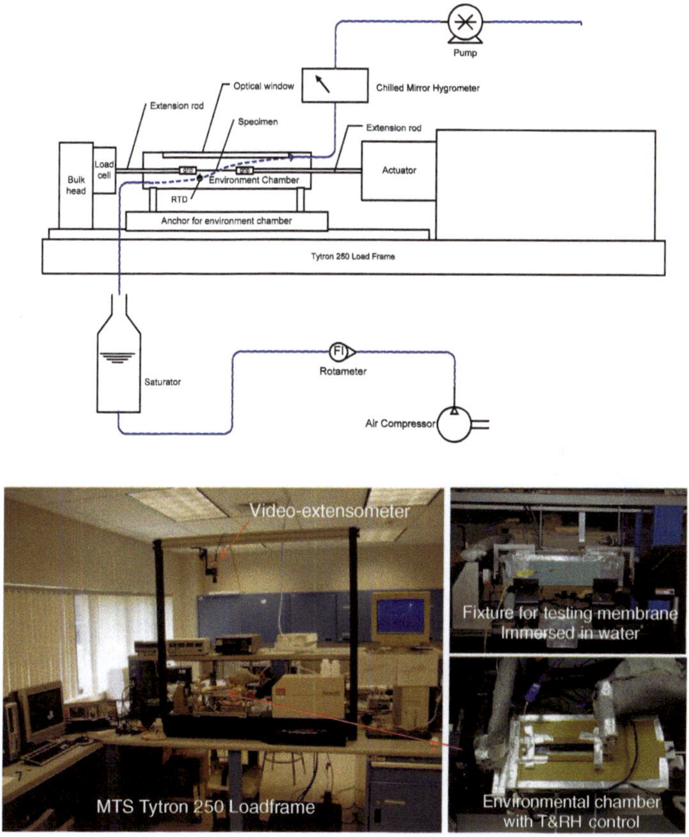

Figure 6. A membrane mechanical testing system consists of a MTS horizontal load frame, a home-built environment chamber with temperature and RH control, and a high-precision video-extensometer.

repeatable, so are the yield points. Strain-to-break shows the largest sample-to-sample variation, typically. To obtain statistically significant information about the strain-to-break, replicate tests are necessary. A scheme the authors used to obtain replicate samples are shown in the insert of Fig. 7. A typical MEA with 5 cm by 5 cm active area is cut into multiple strips about 6 mm

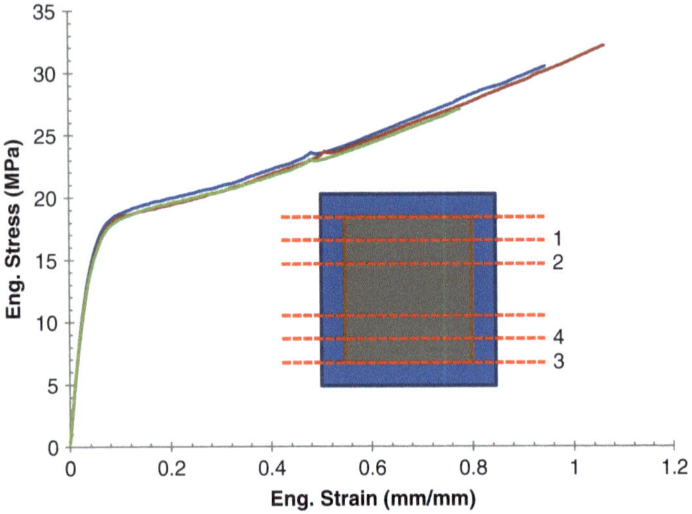

Figure 7. Stress–strain curves of strips cut from an as-fabricated MEA with N-112 membrane; the test was conducted at ambient conditions.

wide and 60 mm long (including the edge seal) and the strips are numbered as shown in Fig. 7. The strip is gripped on both ends by the edge seal and tested in the uniaxial load frame. When the edge seal is not present, the MEAs are gripped directly under a mechanical clamping type grip or a Capstan type roller grip, which is effective in avoiding failure under the grips. Engineering stress and strain are typically calculated based on the recorded load versus displacement data. When calculating engineering stress, a nominal cross section area of the membrane is typically used. This is based on the assumption that porous electrodes do not carry a significant amount of load. This treatment is adequate for characterizing strength decay by comparing failure stress/strain of the degraded samples to the failure strain/stress of control samples. For model validation purpose, true strain and stress based on the true cross section area are preferred. True stress/strain curves can be derived from engineering stress/strain curves. In this manuscript, all stress–strain curves are reported in terms of the engineering stress and strain.

3.3. Chemical Degradation Processes

Early research of ionomer membrane degradation was conducted in the context of PEM electrolyzers. The detection of fluoride and other chain fragments in the condensed effluent water indicates the decomposition of PFSA ionomer and has long been noticed. Baldwin[15] reported the detection of fluoride in the effluent of PEM electrolyzer and believed that it is the result of membrane mechanical failure. Extensive research has been conducted to elucidate the reaction pathways for membrane decomposition. Many controversial results and mechanisms have been reported in the literature, demonstrating the complex nature and the current inadequate understanding of the membrane degradation mechanisms.

The role of H_2O_2 in the degradation of fuel cell membrane is still being debated. Early reports attribute the degradation of membrane to the transition metal contaminants and the H_2O_2, which can be formed in the electrode and diffuse into the membrane. Early theories of the chemical decomposition pathways for the decomposition of PFSA membrane are based on Fenton's reaction mechanism,[16,17] where hydrogen peroxide is catalytically decomposed by transition metal ions, such as Fe^{2+} and Cu^{2+}. Fenton reactions (Eqs. 1, 2, 3, 4, 5, 6) are known to produce hydroxyl (OH^*) and hydroperoxyl radicals (OOH^*) as intermediate reaction products. These radicals are strongly nucleophillic and can decompose PFSA membrane. Pozio et al.[4] postulated the degradation mechanisms as free radical attack of the perfluorinated molecular chains from weak bonds. It is postulated that the free radicals are derived from H_2O_2, which can form a two-electron oxygen reduction at anode or cathode. H_2O_2 and the transition metal ions (Fe^{2+}, and Cu^{2+} from contaminants) induce the Fenton reactions, which generate damaging radicals inside the membrane. Mittal et al.[18] conducted an experimental study and reported that H_2O_2 may not be necessary in the degradation of fuel cell membrane. Mittal et al.[19] also concluded that molecular H_2, O_2 and platinum are three necessary ingredients for the occurrence of PFSA membrane degradation, without revealing any details of the polymer decomposition reactions. Aoki et al.[20] have come to the same conclusion via an ex-situ experimental study.

$$H_2O_2 + Fe^{2+} \rightarrow HO\bullet + OH^- + Fe^{3+} \tag{1}$$

$$Fe^{2+} + HO\bullet \rightarrow Fe^{3+} + OH^- \quad (2)$$

$$H_2O_2 + HO\bullet \rightarrow HO_2\bullet + H_2O \quad (3)$$

$$Fe^{2+} + HO_2\bullet \rightarrow Fe^{3+} + HO_2^- \quad (4)$$

$$Fe^{3+} + HO_2\bullet \rightarrow Fe^{2+} + H^+ + O_2 \quad (5)$$

$$Fe^{3+} + H_2O_2 \rightarrow Fe^{2+} + H^+ + HO_2\bullet \quad (6)$$

Another controversy around membrane degradation in the literature is about the role of platinum. Watanbe et al.[21,22] reported a Pt-impregnated membrane as a self-humidification mechanism; they also suggested that distributed Pt particles in the membrane can reduce gas crossover and can mitigate membrane degradation. Ohma et al.[5] reported that platinum, passively formed in the membrane in the form of a band, has contributed to the accelerated local degradation of membrane near the platinum band. A common feature of all theories is the recognition of free radicals as the damaging species of the PFSA membrane. The controversy resides in the understandings on how the radicals are generated and how the radicals chemically react with PFSA membrane.

Curtin et al.[23] suggest that radical attack starts from the carboxylic end group of a Nafion molecular chain; by eliminating the unstable carboxylic acid end group, the stability of Nafion in Fenton reagent can be enhanced. Zhou et al.[24] studied a group of model compounds which have structures similar to moieties found in PFSAs; their results confirm that the carboxylic acid end group is indeed weak in Fenton's reagent. Escobedo[25] reported a model compound study showing that branched side-chains have much less contribution to membrane instability than the carboxylic acid end group. Hicks[26] suggested that the end group attack cannot be the only degradation mechanism due to the projected finite fluoride emission rate at zero concentration of carboxylic acid end group.

Mittal postulated that radical formation is likely due to the chemical reaction of H_2 and O_2 on Pt surface, this reaction is chemical in nature and shows strong dependence on the surface properties of Pt particles, and the sulfonic acid groups in the PFSA membrane maybe the key to the radical attack mechanisms.[27] Cipollini[28] in

2007 and Coms[29] in 2008 hypothesized detailed pathways for membrane decomposition reactions, with Coms' results partly based on first principle calculations.

Accelerating factors of membrane chemical degradation processes have been reported. Endoh et al.[30] and Inaba et al.[31] reported the open circuit voltage (OCV) test at low RH is more damaging for the decomposition of the electrolyte membrane. Akoi et al.[20] reported a novel accelerated electrolyte degradation test method using a model system of carbon supported platinum catalyst bonded by Nafion resin and dispersed in deionized water. And the reaction is accelerated by temperature and the concentrations of H_2 and O_2. Quintus et al.[32] reported accelerated membrane test in H_2O_2 (without adding any transition metal ions). It was shown that Nafion-112 membrane in gaseous H_2O_2 (30%, balance N2) environment at 70°C suffered from 18% weight loss after 36 h of exposure; similar experiments conducted in liquid H_2O_2 showed no measurable weight loss; and in addition, the membrane degradation in gaseous H_2O_2 was shown to be accelerated by lowering the dew point.

In summary, though there is still no unifying theory of chemical degradation, it appears that reactant gas crossover, hydrogen peroxide formation and movement, recrystallized Pt particles, and transition metal ion contaminants are relevant factors contributing to the chemical decomposition of PEMs. In addition to free radical attack, there are other possible degradation modes due to the platinum in the membrane: the heat generated during recombination of crossover hydrogen and oxygen on the surface of platinum embedded in the PEM may also lead to local hotspots which may lead to thermal decomposition of the membrane; the platinum particles on the platinum band can grow to fairly large size ($\sim$60 nm), and those large mechanical inclusions may lower the mechanical strength of the membrane.

After chemical degradation, membrane thinning is typically observed. Besides thinning, scanning electron microscopy (SEM) images of MEA after OCV test frequently show a characteristic long crack or a band of microcracks along the center of the membrane, as shown in Fig. 8. These cracking features are not observed in the SEM pictures of the control samples prepared with the same epoxy potting and polishing procedure. The crack is believed to be the result of tensile stresses developed in the ionomer membrane during the dry out of the potted sample in the vacuum chamber. There likely

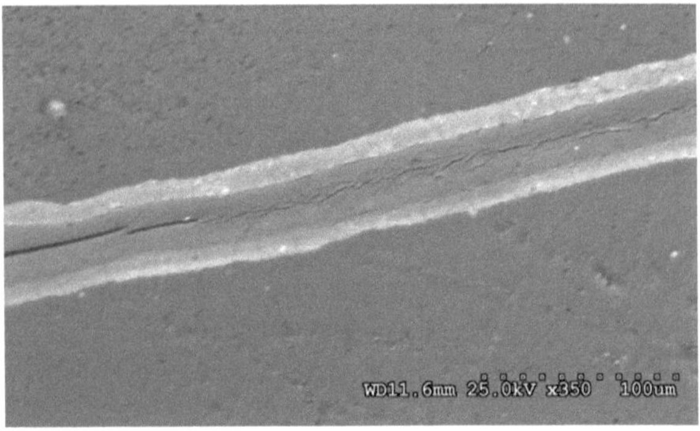

Figure 8. SEM image of Nafion-112 MEA after 100 h of OCV test.

exists a weakened band in the chemically degraded membrane, but not in the control membranes.

3.4. Mechanical Degradation Processes

Pure mechanical effects can result in the degradation of membranes.[33,34] Variations in temperature and humidity inevitably occur during PEM fuel cell operation. Due to the large hygrothermal expansion coefficient of ionomer membrane, stresses and equivalent strains ("mechanical loading") can be induced in the membrane or MEA, which is rigidly constrained by gas diffusion media, the bipolar plate and the edge seal. Cycling of RH can cause cyclic stress in the membrane or MEA. This has been found to be particularly detrimental to the mechanical integrity of the membrane, similar to the fatigue process observed in many structural materials under cyclic stresses.

The authors[35] have reported mechanical behavior of ionomer membrane after RH cycling. Figure 9 shows the typical stress–strain curve of the RH cycled MEA and the strain-to-break as a function of number of RH cycles. The yield stress and strain remain largely unchanged. However, the strain-to-break was found to decrease. Large sample-to-sample variation of strain-to-break is typically observed, a characteristic of defect controlled behavior. The result of multiple strips (from one MEA) test reveals a clear decreasing trend. Especially the minimal strain-to-break among four strips was tested for each MEA.

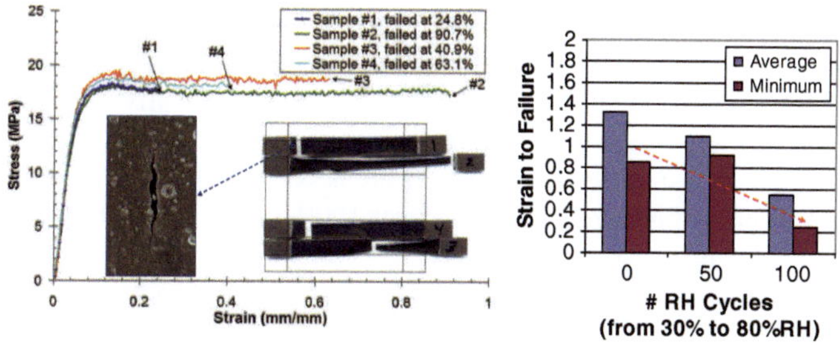

Figure 9. Left: stress–strain curves of MEA after 100 cycles from 30 to 80% RH at 65°C; pictures of broken sample are shown in the inserts; microcracks were observed on the surface of the sample; right: the average and the minimum strain-to-break of MEA as function of the number of RH cycles [reprinted with permission from reference 34].[34]

Our evidence has also shown that defects generated and propagated during RH cycling condition do not contribute much to the gas crossover before the occurrence of the final mechanical breach. Figure 10 shows the RH cycling test results of MEAs cycled from 0 to 100% RH at 80°C. Multiple test runs were conducted, and the sample was taken out and mechanically tested after 200, 400, and

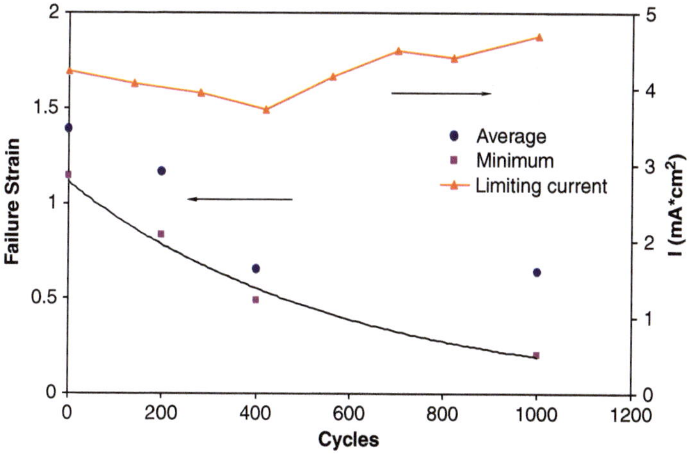

Figure 10. RH cycling results in steady degradation of membrane strain-to-break, but little effect on the limiting current due to hydrogen crossover.[36]

1000 RH cycles. Four to six strips from each MEA were tested. The average and minimal strain-to-break as function of the number of RH cycles are plotted together with the periodically measured hydrogen crossover rate during the course of RH cycling test. A steady decreasing trend of strain-to-break is observed, and the minimal strain-to-break reduced from over 100% to less than 20%; while the hydrogen crossover rate stays between 1.5 and 2 mA/cm^2.

Stress–strain test results also provide insights into the nature of the defects propagated by RH cycling. It was found that the reduction of strain-to-break is a result of the formation and growth of a small number of discrete and localized defects. And these defects are likely also responsible for reactant gas cross over once they penetrate the membrane. These are demonstrated in Figs. 11 and 12. An MEA was subjected to 50 cycles from 0 to 100% RH. The post mechanical test results show that the strain-to-break has significantly reduced for 3 out of 4 strips tested. The fracture locations of the test strips roughly correlate to leak spots identified by a bubble test. Strip #3 which showed high strain-to-break corresponds to the lower portion of the MEA, where no leak spot was observed.

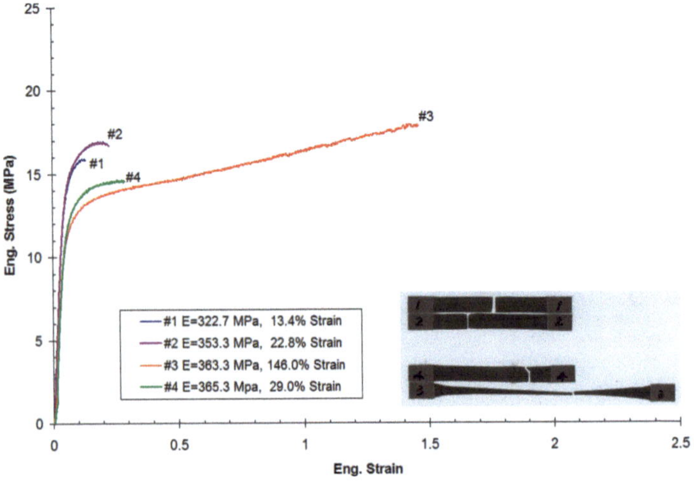

Figure 11. Mechanical stress–strain behavior of four strips cut from a cell after a RH cycling test of 50 cycles from 0–100% RH; the blue spots marked on the MEA are leaking locations identified by a bubble test after the RH cycling; the insert shows the fracture location ofeach strip.

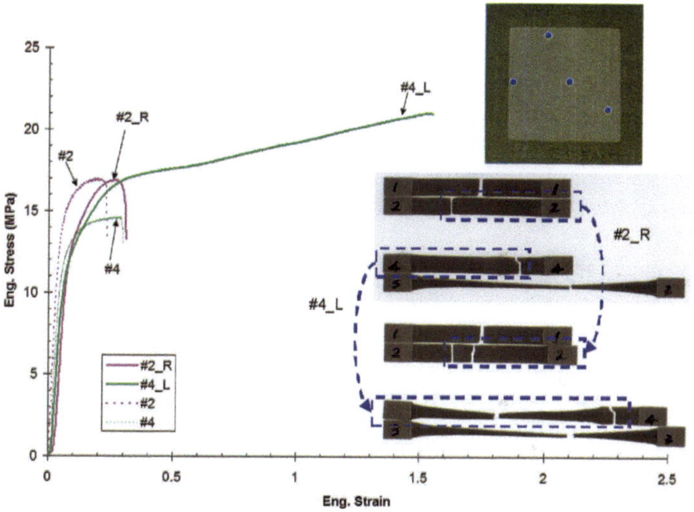

Figure 12. Results of the second stress–strain test for the remaining portion of the strips after the first test.

The remaining portion of the broken strips (#2 and #4) that have shown reduced strain-to-break was mechanically tested again. The results are shown in Fig. 12. Interestingly, the longer portion (#4_L) of the remaining part of the sample #4 indicated strain-to-break greater than 150% in the second test. This is comparable to the strain-to-break of the control samples. The longer portion (#2_R) of the remaining broken part of sample #2 was also tested; the result indicated only a slightly higher strain-to-break as compared to the original one. These results suggest that the largest defect contained in the strip is likely responsible for the observed reduction of strain-to-break. Once it has been consumed in the first test, the remaining portion tends to show higher strain-to-break, because it contains lesser or no defects.

It is important to understand the accelerating factors in RH cycling test. In-cell RH cycling tests are typically carried out by alternating an inert feed gas (N_2) of controlled dew points. The water content of the membrane can be monitored by high-frequency cell resistance measurements. The result of the RH cycling test of a Nafion® based MEA cycled from 0 to 100% RH at 80°C is shown in Fig. 13. It can be seen that with the chosen cycling period (~30 min)

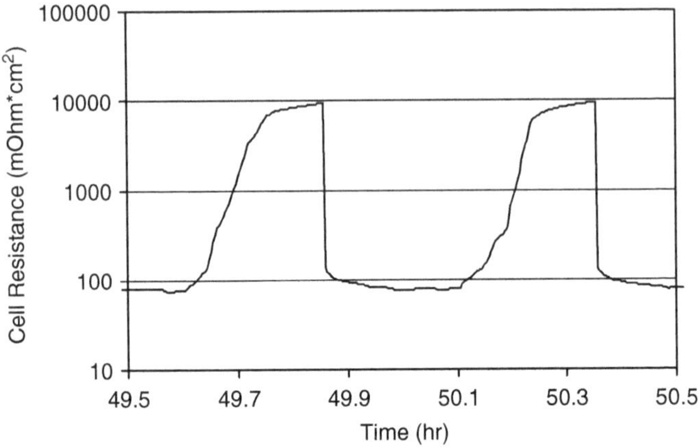

Figure 13. Plot of cell resistance as function of time during a RH cycling test from 0 to 100% RH at 80°C at 0.5 SLPM N_2 flow; data captured in about two 30-min cycles are shown on the plot; the cell resistance was measured by an Agilent® 4338B milliohmmeter.

and flow rate (~0.5 standard litre per minute (SLPM)), MEAs gradually equilibrate with both the wet and dry gas as indicated by the jumps and leveling off trend of high frequency cell resistance. It takes approximately 10–20 min for the cell resistance to rise (losing water) from 10^2 to 10^4 mΩ cm^2, while it takes only10–20 s for the cell resistance to fall (taking in water) from 10^4 to 10^2 mΩ cm^2. It appears that MEA uptakes water much faster than it loses water. If the cycling frequency is increased, the membrane dry-out process may not proceed to the equilibrium level, and the peak tensile stress is expected to be lower than the stress predicted by equilibrium value of water content at the corresponding RH. This is expected to lower the rate of defect growth.

The authors also investigated the dependence of the mechanical degradation rate on the amplitude of RH cycling. Relatively long cycling period (>30 min) is used for the membrane to approach the equilibrium water content. The strain-to-failure data for two types of MEAs as function of RH cycles were compiled. Figure 14 shows the sample-to-sample variation of strain-to-failure at each test condition. Figure 15 shows the average strain-to-failure as a function of RH cycles. Both MEAs have shown faster degradation rate when

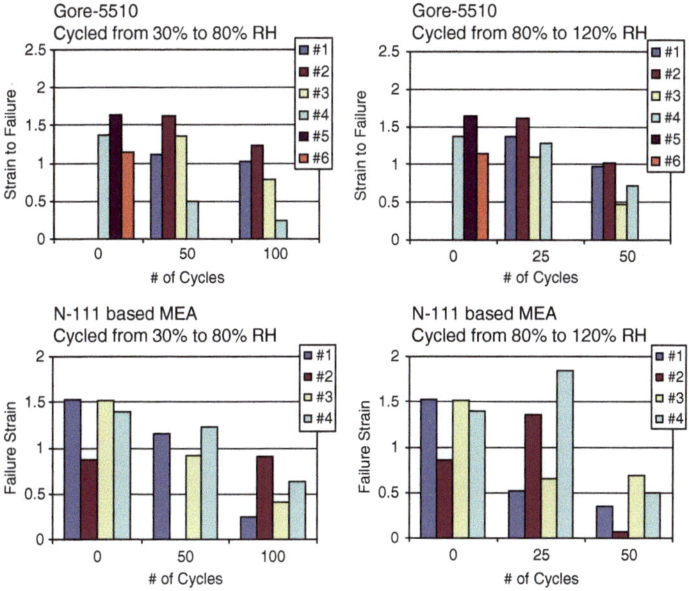

Figure 14. Strain-to-failure versus number of cycles for two types of MEAs: one with a reinforced composite membrane (Gore-5510) and the other with regular PFSA membrane (N-111). The samples are cycled from 30 to 80% RH and from 80 to 120% RH.[37]

cycled from 80 to 120% than cycled from 30 to 80% RH. The water content of PFSA membrane is typically characterized by the number of water molecules per sulfonic groups, defined as λ. It appears that the degradation rate is roughly proportional to the cycling amplitude, defined as the difference of the deformation, $\Delta\varepsilon$, which corresponds to a difference of water content, $\Delta\lambda$, resulted from two different hydration conditions. The swelling induced dimensional change $\Delta\varepsilon$ is approximately proportional to $\Delta\lambda$.[35] Assuming that equilibrium was achieved during the membrane swelling and shrinking process, $\Delta\lambda$ is estimated to be 3–4 for cycling from 30 to 80% RH and around 14 for cycling from 80 to 120% RH (liquid water).

High resolution SEM images of the freeze-fractured cross section of the RH-cycled samples were obtained. Many craze-like defects were observed on the cross section of samples that had been cycled from 80 to 120% RH, as shown in Fig. 16. In comparison,

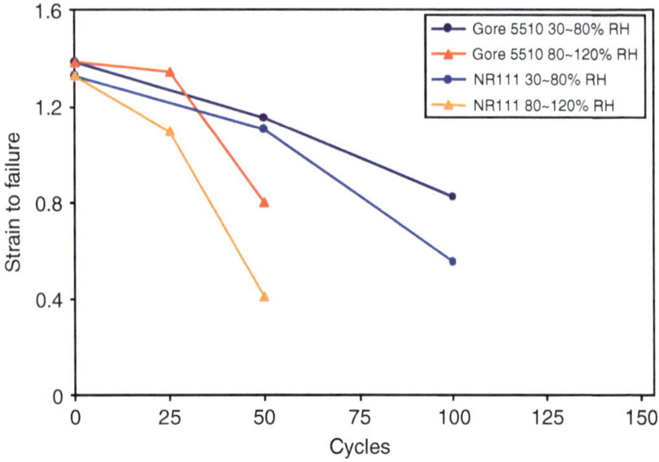

Figure 15. The trend of average strain-to-failure versus number of cycles for two types of MEAs tested.[37]

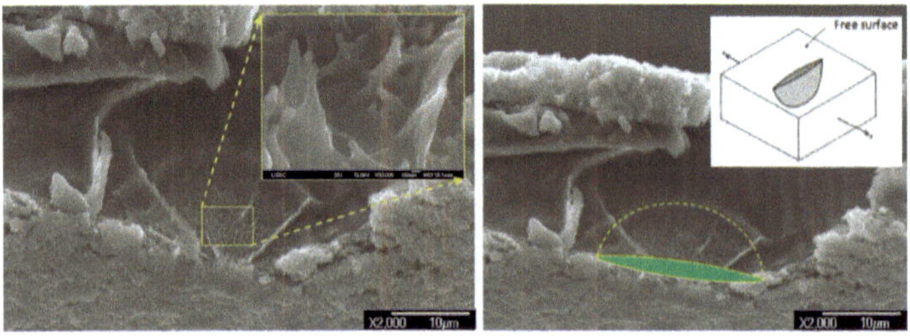

Figure 16. SEM images of the fracture surface of an RH-cycled (80–120% RH) sample, showing crack initiation and growth from a craze site near the anode and membrane interface. A crazing site found on Hi-RH cycled samples. The half-penny shaped craze closely resembles crazes formed on free surfaces.

control samples and chemically degraded samples prepared in the same way (freeze fracture) did not show such craze-like features. The facture morphology clearly shows that RH cycling and accelerated chemical degradation promote different failure modes. The RH cycling seeded the surface with unique "defects" that can later develop into craze-like features upon freeze fracture. The RH-cycled samples showed multiple crazing sites, where (it is believed) the final fracture originates. The chemically degraded sample typically shows brittle fracture without crazing formation. It appears that the RH cycling process generates localized defects; upon mechanical stress cycling, these defects grown larger and result in the crossover or fracture of the membrane.

The onset of crazing in an uniformly stressed polymer is determined by the local stress/strain state. Since crazing is accompanied by significant increase in volume (decrease in apparent density due to void formation and growth) hydrostatic stress/strain is particularly important. For the two-dimensional case, Oxborough and Bowden[37] suggested that crazing occurs when the tensile strain in any direction reaches a critical value ε_1 determined by:

$$E\varepsilon_1 = A + \frac{B}{\sigma_1 + \sigma_2} \tag{7}$$

where E is Young's modulus of the material, σ_1 and σ_2 are the in-plane stresses, and A and B are constants when depends on temperature. This equation clearly shows the dependence of crazing on dilatational stress state.

High dilatational stress that is required for crazing or defect initiation can potentially arise at the electrode/electrolyte interface amplified by the irregularities. Fiber impression is the potential cause of such geometric irregularities in the electrode or membrane layer. Our previous work[38] has shown the mesoscale fluctuation may be the source of defect initiation. An FEM model was generated for modeling a 2D anode/membrane/cathode structure in ABAQUS. The geometry of the model and the interfacial irregularities were extracted from cross section SEM images of actual MEA samples. The boundary conditions are set to allow the three-layer structure to freely expand; no external loads are applied. The only source of internal stress is the property-mismatch of the electrode and electrolyte layer. The material properties for the electrodes were estimated by the rule of mixtures. The electrolyte membrane properties as a function of temperature and humidity were obtained by controlled uniaxial tests. Membrane hygrothermal expansion coefficients were modeled by the ABAQUS user subroutine UEXPAN based on the experimental data of Morris and Sun.[38] The change in RH from 0 to 90% was imposed and the calculated dilatational strain was shown in Fig. 17 with and without considering the interfacial irregularities. It can be seen that the stress/strain during dry out

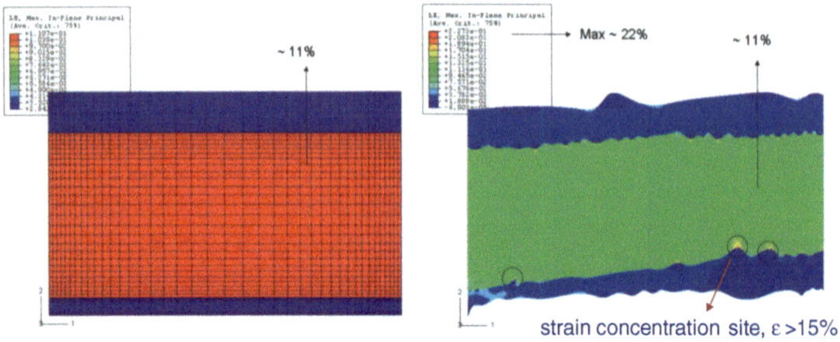

Figure 17. Strain concentrations due to irregularities at the electrode and membrane interfaces. Circled sites on the right figure are locations with maximum principal strain greater than 15%.

of a three-layer MEA is greatly amplified (doubled) by mesoscale geometric irregularity. It is clear that these strain concentration sites occur at the irregularities along the interfaces. The maximum strain at a stress-concentration site can be twice the average strain. These sites are susceptible to crazing initiation.

Based on our observation, a membrane degradation and failure mechanism under the RH cycling, a pure mechanical effect is theorized as the following sequence: electrode-microcracking→ crazing initiation at the electrode/electrolyte interface →crack growth under stress cycling→fast fracture/instability.

3.5. Interactions of Chemical and Mechanical Degradation

A pronounced effect of membrane chemical degradation process is the mechanical weakening of the membrane (Fig. 18). The accelerated chemical aging can reduce the strain-to-failure to values far less than the yield strain of the membrane. Considering the equivalent strain levels estimated in a reasonable RH variation, the mechanical fracture of chemically degraded membrane is very likely, unless appropriate mitigation strategies are implemented. The drastic reduction of strain-to-break after chemical degradation suggests several potential operative degradation mechanisms occurring at the molecular level, such as chain scission, detangling, cross-linking, and crystallization. These changes have recently been suggested by Page et al.[39] to affect ionomer physical properties, such

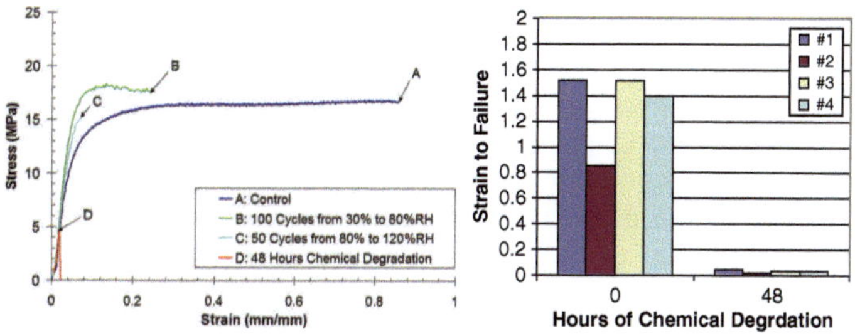

Figure 18. Stress–strain curves showing the effect of RH cycling and chemical degradation on the stress–strain behavior of the MEA (left) and drastic reduction of membrane strain-to-break as a result of chemical [reprinted with permission from reference 34].[34]

as ionic aggregation, glass transition temperature, and mechanical properties.

Experiments indicate that the critical strain-to-failure is also affected by the average molecular weight and by material processing history. McGrath[40] reported strain-to-break of a non-crystalline ionomer (a poly(arylene-ether) random copolymer, biphenyl sulfone in H form, or bi-phenyl sulfone in H form (BPSH)) is proportional to the length of the chain. We found in our laboratory that the casting procedure also affects the strain-to-break of the solution-cast ionomer film. As shown in Fig. 19, a Nafion film cast at near room

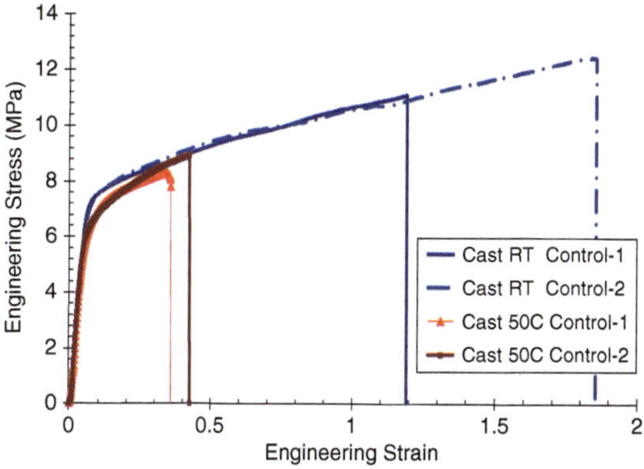

Figure 19. The strain-to-break of solution-cast Nafion (EW = 850) films cast at RH and 50°C.

temperature results in much larger strain-to-break than the film cast at 50°C (faster solvent evaporation). These findings may be rationalized as the entanglement densities may change as a function of molecular chain length and processing history. This suggests that chain shortening due to chemical attack and potential chain disentanglement due to RH cycling may contribute to the observed embrittlement of the ionomers. However, the degree of embrittlement reported/observed is not quite as severe as those observed in our OCV test.

To further probe the root cause of mechanical weakening/embrittlement under chemical degradation conditions, the authors have conducted Fenton degradation experiments on Nafion membranes. The Fenton test conditions (10 ppm Fe^{2+} and 3% H_2O_2 concentration) was controlled so that the membrane decay occurs uniformly throughout the bulk of the sample without forming blisters or other artifacts inside the membrane. Surprisingly, the authors found that the Fenton-tested membrane can be severely degraded (loss of 40% of weight and 30% of F account) but still maintain the same level of ductility, as shown in Fig. 20. The strain-to-break increased from approximately 250% to over 300% in case of Nafion 117 sample. The same Fenton test conducted on NRE-212 sample showed similar behavior. These results indicate that chemical decomposition of PFSA membrane does not necessarily reduce mechanical toughness of the membrane. On the other hand, in the constant current endurance test (Fig. 24), the membrane lost only 2–3% of total F account; however, its ductility has reduced by 87%. These results lead the authors to believe that the loss of mechanical toughness is mainly due to localized or inhomogeneous membrane decay.

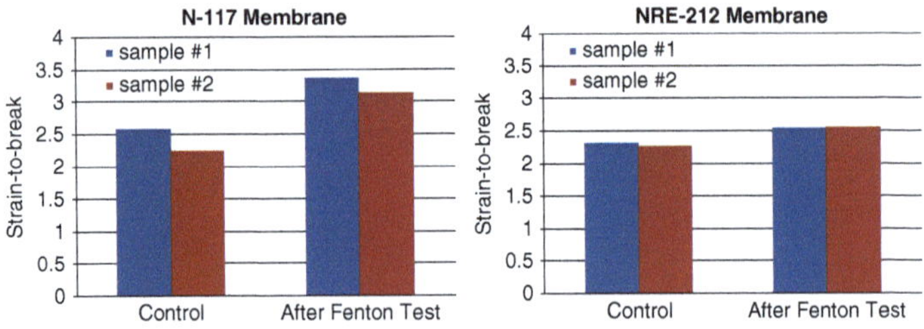

Figure 20. Strain-to-break for Nafion 117 (left) and NRE-212 (right) membrane samples before and after Fenton degradation test.

It is suspected that membrane decomposition occurs non-uniformly through the thickness direction of the MEA, as suggested by the SEM images shown in Fig. 8. Locally decomposed region in the membrane forms mechanical defects, which may be responsible for the drastic reduction of the mechanical toughness of the membrane. To confirm the hypothesis, the authors have recently conducted a two-layer membrane OCV (hydrogen/air) test. After OCV test, the two layers are separated and their mechanical properties are evaluated. The results (Figs. 21 and 22) show that the cathode-side membrane lost mechanical integrity much more than the anode-side membrane. This clearly indicated that the degradation of the membrane is non-uniform through the thickness direction of the membrane. The results suggest that mechanical defects due to membrane decomposition primarily form in one layer of the bi-layer MEA.

Much recent ongoing work has focused on the chemical degradation mechanisms and revealed that radical attack is the root cause of the membrane decomposition. However, there is still a lack of fundamental understanding of the mechanisms for the degradation of mechanical strength, which is related to the membrane failure mode responsible for the "sudden death" behavior of fuel cells.

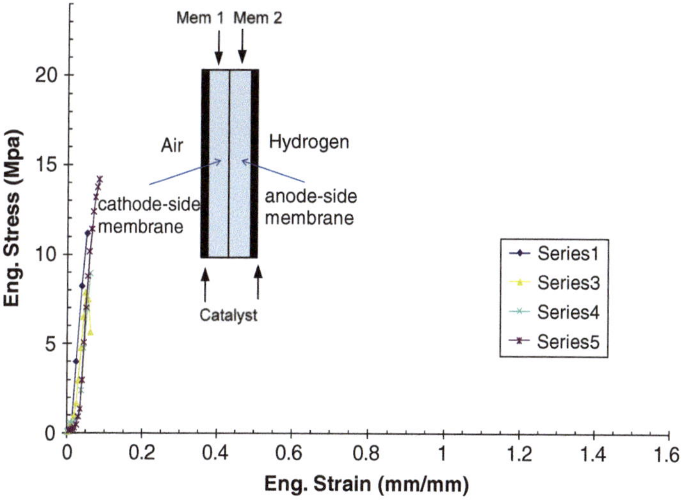

Figure 21. Residual mechanical properties of cathode-side membrane.

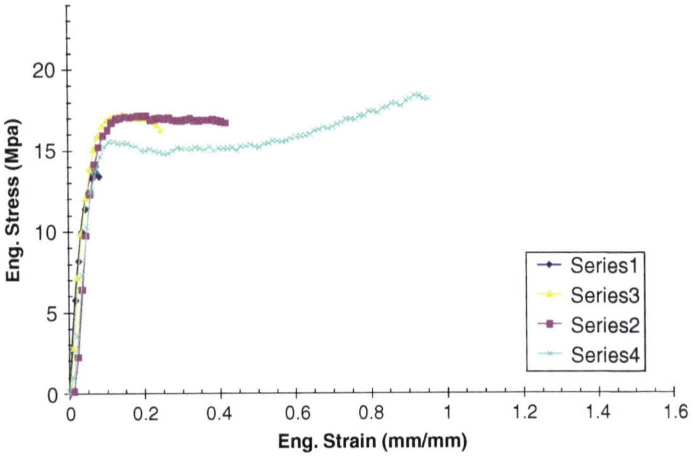

Figure 22. Residual mechanical properties of anode-side membrane.

Based on our observations, we generalize the fuel cell membrane degradation and failure mechanisms as the schematics in Fig. 23. So far, the evidence has shown that defects formation and growth play an important role both in chemical and in mechanical degradation processes. Drawing an analogy with material corrosion behavior,

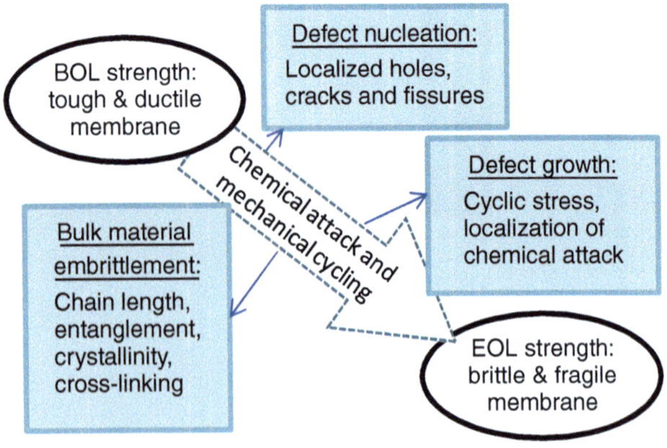

Figure 23. Chemical attack and cyclic mechanical stresses lead to fuel cell membrane mechanical failure.

our evidence and other literature reports have shown that the membrane degradation is somewhat comparable to pitting corrosion, or a mixed pitting and uniform corrosion, rather than uniform corrosion. There is still much to be understood about the fundamental mechanisms of the localized or inhomogeneous membrane degradation phenomena, which is believed to be the major cause of mechanical embrittlement of the membrane.

4. ACCELERATED TESTING AND LIFE PREDICTION

The degradation behavior of the PEM fuel cell is extremely complicated. It encompasses a set of convoluted physical, mechanical, chemical, and electrochemical processes. Several failure modes, such as performance decay due to cathode catalyst dissolution and sintering, carbon corrosion, and catastrophic failure due to membrane mechanical decay, have been studied and quantified to some extent, but clearly are not understood at the fundamental level. Nonuniform decay and damage localization generate another barrier for understanding the degradation and failure of the fuel cell stack. At stack-level, frequently, the stack failure is due to the breach of a single cell, and at cell-level, the failure is frequently due to the formation and propagation of a small number of defects at certain local regions, such as areas near the edge seal or gas inlets/outlets. Endurance testing of stacks is necessary to provide information on the location of weak spots in a given stack design and the relevant degradation and failure modes. Due to time and cost limitation, frequently accelerated test on subscale cells are used to reduce the amount of full-size stack tests. The results of these accelerated tests on sub-scale cells (ex situ study) need to be correlated to the full-size stack tests (in situ). Here we described a combined accelerated degradation test and modeling (mechanistic and statistical) approach to correlate ex situ test results of sub-scale cells to in situ degradation in the stack.

4.1. Accelerated Degradation Testing and Degradation Metrics

Separate Accelerated Stress Test (AST) protocols have been proposed by the US Department of Energy to assess the durability of fuel cell under conditions that promote distinct degradation modes dominated by catalyst dissolution, carbon corrosion, membrane chemical decomposition, and membrane mechanical degradation. In the real degradation process of fuel cell stack in service, these degradation modes are occurring simultaneously, especially under transient and cyclic conditions, such as startup-transient (fuel starvation), shut-down transient, load cycling (catalyst corrosion and membrane stress cycling), and partial power (high voltage and membrane swelling). It is necessary to understand the interactions of these degradation modes, such as mechanical stress and chemical degradation. Alternative AST protocols that induce interactions of the degradation modes are necessary.

A quantifiable degradation indicator or damage metric is needed to determine the acceleration factor and to enable phenomenological modeling of progressive degradation for life prediction purposes. Traditional fuel cell diagnostics techniques include voltage decay rate, fluoride emission (FE), hydrogen crossover rate, and the catalyst surface area (via cyclic voltammetry). Typical measurable quantities in these tests (cell current, voltage, hydrogen crossover, total FE, and electrochemical area) do not reflect the extent of local damage till the end of cell life. For example, the FE represents the overall amount of membrane material loss across the cell/stack planform, but may not accurately reflect the local material degradation, which depends on local temperature, hydration level, potential, etc. It has been shown that the material degradation of an MEA can be very different from one region to another. For Nafion membrane in an electrolyzer application, Stucki[41] reported a pattern of non-uniform thinning of the membrane in a PEM electrolyzer. The cell/stack failure will occur when the weakest spot in the MEA develops a pinhole or a tear. It has been reported that the FE can vary up to one order of magnitude within replicate tests, which demonstrated reproducible membrane life.[36] The voltage decay rate reflects the performance loss in a period of time; the voltage decay rate cannot be used to reliably predict the time it takes for the cell/stack to fail in a "sudden death" mode. A recent report[42] shows a negative voltage decay rate (i.e., stack voltage gradually increase!) over 1000 h test before the membrane failed due to excessive crossover.

Based on our experience, the strength of membrane/MEA, particularly strain-to-failure, measured ex situ has the potential to serve as a metric/indicator for PEMFC life prediction and accelerated testing when the failure mode is membrane crossover. It is an appropriate gauge of membrane integrity for the following reasons: (1) it is a relevant failure criteria for this life-limiting failure mode, (2) it is sensitive to material damage caused by both chemical degradation and RH cycling, (3) it is found to decrease gradually with number of RH cycles, and (4) it is sensitive to localized material degradation. The downside of using mechanical strength/toughness as a damage metrics is the unavoidable destructive testing to obtain the strength/toughness measurements.

Recently, a set of endurance tests on Nafion NR-111 membrane-based MEA have been carried out at the author's lab. The cells were run at a constant current density of 1000 mA/cm^2 under fully humidified conditions. The cell performance gradually decreased over the test period and failed catastrophically in the end. By testing multiple cells and terminating the endurance test at different running time, the author has observed a trend of steady decay of membrane strain-to-break, as shown in Fig. 24. Since regular performance diagnostic tests do not reveal any local properties of the membrane, this steady decay of membrane ductility was largely unnoticed and rarely reported in the literature. These test data suggest that the electrochemical performance degradation and the mechanical degradation of membrane are two parallel competing degradation mechanisms, the second one is likely responsible for the "sudden death" behavior of the cell/stack. The rate of mechanical decay may determine the useful life of PEM cells.

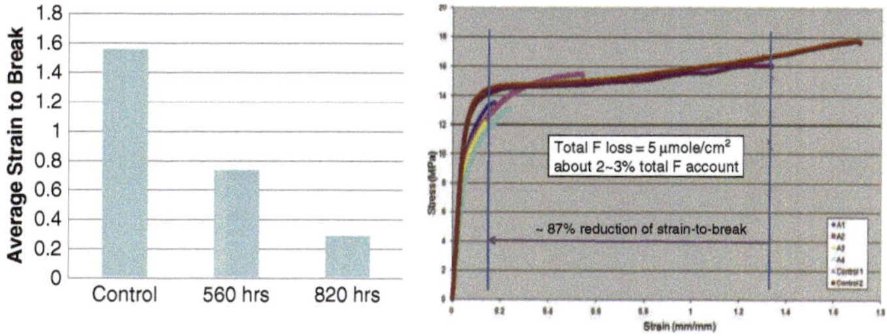

Figure 24. Mechanical decay of MEAs after endurance test at 1000 mA/cm^2; the trend of reduction of membrane ductility is shown on the left figure and the stress–strain behavior of the sample tested for 820 h and the control is shown on the right figure; the total fluorine (F) loss over 820 h accounts for 2–3% of total amount of F in the membrane.

We describe a phenomenological approach that can potentially synthesis our current understandings of various degradation modes and provide a meaningful and verifiable way to correlate ex situ ADT data to in situ real time degradation. With certain prior knowledge of the failure modes of a given plate-to-plate fuel cell package, we attempt to define a proper damage metric. The life of the fuel cell can be predicted by tracking the change of a damage metric which captures the material state change as multiple forms of degradation accumulate and evolve. With such a damage metric, it is possible to quantitatively study the coupling of different failure mechanisms and perform accelerated degradation test by tracking the change of the metric with accelerating parameters, i.e., "stresses" (frequency of certain transients, temperature, RH, and RH cycling), establish the rate of degradation as a function of applied "stress" and then extrapolate the degradation rate to normal "stress" levels for life prediction. A schematic of the concept is shown in Fig. 25.

Comparing to the accelerated life test where only the life distribution under a given "stress" level is recorded, the accelerated degradation test keeps track of the evolution of a damage metric (Fr), which allows the experimental validation of the correlation approach if the selected damage metric is observable. For example, the projected life distribution under normal "stress" can sometime

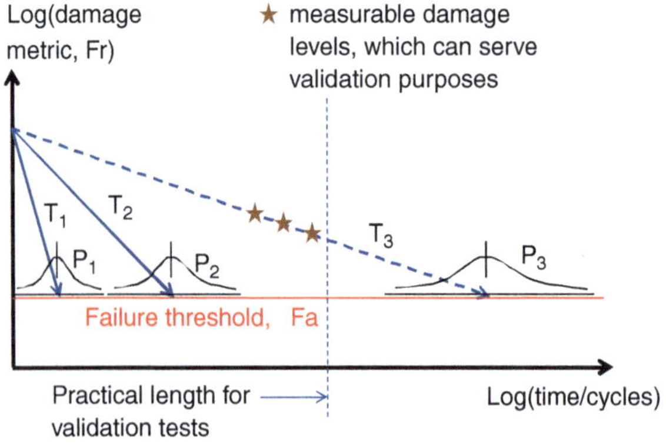

Figure 25. Concept of accelerated life and accelerated degradation test; T1 and T2 are two stress levels and T3 is the normal stress level; P1, P2, and P3 are the life distribution under stress level T1, T2, and T3.

be difficult to obtain due to practical and economical constraints. In accelerated degradation test, one can project the locus of the degradation metric as a function of operation history within a practical period of time and compare the experimentally measured damage metric to the projected value.

As explained, the degradation of fuel cell is driven by many local intensive variables, such as temperature, mechanical stress, humidity, voltage, current density, concentration of gas species, etc. In a real degradation process, such as a driving profile, the time history of these intensive variables will be stochastic in nature. A key question, and also a current knowledge gap, is how to use the ADT data to predict fuel cell life under realistic stochastic operating history. One approach is to take advantage of the inherent correlations among the intensive variables that drive fuel cell degradation. For example, the partial load (high voltage, low current, and low power) results in decreasing temperature and rising RH (water condensing and membrane swelling), and the full load (low voltage, high current, and high power) results in increasing temperature and lowering RH (membrane shrinking). Stack integrators typically collect such data set for their fleet of fuel cell engines; statistical theory and data mining method can help analyze these data and discover inherent correlations between variables. For example, multifactor dimensionality reduction (MDR) can detect and characterize the combination of these intensive variables that interact to influence a dependent variable. These findings can guide the design of alternative ADTs that can capture the degradation characteristic of fuel cells under realistic conditions. An alternative approach is to use a damage accumulation scheme to account for complex operation history of fuel cells. The applicability of any damage accumulation scheme needs to be experimentally validated; the next section describes a progressive degradation modeling approach that can potentially be used to combine multiple degradation effects, such as the RH cycling and the chemical degradation.

4.2. Progressive Degradation Model of Combined Effects

We will construct an example to demonstrate a progressive degradation (damage accumulation) modeling approach to combine multiple degradation effects. We consider chemical degradation due to, e.g., exposure to Fenton's reagent and mechanical degradation due to RH cycling. Hypothetical data of RH cycling and chemically induced membrane degradation will be used to illustrate how the two degradation effects can be combined and the approach we will use to estimate life under combined effects.

For example, assume a mechanical strength indicator, in this case the strain-to-failure of membrane, decreases from about 100% for the control samples to about 4% after chemical degradation. The exposure time is taken to be about 48 h. For this degradation mechanism, we assume that the degradation is linear with time as a first-order approximation and write the strain-to-failure (ε_f) as

$$\varepsilon_f(t) = (1 - D \cdot t)\,\frac{C}{C_0} \cdot \frac{T - T_g}{T_0} \tag{8}$$

where C_0 is the concentration of the Fenton reagent used for the present data, C is the concentration of the reagent for any other situation to be represented, T is temperature, T_g is the glass transition temperature of the membrane, T_0 is a material constant (all with units of °C), and t is time in hours. In this example, we take the last two ratios in Eq. (8) to have unit value. This equation represents the data as described, since at 48 h the strain-to-failure changes from 1 (or 100%) to 0.04 (or 4%) in the prescribed time of degradation; we set $D = 0.02$. With actual experimental data sets, we can establish the other constants, e.g., C_0 and T_0 in Eq. (8) and correct their dependence on temperature T, if needed. For now, we use it to represent the degradation kinetics of strain-to-failure as a result of chemical attack.

For the RH cycling effects, we use two hypothetical test data sets as follows:

1. After 100 cycles from 30% RH to 80% RH, the failure strain decreases to 40% (0.4).
2. After 50 cycles from 80% RH to 120% RH, the failure strain decreases to 34% (0.34).

Clearly, the number of RH cycling and the amplitude of the RH cycle (ΔRH) are important to this relationship, so we use the following representation:

$$\mathrm{Fa} = \frac{\mathrm{RH\,max}}{120} \cdot \frac{\Delta\varepsilon_{\mathrm{hydration}}}{\varepsilon_{\mathrm{f}}(t)} = A - B \cdot \log(N(t)) \tag{9}$$

where Fa is a failure function defined as the ratio of an applied condition to a material strength indicator $(\varepsilon_{\mathrm{f}})$, RHmax is the maximum value of RH used for the cyclic hydration test, 120 is the maximum %RH represented in the data, $\Delta\varepsilon_{\mathrm{hydration}}$ is the range of strain caused by the range of hydration used in the test (which is assumed to be 0.24 for case 1 and 0.192 for case 2), N is the life (cycles-to-failure) of the specimen, and A and B are material constants. By proper scaling of the parameters in the above representation, the failure function (Fa) can be set equal to 1 when failure occurs and less than 1 for general applied conditions. If the cycles-to-failure point is defined as the cycles required to reduce the strain-to-failure to the range of strain induced by the hydration cycles, then (from linear extrapolation) we estimate that $N = 126$ cycles-to-failure for data set 1 and 61 cycles-to-failure for data set 2. Then for the data discussed above, we find that $A = 0.374$ and $B = -0.102$.

To combine these two effects, we appeal to the damage mechanics formulation of Reifsnider and Case.[43] The remaining strength for any metric of strength, Fr, is written as

$$\mathrm{Fr}(t) = 1 - \int_0^t (1 - \mathrm{Fa}(t)) \cdot j \cdot \frac{a^{j-1}}{N(a)} \cdot \mathrm{fq}^j \cdot \left(1 - \frac{t \cdot 2.3 \cdot C}{B \cdot \varepsilon_{\mathrm{f}}^2(a)}\right) \mathrm{d}a \tag{10}$$

where Fa is defined by Eq. (8), fq is the frequency of strain cycling (here, in cycles per hour for convenience), and j is a material constant taken to be a typical value of 1.1 for this calculation. When the remaining value of the metric (left hand side of Eq. (10)) is just equal to the instantaneous value of the applied conditions (the current value of Fa(t) in the kernel of the integral), then the end-of-life is predicted at that instant, since the remaining "resistance" of the material to the applied conditions is just reduced to the current value of the applied conditions (e.g., material strength is reduced to the level of applied load, etc.).

We can check our representations by considering the two degradation processes separately, which can be done by changing the frequency of cycling (which applies to the hydration cycles). If the frequency of cycling is taken to be a very low value, then there is essentially no damage due to RH cycling, and one would expect the remaining strength to drop to the applied value (to cause failure) in somewhat less than 48 h, since primarily chemical attack is reducing the strength, as represented by Eq. (8).

The other extreme used as an input is cyclic hydration alone, with no chemical degradation. For that case, we can set the chemical degradation rate, *D*, in Eq. (8) to zero, so that no loss of strain-to-failure is occurring due to chemical attack. Then if we set the frequency of cyclic hydration to 2 cycles per hour and take the case 1 of our data (30% RH to 80% RH), then Eq. (10) produces the result shown below in Fig. 26, i.e., predicts failure in about 25 h at 50 cycles.

Finally, suppose we allow chemical attack and cyclic hydration. As shown in Fig. 26, the prediction is that failure would occur in about 7 h, which is about 14 cycles of hydration combined with 7 h of chemical degradation. Of course, this result can be checked in the laboratory. That is how this philosophy was validated in earlier publications for other combined effects.[44] It is also possible to evaluate what happens when one type of degradation (alone) is followed by another one in a sequence and to check to see if reversing the sequence changes the result.

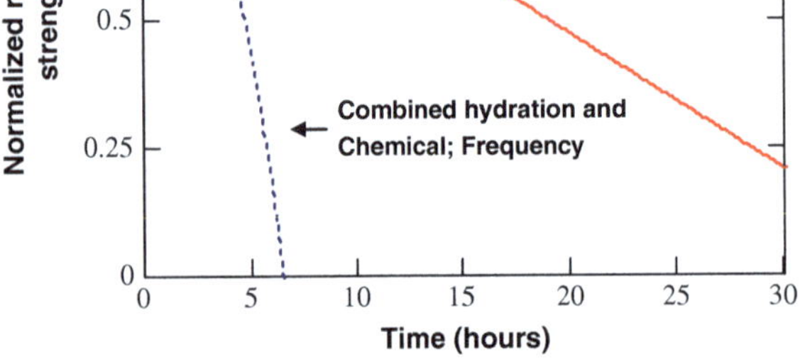

Figure 26. Normalized remaining strength for two conditions, cyclic hydration alone (top) and combined cyclic hydration and chemical attack (bottom).

This is an illustrative example with hypothetical inputs. Actual experimental data are needed to properly set the Eqs. (8) and (9), and it may be necessary to select a different strength metric (rather than the strain-to-failure used here). But the benefit of constructing such a model for estimating life, accelerated characterization, and conducting design studies without the cost and delay of trying all cases of interest has been shown to be great in the turbine and aircraft communities where this philosophy and methodology have previously been used successfully. In the present context, the example demonstrates a more general principle wherein a metric which defines the changes in a physical variable that defines the state of the material (e.g., strength, strain-to-break, conductivity, etc.) can be constructed in terms of the external variables that drive the processes that cause those changes. In general, that metric will depend on multiple variables, such as applied mechanical loading, temperature, chemical environments, etc. However, if the metric is constructed in such a way that those influences are correctly represented, as in the integral metric in Eq. (10), then it is possible to use the metric to correctly assess the effects of combined effects as a foundation for predictive simulations of future performance.

5. MITIGATION

Mitigation approaches to membrane degradation can be categorized as methods at material-level, cell- or device-level, system- and operation-level. Material-level mitigation approaches include the development of new and better membrane materials that show higher resistance to degradation. Cell-level approaches include the design and configuration of cell components (edge seal, gas diffusion media, flow field and flow plates, etc.) that closely interacts with MEA. At system- and operation-level, the proper control of the cell/stack operation conditions, such as temperature, humidity, liquid water removal, potential and potential cycling, etc., can also greatly benefit the durability of fuel cell membrane. A recent review article[45] briefly discussed the development of various new membranes that are more resistant to degradation. Most of the effective membrane degradation mitigation methods, particularly those at device- and system-level, are either published in the patent literatures or remain as proprietary knowhow of the leading industrial developers. In the following text, representative examples of the mitigation approaches will be discussed. These are not intended to be a comprehensive survey of currently available mitigation approaches.

It has been reported that certain additives into PFSA membrane, such as Ce^{+}, Mn^{+}, CeO_2, and MnO_2 can greatly reduce FE rate at OCV conditions.[44,46,47] These transition metals in ionic or oxide form were found to have the capability of quenching radicals generated inside membrane.[48] Membrane structural reinforcement, such as expanded PTFE, has shown benefits to mechanical strength and resistance to both OCV aging[49] and RH cycling. Gore reported the improvement to RH cycling life by reinforcing PFSA type membrane.[36]

At the device-level, cell design can help improve the durability of PEM fuel cell. Particularly the edge and the edge seal configuration of the cell have been found to influence the local membrane degradation behavior. Evidence of highly localized chemical decomposition occurring near the edge of the MEA and within the MEA have been reported by Ohma et al.[26] Sompalli et al.[50] recently published a paper discussing membrane degradation at catalyst edge in PEMFC. Their results show a fast failure of membrane in the vicinity of the electrode perimeter in the case that cathode electrocatalyst layer extends over the anode electrocatalyst layer, i.e., cathode overlap. The cathode overlap causes a local potential redistribution of membrane, which results in the high potential of cathode catalyst layer relative to the electrolyte membrane potential. This leads to O_2 evolution and carbon corrosion process unintentionally in the cathode side, causing membrane thinning and failure eventually. Gas diffusion media that come into close interaction with membrane can also impact the durability of the MEA. Ballard has patented a technique of abrading the GDM, which has been found to reduce the likelihood of pinhole formation.[51]

It has been shown that temperature, water management, potential/potential cycling, and contaminants can impact the degradation behavior of the membrane. While higher temperature and lower humidity operation of PEM fuel cell is highly desirable for a number of system benefits, such as improved heat rejection efficiency and better tolerance to certain contaminants. Major efforts have been focused on the development of membrane with adequate proton conductivity. Both high cell operating temperature and low humidity operation tend to drastically accelerate membrane degradation. It is equally important to ensure the membrane being developed has adequate resistance to chemical and mechanical degradation.

High-potential hold and potential cycling are detrimental to electrocatalyst and catalyst support via platinum dissolution[6] and carbon corrosion, which can also accelerate membrane degradation. It is believed that the localized membrane decay is a result of a chain of self-amplifying events that start with local flooding in the gas diffusion media and eventually leads to locally reduced hydrophobicity that reinforces and stabilizes the local flooding. This chain of events includes local air or fuel starvation, perturbated membrane electrochemical potential distribution which leads to locally accelerated carbon corrosion and Pt dissolution, and inhomogeneous membrane decay. Improving the stability of the Pt-based catalysts in the electrodes should also benefit the endurance of the PEM. A recent discovery[52] showed that small gold clusters formed on the surface of Pt catalyst particles have a surprisingly pronounced stabilizing effect against Pt dissolution under potential cycling conditions. At the system level, it is possible to eliminate or at least to reduce potential cycling through system-level control and/or hybridization techniques with batteries or super capacitors. System level cell potential control has been found effective in reducing carbon corrosion and platinum dissolution.[3]

The author[53] has found that water management scheme can greatly affect the membrane decay modes: homogeneous/uniform or localized/inhomogeneous. In an endurance test conducted at constant current density (1000 mA/cm^2) using a solid-plate cell hardware, highly non-uniform membrane thinning was observed in the MEA in the post-mortem analysis. However, the same type of MEA tested at the same running conditions using a cell hardware with an active liquid water removal technology[54] did not show highly localized membrane degradation. Uniform membrane hydration tends to preserve the membrane mechanical properties better.

The most effective mitigation approach is likely a combined system-, device- and material-level approach for a particular application. Fundamental understanding of the degradation mechanisms of the PEM at the nano- and microscale is essential to provide a linkage of the key processes occurring at that scale to the device- and system-level engineering strategies. It will provide the most probable pathway to break the current technical barriers in the PEM fuel cell development.

ACKNOWLEDGMENTS

In the past 7 years or so, our work on the membrane degradation has been sponsored by National Science Foundation (CMS-0408807, CBET-0829082), UTC Power and UTRC, and Connecticut Innovation's Yankee Ingenuity Program. A number of current and former students have contributed to various extents in carrying out the experimental and analytical work. The authors wish to thank Matthew Feshler, Yue Zou, Roham Solasi, Hongying Zhao, Xiaofeng Wang, Wonseok Yoon, and William Rigdon.

REFERENCES

[1] R. Borup et al., Chem. Rev. **107**, 3904 (2007)
[2] T.W. Patterson et al., US Patent 7, **442**, 453 B1, (2008)
[3] T.A. Bekkedahl et al., US Patent 6, **913**, 845, (2004)
[4] A. Pozio et al., Electrochem. Acta **48**, 1543, (2003)
[5] A. Ohma, S. Yamamoto, and K. Shinohara, J. Power Sources **182**(1), 39 (2008)
[6] P.J. Ferreira et al., J. Electrochem. Soc. **152**, A2256 (2005)
[7] K. Yasuda et al., Z. Phys. Chem. Chem. Phys. **8**, 746 (2006)
[8] E. Guilminot et al., J. Electrochem. Soc. **154**, B1106 (2007)
[9] C. Iojoiu et al., J. Electrochem. Soc. **154**, B1115 (2007)
[10] K. Matsuoka et al., J. Power Sources **179**, 560 (2008)
[11] S. Sugawara et al., J. Power Sources, **187**, 324 (2009)
[12] T.W. Patterson, R.M. Darling, Electrochem Solid-State Lett **9**, A183 (2006)
[13] G. Gebel, P. Aldebert, M. Pineri, Polymer **34**, 333 (1993)
[14] G. Gebel, Polymer **41**, 5829 (2000)
[15] R. Baldwin et al., J. Power Sources **29**, 399 (1990)
[16] A. Bosnjakovic, S. Schlick, J. Phys. Chem. B **108**, 4332 (2004)
[17] M.L. Kremer, J. Phys. Chem. A. **107**, 1734 (2003)
[18] V.O. Mittal, H.R. Kunz, J.M. Fenton, Electrochem Solid-State Lett **9**(6), A299 (2006)
[19] V.O. Mittal, PhD Dissertation, Chemical Engineering Department, University of Connecticut: Storrs, Connecticut, 2006
[20] M. Aoki, H. Uchida, M. Watanabe, Electrochem. Comm. **7**, 1434 (2005)
[21] M. Watanabe et al., J. Electrochem. Soc. **143**(12), 3847 (1996)
[22] M. Watanabe, H. Uchida, M. Emori, J. Electrochem. Soc. **145**(4), 1137 (1998)
[23] D.E. Curtin et al., J. Power Sources **131**, 41 (2004)
[24] C. Zhou, M.A. Guerra, Z.M. Qiu, etc., Macromolecules, **40**, 8695 (2007)
[25] G. Escobedo, Project ID# FC5, 2006 DOE Hydrogen Program Merit Review, available in CD and online.
[26] M. Hicks, Project ID# FC8, 2006 DOE Hydrogen Program Merit Review, available in CD and online.
[27] V.O. Mittal, H.R. Kunz, J.M. Fenton, ECS Trans. **1**, 275 (2006)
[28] N.E. Cipollini, ECS Trans. **11**, 1071 (2007)
[29] F.D. Coms, ECS Trans. **16**, 235 (2008)

[30] E. Endoh et al., Electrochem. Solid-State Lett. **7**, A209 (2004)
[31] M. Inaba et al., Electrochim. Acta **51**, 5746 (2006)
[32] M. Quintus et al., Chemical membrane degradation in automotive fuel cell – Mechanisms and mitigation, *2nd Annual International Symposium on Fuel Cell Durability & Performance*, Miami Beach, FL,7–8 Dec 2006
[33] W. Liu, K. Ruth, G. Rusch, J. New Mater. Electrochem. Syst. **4**, 227 (2001)
[34] X. Huang et al., J. Polym. Sci., Part B: Polym. Phys. **44**, 2346 (2006)
[35] D.R., Morris, X. Sun, J Appl Polym Sci. **50**, 1445 (1993)
[36] Y. Zou, Master Thesis, Mechanical Engineering Department, University of Connecticut, 2007
[37] P.B. Bowden, R.J. Oxborough, Phil. Mag. **28**, 547 (1973)
[38] R. Solasi et al., Mechanical response of 3-layreed MEA during RH and T variation based on mechanical properties measured under controlled T and RH, *Proceedings of 4th International Conference on Fuel Cell Science, Engineering and Technology*, Irvine, CA, 19–21 June 2006
[39] K.A. Page, K.M., Cable, R.B., Moore, Macromolecules **38**, 6472 (2005)
[40] J.E., McGrath, Advanced materials for proton exchange membranes, DOE Hydrogen Program Merit Review Presentation, 2006
[41] S. Stucki, G.G. Scherer, J. Appl. Electrochem. **28**, 1041 (1998)
[42] R. Abouatallah, Hydrogenics' fuel cell stack durability at non-humidified conditions, *2nd Annual International Symposium on Fuel Cell Durability & Performance*, Miami Beach, FL,7–8 Dec 2006.
[43] K.L. Reifsnider S. Case, *Damage Tolerance and Durability of Material Systems* (Wiley, New York, NY, 2003)
[44] E. Endoh, ECS Trans. **3**, 9 (2006)
[45] J. Wu et al., J. Power Sources **184**(1), 104 (2008)
[46] E. Endoh, US Patent Application 0111076, (2007)
[47] E. Endoh, US Patent Application 0104994, (2007)
[48] E. Endoh, ECS Trans **16**, 1229 (2008)
[49] X. Huang et al., ECS Trans. **16**, 1573 (2008)
[50] B. Sompalli et al., J. Electrochem. Soc. **154**, B1349 (2007)
[51] V. Peinecke et al. US Patent 6,716,551, (2004)
[52] J. Zhang et al., Science **315**(5809), 220 (2007)
[53] X. Huang et al., ECS Trans. **16**, 1697 (2008)
[54] A.Z. Weber, R.M. Darling, J. Power Sources **168**(1), 191 (2007)

2

Modeling of Membrane-Electrode-Assembly Degradation in Proton-Exchange-Membrane Fuel Cells – Local H_2 Starvation and Start–Stop Induced Carbon-Support Corrosion

Wenbin Gu,[1] Paul T. Yu,[1] Robert N. Carter,[1] Rohit Makharia,[1] and Hubert A. Gasteiger[2]

[1] *General Motors Research and Development, Electrochemical Energy Research Laboratory, 10 Carriage Street, Honeoye Falls, NY 14472, USA*
[2] *Department of Chemistry, Technische Universität München, Lichtenbergstrasse 4, D-85747, Garching, Germany*

1. INTRODUCTION

Hydrogen-powered fuel cell vehicles are perceived as the future of transportation to replace current gasoline-fueled automobiles.[1] Proton-exchange-membrane (PEM) fuel cells, operating at relatively low temperatures with zero emission and high efficiency, are being developed to make automotive fuel cell power systems competitive to current internal-combustion engines. As a durable membrane-electrode-assembly (MEA) is essential for hydrogen-powered fuel cell vehicles to be commercially viable, understanding MEA degradation mechanisms and improving MEA durability are increasingly critical to the successful commercialization of fuel cell vehicles. There are many factors influencing MEA durability.[1–7] Electrode degradation is one of the most challenging

U. Pasaogullari, C.-Y. Wang (eds.), *Modeling and Diagnostics of Polymer Electrolyte Fuel Cells*, Modern Aspects of Electrochemistry 49,
DOI 10.1007/978-0-387-98068-3_2,

durability issues for automotive applications that involve dynamic driving cycles with frequent load changes and start–stop events. An electrode is normally composed of platinum (Pt) catalyst supported by high surface-area carbon, such as Vulcan or Ketjen Black (KB). Cathode carbon-support corrosion, which can occur under certain automotive operating conditions, causes electrode structure damage and thus electrode degradation;[8–13] and therefore it is of great concern to MEA durability.

Figure 1 shows the schematic of a PEM fuel cell where there is an H_2/O_2 front dividing the cell into two portions: one is H_2-rich and generates electric power as a power source, and the other is H_2-free and becomes a load driven by the power source. The existence of the H_2/O_2 front results from two scenarios: (a) start–stop events frequently seen in automotive application and (b) local H_2 starvation caused by local blockage of the H_2 fuel supply, e.g., part of anode flow-field is filled with liquid water. As shown in Fig. 1,

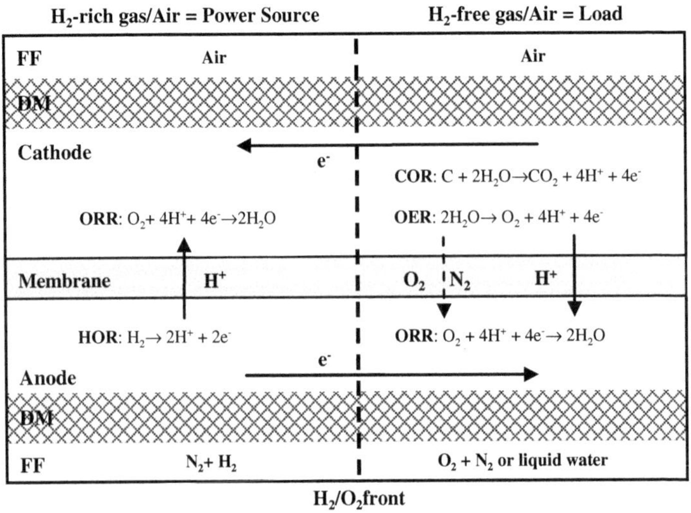

Figure 1. The schematic of a PEMFC having the H_2/O_2 front in the anode and major electrochemical reactions considered in the analysis. The H_2/O_2 front divides the fuel cell into the power source and the load. In the load portion, anode flow-field (FF) is occupied by air (or O_2+N_2) in the case of start–stop, while it is filled with liquid water in the case of local H_2 starvation.

normal electrode reactions, HOR (hydrogen oxidation reaction) and ORR (oxygen reduction reaction), occur at the anode and cathode of the H_2-rich region, respectively. Protons are transported from the anode to the cathode. In the H_2-free region, however, ORR takes place at the anode, and to get the needed protons it induces OER (oxygen evolution reaction) and COR (carbon oxidation reaction or carbon corrosion) at the cathode, resulting in a so-called *reverse current* of protons,[8] i.e., protons flowing from the cathode to the anode side. Electrons are transported between H_2-rich and H_2-free regions via an internal path at open-circuit voltage (in the case of a normal start–stop process) and via an external path when a load is applied to the fuel cell (in the case of local H_2 starvation). During a start process, hydrogen gas flows into the anode which initially was filled with air. During an extended stop process, air flows into the anode that had been filled with hydrogen gas. In both cases, an H_2/air-front passes through the anode flow-field, inducing carbon-support corrosion at the cathode where oxygen is present on the corresponding anode side. Start–stop events, where H_2/air-fronts move through the anode flow-field with residence times in the order of 0.1–1 s, lead to cathode carbon-support corrosion and rapid MEA performance degradation.[8,10,12] The analogous effect occurs when H_2 access is locally blocked by water droplets or anode diffusion medium flooding, leading to a buildup of N_2 and O_2 via permeation through the membrane from the cathode to the anode side ("crossover"), and thus leading to localized cathode carbon-support corrosion referred to as local H_2 starvation.[9,11,14]

Much effort has been made to understand carbon-support corrosion mechanisms in recent years. Experiments to simulate start–stop events and local H_2 starvation were conducted with a single cell having two segmented gas flows on the anode as well as with two cells electrically connected in parallel based on the reverse-current mechanism[8] to observe fuel cell performance decay associated with loss of electrochemically active area and electrode thinning.[8,10] To quantify the rate of carbon corrosion, CO_2 (and CO) evolution were monitored at the cathode exit,[15–19] and in-situ electrode potential measurements were made with a reference electrode.[17,20,21] It has been found that H_2/air-front-induced voltage degradation during start–stop events can be correlated to the extent of cathode carbon-support weight loss normalized to carbon loading, independent of whether conventional or graphitized carbon-supports are used.[11,12]

Consequently, since graphitized carbon-supports have lower carbon corrosion rates, the use of cathode catalysts with graphitized supports significantly reduces H_2/air-front start–stop damage.[12,22] Furthermore, if the ORR activity of the anode electrode is reduced by lowering anode Pt loading, H_2/air-front start–stop degradation is decreased.[22,23]

Mathematical models have been developed to understand experimental observations and to further explore mitigation methods.[8,12,14,15,24–32] The complexity of the models varies with model application and physics incorporated. For example, a one-dimensional model for the electrolyte potential profile along the gas-flow direction was developed to explain the reverse-current mechanism.[8] Hydrogen and oxygen transport were incorporated to describe how cathode carbon corrosion can be induced by a maldistribution of H_2 across the anode.[24] A two-dimensional, steady-state model for the electrolyte potential distribution along the gas-flow direction and across MEA thickness was developed to investigate the effect of cell voltage, ionomer proton conductivity, H_2-starved region size, carbon corrosion kinetics, and oxygen permeability through the membrane on the carbon corrosion rate.[25] Later, the model was extended to simulate a transient process with quasi-steady-state assumption and investigate the influence of MEA design parameters, such as ORR exchange current density on the anode and the thickness of the membrane.[26,27] CFD (Computational-Fluid-Dynamics) based models have also been developed with electrode kinetics coupled with transport for two-dimensional, steady-state simulation,[28,29] and two- and three-dimensional, transient simulations.[30] Most recently, attempts were made to couple carbon-support corrosion with voltage degradation via multi-scale modeling.[31,32]

In this chapter, we will review the fundamental models that we developed to predict cathode carbon-support corrosion induced by local H_2 starvation and start–stop in a PEM fuel cell, and show how we used them to understand experiments and provide guidelines for developing strategies to mitigate carbon corrosion. We will discuss the kinetic model,[12] coupled kinetic and transport model,[14] and pseudo-capacitance model[15] sequentially in the three sections that follow. Given the measured electrode kinetics for the electrochemical reactions appearing in Fig. 1, we will describe a model, compare the model results with available experimental data, and then present the implications of the model predictions. Finally, we will summarize the applicability and usefulness of each model and provide an outlook on further model improvement for better fundamental understanding of MEA degradation associated with carbon-support corrosion.

2. KINETIC MODEL

2.1. Electrode Kinetics

A kinetic model is built based on the electrode kinetics of all electrochemical reactions involved in Fig. 1. We have measured the kinetics of HOR[33] and ORR[34] for the power source, as well as the kinetics of OER[35] and COR[12] for the load. The electrode reaction currents are governed by

$$i_{\mathrm{HOR}} = a_{\mathrm{Pt}} L_{\mathrm{Pt}} i_{o,\mathrm{HOR}(T,p_{\mathrm{H_2}})} \frac{2F}{RT} \eta_{\mathrm{HOR}} \tag{1}$$

$$i_{\mathrm{ORR}} = a_{\mathrm{Pt}} L_{\mathrm{Pt}} i_{o,\mathrm{ORR}(T,p_{\mathrm{O_2}})} \exp\left(-\frac{\alpha_{c,\mathrm{ORR}} F}{RT} \eta_{\mathrm{ORR}}\right) \tag{2}$$

$$i_{\mathrm{OER}} = a_{\mathrm{Pt}} L_{\mathrm{Pt}} i_{o,\mathrm{OER}(T,p_{\mathrm{H_2O}})} \exp\left(\frac{\alpha_{a,\mathrm{OER}} F}{RT} \eta_{\mathrm{OER}}\right) \tag{3}$$

and

$$i_{\mathrm{COR}} = a_{\mathrm{C}} L_{\mathrm{C}} (1-\theta)^m i_{o,\mathrm{COR}(T,p_{\mathrm{H_2O}})} \exp\left(\frac{\alpha_{a,\mathrm{COR}} F}{RT} \eta_{\mathrm{COR}}\right) \tag{4}$$

where a and L are the electrochemically active surface area (in $\mathrm{cm^2/mg}$) and loading of an ingredient in an electrode (in $\mathrm{mg/cm^2}$), with subscripts Pt and C representing platinum and carbon, respectively. L_{C} relates to L_{Pt} via $r_{\mathrm{Pt/C}}$, the weight ratio of platinum over carbon used to fabricate C-supported Pt catalyst. θ is the mass fraction of carbon support that has been lost. R is the universal gas constant (8.314 J/mol/K), F is the Faraday constant (96487 C/equiv), and T is temperature (in Kelvin). The reaction-specific exchange current density, $i_{o,j(T,p_i)}$ in units of $\mathrm{A/cm^2}_{\mathrm{Pt\ or\ C}}$, is dependent on temperature and gaseous reactant partial pressure:

$$i_{o,j(T,p_i)} = i_{o,j}^{\text{ref}} \cdot \left(\frac{p_i}{p_i^{\text{ref}}}\right)^{\gamma_i} \exp\left[-\frac{E_j^{\text{rev}}}{RT}\left(1 - \frac{T}{T^{\text{ref}}}\right)\right] \tag{5}$$

where $i_{o,j}^{\text{ref}}$ is the specific exchange current density normalized to the reference gaseous reactant partial pressure (p_i^{ref}) of 101.3 kPa and a reference temperature (T^{ref}) of 353 K (80 °C), γ_i is the apparent reaction order of species i in an electrode reaction, and E_j^{rev} is the activation energy of the electrode reaction j at the reversible cell potential (zero overpotential).

In Eqs. (1–4), the surface overpotential (driving force for an electrode reaction) is defined as

$$\eta_j = \phi_s - \phi_e - E_{j(T)}^o \tag{6}$$

where j = HOR, ORR, OER, and COR. ϕ stands for the electrical potential of a phase, with subscript s and e representing the electron-conducting solid phase and the proton-conducting electrolyte (or ionomer) phase, respectively. $E_{j(T)}^o$ is the standard equilibrium potential of an electrode reaction versus RHE (reversible hydrogen electrode):[36]

$$E_{j(T)}^o = E_{j(298.15\text{K})}^o + \frac{dE_j^o}{dT}(T - 298.15) \tag{7}$$

As COR and OER occur simultaneously in the cathode, their kinetics are particularly important in evaluating carbon-support corrosion. The kinetics of OER is material-specific, dependent on catalyst composition and electrode fabrication.[35,37–39] A number of OER kinetics studies were done on Pt metal electrodes.[37–39] However, there is a lack of OER kinetics data on electrodes made of Pt nano-particles dispersed on carbon supports. Figure 2 shows the measured OER current density with respect to the overpotential defined by Eq. (6).[35] The O_2 concentration was measured at the exit of a 50-cm^2 cell using a gas chromatograph (GC). The O_2 evolution rate (= O_2 concentration × cathode flow rate) was then converted to the OER current density, assuming 4e$^-$/O_2 molecule. Diluted H_2 (10%) and a thicker membrane (50 μm) were used in the measurement to minimize H_2 crossover from anode to cathode, because H_2 would react with O_2 evolved at the cathode and incur inaccuracy in the measured OER current density. Figure 2 indicates that the OER

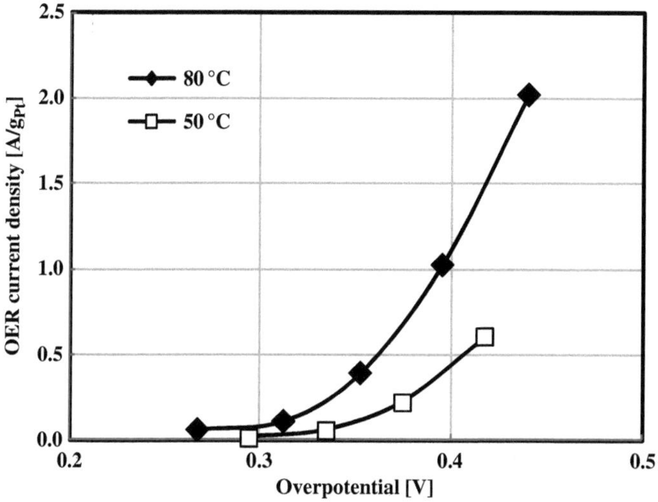

Figure 2. Oxygen evolution reaction current density (in units of A/g_{Pt}) with respect to the overpotential applied to the cathode electrode. The OER current density is based on the measured O_2 concentration at the exit of a 50-cm^2 cell using a GC, assuming $4e^-/O_2$ molecule. The cell is operating with 10% H_2/He (150 kPa_{abs}, varying T and RH) at various potentiostatic conditions.

current density follows the Tafel equation and is a strong function of temperature. The OER kinetic parameter values were determined by fitting the OER current densities to Eq. (3).[35]

The carbon corrosion rate depends on the potential, temperature, and time for a specific carbon material.[1] Figure 3 displays carbon corrosion rate versus carbon weight loss based on the measured CO_2 concentration at the exit of a 50-cm^2 cell using a GC when the cell was held at 1.2 V versus RHE.[12] It shows that the corrosion rate of the conventional carbon support (KB) decreases with time (reflected by the carbon weight loss), whereas the corrosion rate of the graphitized carbon support (Gr-KB) remains nearly constant. The time dependence of the carbon-support corrosion rate is most likely related to the inhomogeneous graphitization of commonly used carbon supports, which generally consist of primary particles with diminishing graphitic order toward the particle center.[22] As depicted in Fig. 3, a carbon particle is made up of an amorphous core with a graphene-like shell.[40] Since the corrosion

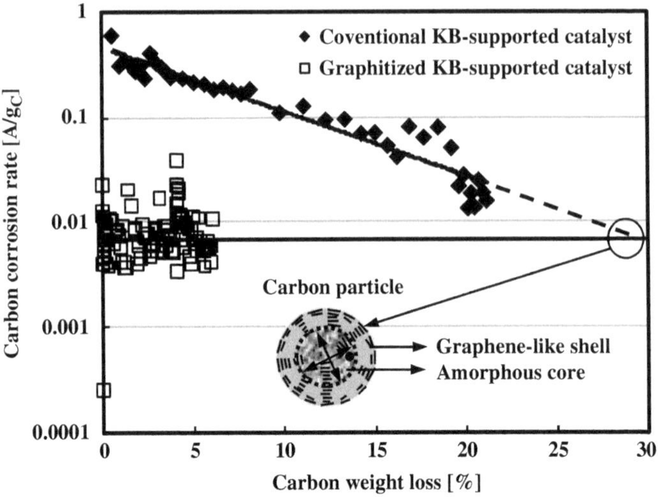

Figure 3. Carbon corrosion rate versus carbon weight loss for both conventional and graphitized KB-supported Pt catalysts. The carbon corrosion rate (in units of A/g_C) is based on the measured CO_2 concentration at the exit of a 50 cm^2 cell using a GC, assuming $4e^-/CO_2$ molecule. The carbon weight loss is obtained by integrating the measured CO_2 evolution rate over time. The cell is operating on neat H_2/N_2 (95 °C, 80% RH_{in}, and 120 kPa_{abs}) with potential held at 1.2 Volts versus RHE.

resistance of carbon within a graphitic plane is approximately one to two orders of magnitude higher than that of carbon along the edges of graphitic planes,[41] the corrosion of conventional carbon supports like KB should begin in the center region of the primary carbon-black particles (high density of graphitic plane edge sites and/or amorphous carbon), and progress over time toward the more graphene-like (and thus more corrosion resistant) shell region that has a lower density of graphitic plane edge sites.[22] The most recent electron microscopy observations of the centered-hollowed carbon particles and interconnected voids concluded that carbon corrosion indeed takes place in an inside-out mode, with the carbon aggregate, rather than the carbon primary particle, as a basic corrosion unit.[13] In Eq. (4), the time dependence is taken into account via the index m by fitting to the measured carbon corrosion rates.

All thermodynamic and kinetic parameter values needed in calculating the electrode reaction current densities are listed in Table 1.

Table 1
Thermodynamic and Kinetic Parameter Values. Values of Standard Electrode Potential (at 25 °C) and Exchange Current Density (at 80 °C) are Referenced to 101.3 kPa Gaseous Reactant Partial Pressure.

Parameter	Unit	Value
HOR		
Standard electrode potential	V	0
Temperature coefficient	V/K	0
Exchange current density	$A/cm^2{}_{Pt}$	0.3
Activation energy at zero overpotential	J/mol	10×10^3
Hydrogen reaction order		1
ORR		
Standard electrode potential	V	1.2291
Temperature coefficient	V/K	-0.8456×10^{-3}
Exchange current density	$A/cm^2{}_{Pt}$	2.47×10^{-8}
Activation energy at zero overpotential	J/mol	67×10^3
Cathodic transfer coefficient		1
Oxygen reaction order		0.79
OER		
Standard electrode potential	V	1.2291
Temperature coefficient	V/K	-0.8456×10^{-3}
Exchange current density	$A/cm^2{}_{Pt}$	1.89×10^{-9}
Activation energy at zero overpotential	J/mol	38×10^3
Anodic transfer coefficient		0.65
Water vapor reaction order		0.88
COR		Vulcan/Gr-Vulcan
Standard electrode potential	V	0.2073
Temperature coefficient	V/K	-0.8530×10^{-3}
Exchange current density	$A/cm^2{}_{C}$	$1.03\times10^{-18}/2.17\times10^{-19}$
Activation energy at zero overpotential	J/mol	$134\times10^3/140\times10^3$
Anodic transfer coefficient		0.67/0.65
Index for carbon weight loss dependence		10.4/0

The standard electrode potential and its temperature coefficient are found in the literature.[36] Kinetic parameter values were measured in-house for HOR,[33] ORR,[34] OER,[35] and COR.[12,22] Table 2 gives cell component materials and transport properties. The membrane and electrode proton conductivity in Table 2 are based on the measured membrane and electrode resistance,[42,43] which is a strong function of relative humidity (RH). In what follows next, we will describe the

Table 2
Cell Component Materials Properties.

Parameter	Unit	Value/equation
Geometric		
Membrane thickness	cm	18×10^{-4}
Electrode thickness ($r_{Pt/C} = 1$)	cm/(mg$_{Pt}$/cm^2)	30×10^{-4}
Electrode Pt loading (anode/cathode)	mg$_{Pt}$/cm^2	0.4/0.4
DM thickness (under compression)	cm	150×10^{-4}
Electrode porosity		0.5
Electrode pore diameter	cm	50×10^{-7}
DM porosity (under compression)		0.6
DM tortuosity		2.4
Specific Pt surface area	cm$^2{}_{Pt}$/mg$_{Pt}$	600
Specific carbon surface area (Vulcan/Gr-Vulcan)	cm$^2{}_{C}$/mg$_C$	2400/800
Transport		
Membrane proton conductivity at 80 °C	S/cm	$1.84\times10^{-5}\times(\text{RH}\%)^{1.84}$
Electrode proton conductivity at 80 °C	S/cm	$4.66\times10^{-8}\times(\text{RH}\%)^{2.84}$
Activation energy for proton conduction	J/mol	6×10^3
Electrode electrical conductivity	S/cm	1
DM electrical conductivity	S/cm	100
FF electrical conductivity	S/cm	10^4
DM/FF electrical contact resistance	Ω cm^2	0.03
Electrode permeability	cm^2	7×10^{-14}
DM permeability	cm^2	10^{-8}

kinetic models for local H_2 starvation and start–stop, respectively, and use the models to evaluate carbon corrosion rate and compare the model results with available experimental data. Furthermore, we will discuss the implications of the model predictions.

2.2. Local H_2 Starvation Model

In the fully developed H_2 starvation region, where H_2 is completely depleted, the cathode electrode potential reaches its maximum value as determined by the mixed potential of all electrochemical reactions involved at the cathode, namely, COR and OER. The reaction current therein is determined by O_2 crossover (i_{x,O_2}) from cathode through membrane to anode, where only ORR takes place in the absence of H_2. Charge balance yields

$$i_{x,O_2} = i_{COR,cath} + i_{OER,cath} \tag{8}$$

Equation (8) states the fact that the sum of carbon corrosion current ($i_{CO_2} = i_{COR,cath}$) and oxygen evolution current ($i_{O_2} = i_{OER,cath}$) on the cathode must be equal to the oxygen reduction current on the anode determined by the O_2 crossover rate:

$$i_{x,O_2} = 4FK^m_{O_2}\frac{p_{O_2,cath}}{\delta_{mem}} \tag{9}$$

Here, the gas permeability through the membrane is calculated by the semiempirical correlation[44]

$$K^m = k^{dry}\exp\left(-\frac{E^{dry}}{RT}\right) + \nu_{H_2O}k^{wet}\exp\left(-\frac{E^{wet}}{RT}\right) \tag{10}$$

where ν_{H_2O} is the volume fraction of water, which is dependent on membrane equivalent weight, density (at dry state), and water content. k and E are the Arrhenius pre-exponential and activation energy (with superscripts dry and wet denoting the phase state). The values of the various constants can be found in the reference[44] for each species of interest.

Figure 4 plots polarization curves of COR and OER, together with O_2 crossover current density that controls the total amount of current at the cathode based on the local H_2 starvation model, Eq. (8). The cell operates at 80 °C, 150 kPa_{abs}, 100% RH_{in}, and 2/2 stoichiometric flows using neat H_2/air. Under these conditions, the O_2 crossover current through the membrane is 0.48 mA/cm^2 based on measured membrane permeability.[44] Figure 4 illustrates that the relative rate of carbon corrosion to oxygen evolution has significant implications for mitigating carbon-support corrosion: either making the carbon-support more corrosion-resistant via graphitization or enhancing the oxygen evolution reaction via addition of O_2 evolution catalysts can greatly reduce carbon-support loss, since in these cases the O_2 evolution reaction will more effectively balance the permeation-limited O_2 reduction reaction on the anode.

The kinetic model described by Eq. (8) can be used to evaluate the material impact on the carbon corrosion rate. As shown for 50% Pt/Vulcan in Fig. 4 and Table 3 (first row), its carbon corrosion rate is essentially equal to the O_2 crossover current. Since the

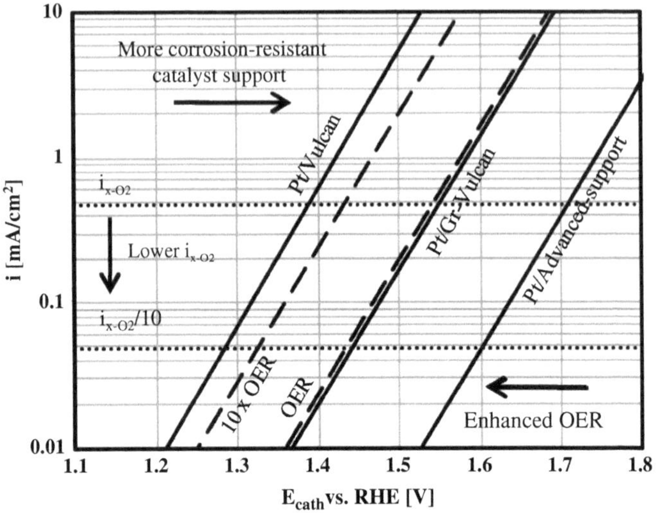

Figure 4. Polarization curves of carbon corrosion and oxygen evolution reactions based on measured carbon corrosion kinetics for Pt/Vulcan and Pt/Graphitized-Vulcan and oxygen evolution kinetics for Pt/C catalysts. The upper horizontal dotted line denotes a current density equivalent to oxygen crossover through membrane from cathode to anode.

Table 3

Catalyst, Catalyst-Support, and Membrane Material Impact on Maximum Catalyst-Support Corrosion Rate Under Fully Developed H_2 Starvation Conditions. Here, *Advanced-Support* is a Hypothetical Support with a 30-Fold Lower Corrosion Rate than Graphitized Vulcan Carbon and *Membrane-X* Refers to a Hypothetical Membrane with a 10-Fold Lower O_2 Permeability.

Catalyst	Membrane	$i_{CO_2,max}$ [mA/cm^2]	Gain
50% Pt/Vulcan	PFSA (18 μm)	46×10^{-2}	1×
50% Pt/Gr-Vulcan	PFSA (18 μm)	26×10^{-2}	1.8×
50% Pt/advanced-support	PFSA (18 μm)	1.8×10^{-2}	26×
50% Pt/Gr-Vulcan + 10× OER	PFSA (18 μm)	4.8×10^{-2}	9.5×
50% Pt/Vulcan	*Membrane-X*	4.6×10^{-2}	10×

electrode performance drops rapidly once ≈10%wt. of the carbon-support is corroded,[6,11] it takes only ≈3000 s under these conditions to destroy the electrode. As shown in Fig. 4 and Table 3 (second row), the carbon corrosion rate is reduced to ≈57% of O_2 crossover current by the use of graphitized Vulcan carbon-support, indicating the importance of the relative rate of carbon corrosion to oxygen evolution. Furthermore, if one was able to develop a high surface area catalyst support with 30-fold higher corrosion resistance than graphitized Vulcan (Table 3, third row), the catalyst-support corrosion damage would be reduced by ≈26-fold (possible candidates are, e.g., boron-doped diamond supports[45,46]). An alternative route toward improved carbon-support corrosion stability during local H_2 starvation is the addition of a more effective O_2 evolution catalyst to the cathode electrode, analogous to what was proposed previously in the context of carbon-support corrosion damage during cell reversal.[47,48] If oxygen evolution kinetics are enhanced by a factor of 10 over Pt/C (e.g., by means of addition of IrO_2 and RuO_2 catalysts,[48] etc.), the carbon corrosion rate would be reduced by a factor of 10 if graphitized Vulcan is used as catalyst-support (Table 3, fourth row). Finally, as the total current is controlled by the O_2 crossover, the use of a hypothetical membrane with a 10-fold lower O_2 permeability would result in a proportionally lower carbon corrosion rate (Table 3, fifth row).

2.3. Start–Stop Model

Considering that at a time t, the H_2/air-front, which divides the fuel cell into the power source (the H_2/air cell) and the load (the air/air cell), is located at x (normalized anode channel length), conservation of charge yields:

$$i_{\text{Power}} = [i_{\text{HOR,an}} = i_{\text{ORR,cath}}]_{\text{H}_2\text{ - rich gas/Air}} \tag{11}$$

for the power source,

$$i_{\text{Load}} = [i_{\text{ORR,an}} = i_{\text{COR,cath}} + i_{\text{OER,cath}}]_{\text{H}_2\text{ - free gas/Air}} \tag{12}$$

for the load, and

$$i_{\text{Power}} x = i_{\text{Load}}(1 - x) \tag{13}$$

as the equal amount of current flows through both H_2/air and air/air cells.

Equations (11), (12), and (13) can be solved for three unknowns $\phi_{e,Power}$, $\phi_{e,Load}$, and $\phi_{s,cath}$ with reference to $\phi_{s,an} = 0$ (set arbitrarily) when the difference in electrolyte phase potential between anode and cathode electrodes is negligible, which is a good assumption at a very low current density. However, anode and cathode electrolyte phase potentials can be differentiated by adding two more equations, namely

$$\begin{cases} \phi_{e,cath} = \phi_{e,an} + i_{Power} R_{H^+} \text{ for the power source portion} \\ \phi_{e,cath} = \phi_{e,an} - i_{Load} R_{H^+} \text{ for the load portion} \end{cases} \tag{14}$$

where the protonic resistance is estimated by[42,43]

$$R_{H^+} \approx R_{H^+,\,mem} + R_{H^+,\,cath}/3 \tag{15}$$

which is the sum of membrane resistance and the effective cathode electrode resistance to proton transport.

Figure 5 gives a graphic solution of the start–stop model with the H_2/O_2 front passing through the center of the anode flow-field, representative of the average carbon corrosion rate. The 50-cm^2 cell undergoes start–stop cycles at 80 °C, 66% RH_{in}, and 150 kPa_{abs} with a residence time of 1.5 s based on anode void volume (including flow-field and diffusion medium). Figure 5 plots polarization curves for both the H_2/air cell (power source) and the air/air cell (load). The intercept point gives the balanced current density ($i_{OER}+i_{COR}$) that flows through the load and therefore causes carbon corrosion accompanied by oxygen evolution. For the base case (0.40 mg_{Pt}/cm^2 on Vulcan), the balanced current density is 0.125 A/cm^2, nearly all of which goes to carbon corrosion. In case (a), replacing Vulcan with Gr-Vulcan significantly increases the resistance to COR and reduces the balanced current density to 0.040 A/cm^2, more than 1/3 of which goes to oxygen evolution reaction. In case (b), lowering anode Pt loading to 0.05 mg_{Pt}/cm^2 on Vulcan increases the resistance of the load to ORR and thus decreases the balanced current density to 0.072 A/cm^2, but only 3% of it is taken by oxygen evolution. The modeled gain (defined as the ratio of the base case carbon corrosion current density over the carbon corrosion current density

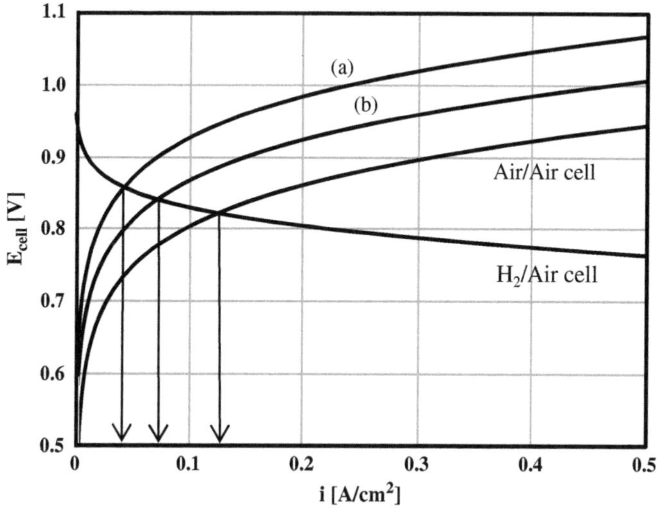

Figure 5. Graph illustrating reduction of cathode carbon corrosion during start–stop with (a) corrosion-resistant carbon support (Gr-Vulcan) and (b) lower anode catalyst loading (0.05 mg$_{Pt}$/cm^2). The base case is Vulcan and 0.40 mg$_{Pt}$/cm^2 anode loading.

of the case of interest) agrees well with the measured start–stop degradation rates at 1.2 A/cm^2, as shown in Table 4.

Carbon corrosion current density can be further reduced by controlling the cell voltage at a low level during the start–stop process.[25,27] Neglecting i_{OER}, Eq. (12) is simplified to

Table 4
Measured Start–Stop Degradation Rates at 1.2 A/cm^2 Normalized to H_2/O_2 Front Residence Time on a 50 cm^2 Cell at 80 °C, 66% RH$_{in}$, 150 kPa$_{abs}$, Compared to the Modeled $i_{CO_{2,base}}/i_{CO_2}$ from Fig. 5.

Anode Pt loading (mg$_{Pt}$/cm^2)	Carbon support	Degradation rate (mV/cycle/s)	Gain	
			Experiment	Model
0.40	Vulcan	1.3	1×	1×
0.40	Gr-Vulcan	0.25	5.2×	4.9×
0.05	Vulcan	0.75	1.7×	1.8×

$$i_{\text{ORR,an}} = i_{\text{COR,cath}} \tag{16}$$

Using Eqs. (2), (4), and (6) in Eq. (16), one finds the surface overpotential that drives COR at the cathode[27] is

$$\eta_{\text{COR,cath}} = \frac{\alpha_{\text{c,ORR}}}{\alpha_{\text{a,COR}} + \alpha_{\text{c,ORR}}}(\phi_{\text{s,cath}} - \phi_{\text{s,an}}) + \frac{RT}{F(\alpha_{\text{a,COR}} + \alpha_{\text{c,ORR}})} \left[\ln\left(\frac{a_{\text{Pt}}L_{\text{Pt}}i_{o,\text{ORR}}}{a_{\text{C}}L_{\text{C}}(1-\theta)^m i_{o,\text{COR}}}\right) + \frac{\alpha_{\text{c,ORR}}F}{RT}\left(E^o_{\text{ORR}} - E^o_{\text{COR}}\right)\right] \tag{17}$$

With the given MEA configuration and operating condition, the second term on the right hand side of Eq. (17) is constant. Therefore, η_{COR} is linearly proportional to the IR-corrected cell voltage ($\phi_{\text{s,cath}} - \phi_{\text{s,an}}$), meaning that i_{COR} can be significantly reduced if the cell potential is controlled to a low enough value. Physically, the current generated from the H_2-rich portion of the fuel cell is forced to be applied to an external load when controlling the cell potential, thus reducing the current flowing through the H_2-free portion of the cell and causing less carbon-support corrosion.

3. COUPLED KINETIC AND TRANSPORT MODEL

3.1. Model Description

A two-dimensional, electrochemical, and transport-coupled transient fuel cell model is developed based on the laws of conservation of species and charge[14]:

$$\frac{\partial(\varepsilon c_i)}{\partial t} + \nabla\cdot(c_i\mathbf{v}) = \nabla\cdot(D_i\nabla c_i) + \frac{s_i i_j}{n_j F\delta_{CL}} \tag{18}$$

$$\nabla\cdot(\sigma\nabla\phi_s) - \sum_j i_j/\delta_{CL} = 0 \tag{19}$$

$$\nabla(\kappa\nabla\phi_e) + \sum_j i_j/\delta_{\text{CL}} = 0 \tag{20}$$

where c_i is the molar concentration of species i ($i = H_2$, O_2, and N_2), ε is the porosity of a porous medium such as DM (diffusion medium) and CL (catalyst layer), D_i is the effective diffusivity of species i, σ is the effective electronic conductivity of the solid phase, κ is the effective proton conductivity of the electrolyte phase, δ_{CL} is the thickness of an electrode, s_i is the stoichiometry of species i in reaction j, and n_j is the number of electrons involved in the electrode reaction. We assume that Darcy's law applies to the gas phase in porous media for the superficial gas velocity:

$$\mathbf{v} = -\frac{K^g}{\mu}\nabla p \tag{21}$$

where K^g is the permeability of a porous medium and μ is the gas viscosity. The gas pressure is calculated based on ideal gas law and the assumption of equilibrium water vapor pressure:

$$p = RT\sum_i c_i + p_{H_2O}^{sat}(T) \tag{22}$$

with the saturated water vapor pressure as a function of T in degree Celsius

$$\log p_{H_2O}^{sat}(\mathrm{atm}) = -2.1794 + 2.9535 \times 10^{-2}T - 9.1838 \times 10^{-5}T^2 + 1.4454 \times 10^{-7}T^3 \tag{23}$$

The permeability of an electrode can be estimated from the electrode porosity (ε) and pore diameter (d) by the Kozeny–Carmen relation

$$K^g = \frac{\varepsilon^3 d^2}{180(1-\varepsilon)^2} \tag{24}$$

The typical value of DM permeability varies from 1 to 10 Darcy (1 Darcy $\equiv 10^{-8}$ cm^2) from Gurley measurement.

The gas viscosity can be calculated by the semiempirical formula of Wilke[49]:

$$\mu_{mix} = \sum_{i=1}^{n} \frac{x_i \mu_i}{\sum_j x_j \Phi_{ij}} \tag{25}$$

in which the dimensionless quantities Φ_{ij} are

$$\Phi_{ij} = \frac{1}{\sqrt{8}}\left(1+\frac{M_i}{M_j}\right)^{-1/2}\left[1+\left(\frac{\mu_i}{\mu_j}\right)^{1/2}\left(\frac{M_j}{M_i}\right)^{1/4}\right]^2 \tag{26}$$

Here x_i and M_i are the mole fraction and molecular weight of species i in the gas mixture of n species and μ_i is the viscosity of pure species i at the system temperature and pressure.

The molecular diffusion coefficient of species i in a multi-component mixture can be calculated by the Wilke formula:[50]

$$D_{i,\text{mix}} = \frac{1-x_i}{\sum\limits_{j=1,j\neq i}^{n} x_j/D_{i,j}} \tag{27}$$

Here $D_{i,j}$ is the binary diffusion coefficient. In DM, the effective diffusivity of species i is calculated by

$$D_i = D_{i,\text{mix}}\frac{\varepsilon}{\tau} \tag{28}$$

in which the effect of porosity (ε) and tortuosity (τ) is accounted for. In an electrode with small pores, Knudsen diffusion (molecule-to-wall collision) contributes to the effective diffusivity of species i:

$$D_i = \left(\frac{1}{D_{K,i}}+\frac{1}{D_{i,\text{mix}}}\right)^{-1}\varepsilon^{1.5} \tag{29}$$

where the Knudsen diffusion coefficient is calculated based on pore diameter (d in cm), species molecular weight (M_i in g/mole), and system temperature (T in Kelvin) by

$$D_{K,i} = 4850d\left(\frac{T}{M_i}\right)^{1/2} \tag{30}$$

Notice that the Bruggeman correlation is employed for tortuosity of the electrode in Eq. (29). The porosity-to-tortuosity of DM in Eq. (28) can be obtained by the limiting current method.[51,52]

In the membrane, the effective diffusivity of species i is calculated by

$$D_i = K_i^m RT \times 10^3 \tag{31}$$

where the gas permeability through membrane is calculated by Eq. (10).

Equations (18–20) are discretized by the control volume method[53] and solved numerically to obtain distributions of species (H_2, O_2, and N_2) concentration, phase potential (solid and electrolyte), and the current resulting from each electrode reaction, in particular, carbon corrosion and oxygen evolution currents at the cathode catalyst layer, with the following initial and boundary conditions:

At $$t = 0,$$

$$c_i = c_i^o \tag{32}$$

At $$t > 0,$$

$$c_i = c_{i,\text{in}} \quad \text{at inlets based on gas compositions,} \tag{33a}$$

$$\frac{dc_i}{dx} = 0 \quad \text{at outlets,} \tag{33b}$$

$$-\sigma \frac{\partial \phi_s}{\partial y} = i_{\text{app}} \quad \text{at outer boundaries of flow-fields,} \tag{33c}$$

$$\frac{\partial \phi_e}{\partial y} = 0 \quad \text{at CL/DM interfaces,} \tag{33d}$$

$$\mathbf{v} = \mathbf{v}_{\text{in}} \quad \text{at inlets based on given flow rates.} \tag{33e}$$

The model applies to both Cartesian and cylindrical coordinates. The x-coordinate denotes in-plane direction and the y-coordinate refers to thru-plane direction.

3.2. Local H_2 Starvation Simulation

We use the coupled kinetic and transport model to predict when local H_2 starvation occurs and how it affects carbon corrosion rate.

Figure 6 shows in a cylindrical coordinate (radial/thru-plane) how H_2 depletes and N_2 pressure builds up with time. At the center of the H_2-starved region, the N_2 pressure becomes higher than 80 kPa_{abs} after 1 s. The radius of this N_2 bubble grows to 7 mm after 10 s and covers 80% of the liquid-water-blocked region at 100 s (near steady-state). The H_2 pressure drops quickly to zero at the edge of the N_2 bubble, within which the cathode potential rises sharply, and the carbon corrosion rate starts to increase, as shown in Fig. 7. Conventional DM has a value of permeability ranging from 1 to 10 Darcy. There is no impact of DM permeability within its realistic range.[14] N_2 crossover through the membrane results in N_2 pressure build-up in the H_2-starved anode region. As a result, convective

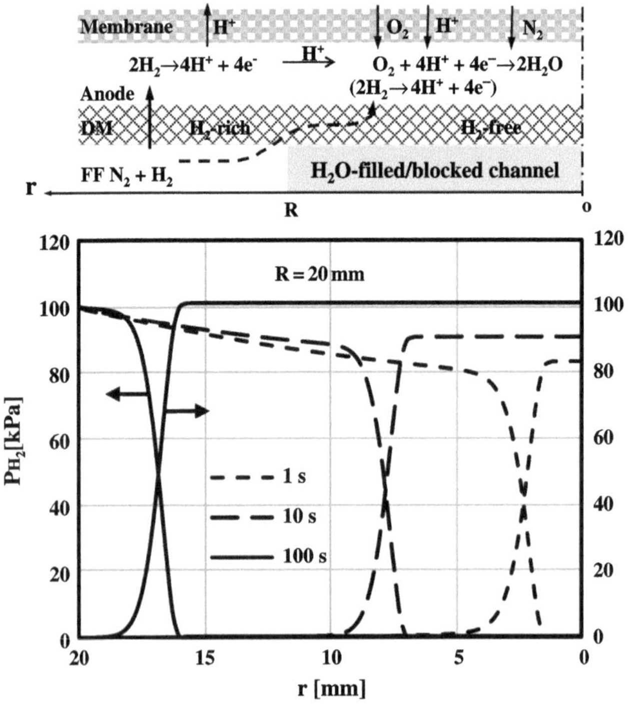

Figure 6. H_2 depletion and N_2 bubble buildup at the center of H_2-starved region resulting from H_2O-filled/blocked channels with a radius of 20 mm as shown above on the plot. The cell operates on neat H_2/air at $\approx$1.5 A/cm^2 (80 °C, 150 kPa_{abs}, 100% RH_{in}). There are only H_2 and water vapor in the anode and DM at $t = 0$.

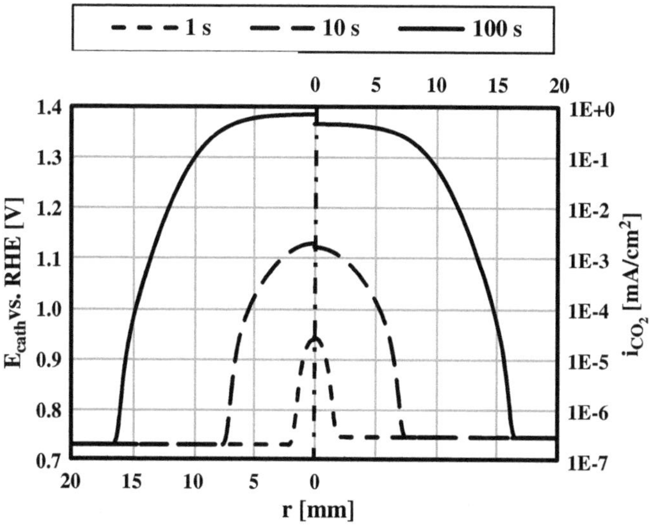

Figure 7. Evolution of cathode potential (defined as $\phi_S - \phi_e$ at cathode electrode) and carbon corrosion current distributions corresponding to H_2 depletion and N_2 bubble buildup as shown in Fig. 6.

flows become insignificant at steady-state, rendering both potential and current distributions that are solely determined by diffusion.

Figure 8 compares the predicted residual carbon support in the cathode with measured electrode thinning after a controlled local H_2 starvation test.[54] Above the plot a schematic is shown. The anode flow-field is partially blocked, creating a region of 60 mm in width where there is no channel. The cathode electrode thickness was measured at the four locations marked by dashed lines after a 13-h run at 1.2 A/cm^2. The model simulation was carried out on a Cartesian coordinate (two-dimensional across the blocked region – through cell thickness). Figure 8 shows that the modeled residual carbon support agrees qualitatively with the measured electrode thickness across the blocked region.

Information on the length scales beyond which H_2 depletes and carbon corrosion takes place is valuable for developing system mitigation strategies. Figure 9 shows that the length scale for H_2 to deplete is independent of the length scale of the H_2-starved region, while the larger the H_2-starved region, the higher the carbon corrosion at the center of the H_2-starved region. The insert in Fig. 9 (top) illustrates that the length scale is determined by letting $i_{HOR} - i_{x,O_2}$

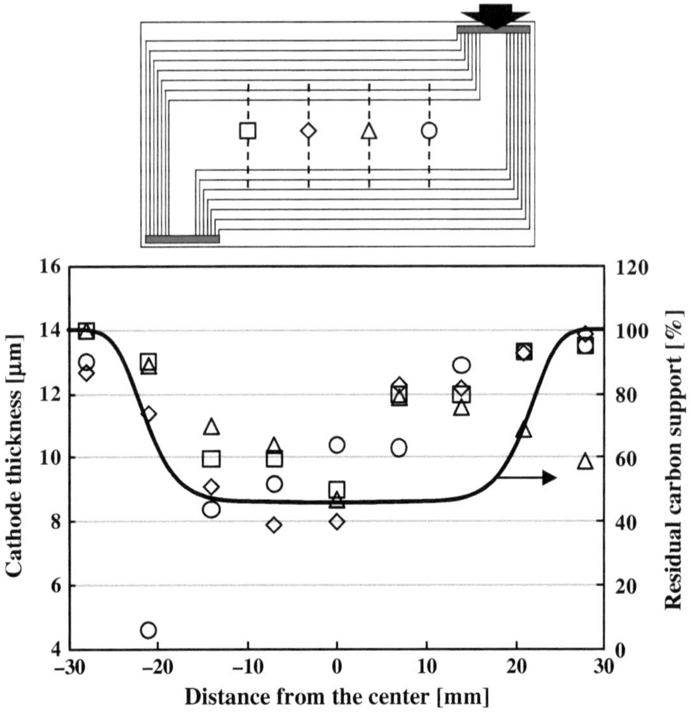

Figure 8. Comparison of the predicted residual carbon support in the cathode with the measured cathode thickness after 13 h run at 1.2 A/cm^2 (70%H_2/air counter-flow, 80 °C, 150 kPa$_{abs}$, 40%/60% RH$_{in}$) with partially filled flow channels as shown above the plot.

be zero. Figure 9 (bottom) shows that a radial length scale of more than 20 mm (40 mm in diameter) is required to allow for a fully developed H_2 starvation region, indicating that a large number of adjacent channels need to be filled with water or blocked by water droplets. The cathode potential (reflected by normalized carbon corrosion current density) is substantially lower in the region where a local H_2 starvation region is not fully developed.

The length scale for H_2 to deplete depends strongly on applied current density, as shown in Fig. 10. The length scale increases as current density decreases. As a result, an H_2-starved region is more likely to be fully developed at a high current density, whereas it tends to be underdeveloped at a low current density. Figure 11 gives the length scales beyond which H_2 would deplete, and the carbon

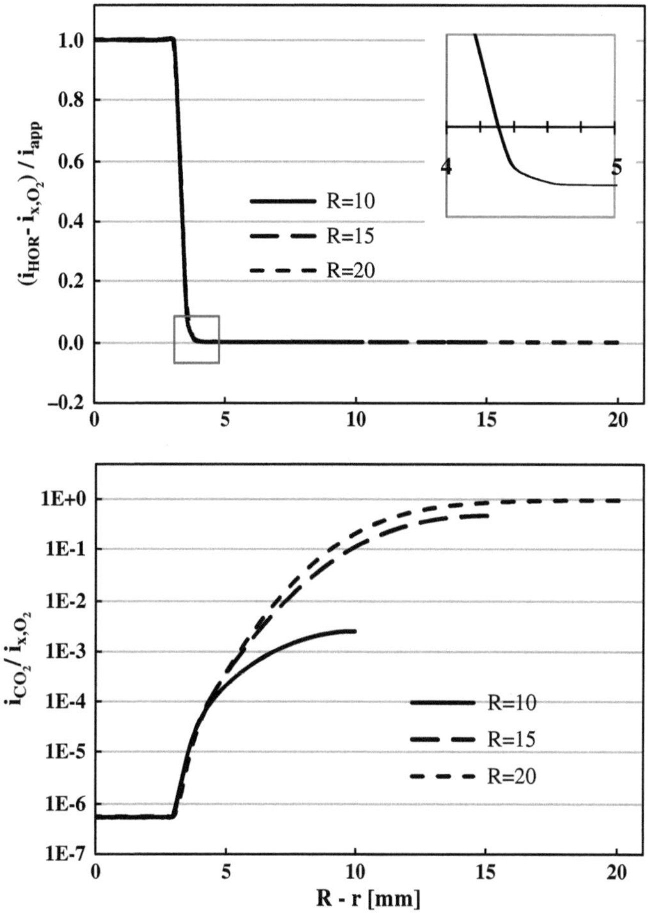

Figure 9. Top: HOR (less O_2 crossover) current density distribution with respect to length scales of a local H_2-starved region determines where H_2 depletes. Bottom: carbon corrosion current distribution with respect to length scales of a local H_2-starved region shows how a fully starved region develops. The cell operates on neat H_2/air at ≈1.5 A/cm^2 (80 °C, 150 kPa$_{abs}$, 100% RH$_{in}$).

corrosion current density would exceed 10 and 50%, respectively, of i_{x,O_2} at the center of the H_2-starved region, with respect to applied current density. It shows that the radial length scale for 10% i_{x,O_2} carbon corrosion current density is 6 mm larger than the radial length scale for H_2 depletion.

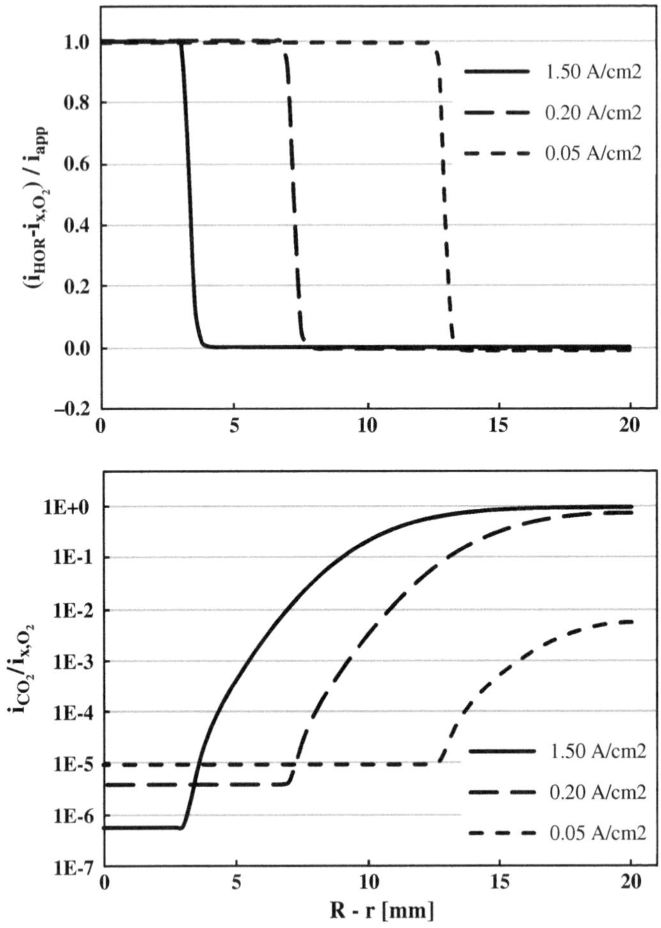

Figure 10. The scale of H_2 depletion (top) and corresponding carbon corrosion current distribution (bottom) depends on the applied current density. The cell operates on neat H_2/air (80 °C, 150 kPa_{abs}, 100% RH_{in}) with a local H_2-starved region of 20 mm in radius.

Figure 12 shows the time scales at which the carbon corrosion current reaches 50% of its maximum value at the center of the H_2-starved region. Local H_2 starvation takes place within ~2–40 s of anode channel flooding/blocking at a high current density,

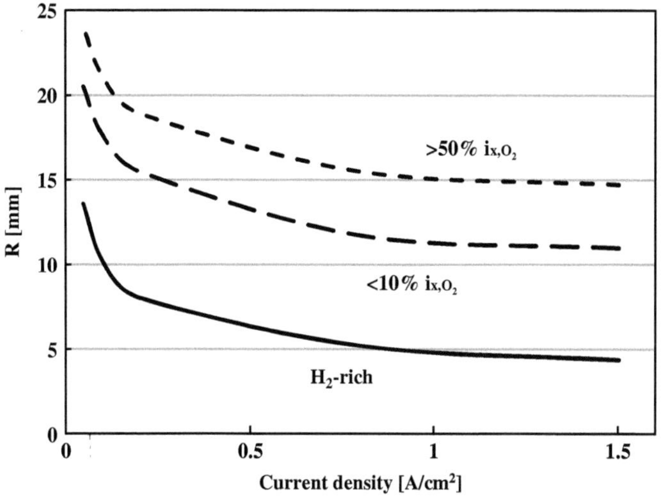

Figure 11. Calculated length scales with respect to applied current density for a cell operating on neat H_2/air (80 °C, 150 kPa_{abs}, 100% RH_{in}). The solid line represents the length scale beyond which H_2 depletes. The long and short dashed lines denote the length scales beyond which the maximum carbon corrosion current density would exceed 10% and 50% of O_2 crossover current density, respectively.

depending on the length scale of the blocked flow-field region. It takes longer for a shorter length scale of the flooded/blocked area. The time scale correlates closely to the N_2 pressure buildup, which in turn depends on the membrane's N_2 permeability. Longer times to reach steady-state are required for membranes with lower N_2 permeability. Figure 12 also shows that the time scales are strongly dependent on the applied current density.

Diluting H_2 with N_2 has a drastic impact on length and time scales. Figure 13 shows the comparisons between 100% H_2 and 70% H_2 diluted with 30% N_2. When H_2 is consumed in the blocked area, a vacuum is created, inducing convective flow. When there is N_2 in the anode gas stream, such convection helps to build up N_2 pressure much more quickly than N_2 permeation from cathode to anode. Figure 13(a) shows that the length scale for H_2 depletion decreases greatly with dilution of H_2. More strikingly, the H_2-starved region becomes fully developed even at 0.05 A/cm^2, which yields only 1% i_{x,O_2} with 100% H_2. Figure 13(b) shows that the time scales

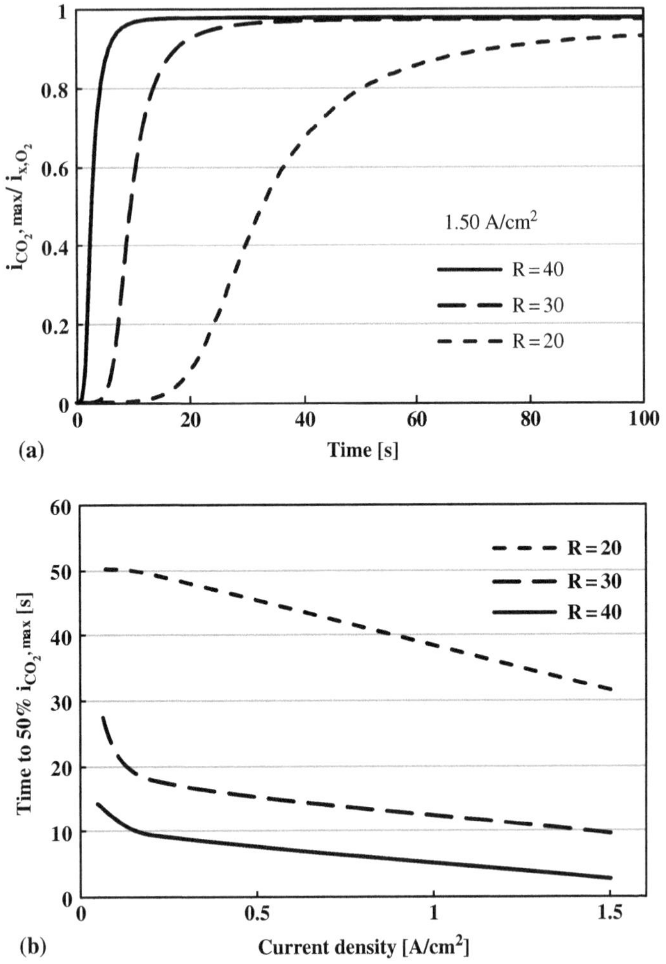

Figure 12. Evolution of the maximum carbon corrosion current density at the center of a local H_2-starved region and calculated time constants (defined as the time to 50% of the maximum carbon corrosion current density) with respect to length scales of a local H_2-starved region for a cell operating on neat H_2/air (80 °C, 150 kPa_{abs}, 100% RH_{in}).

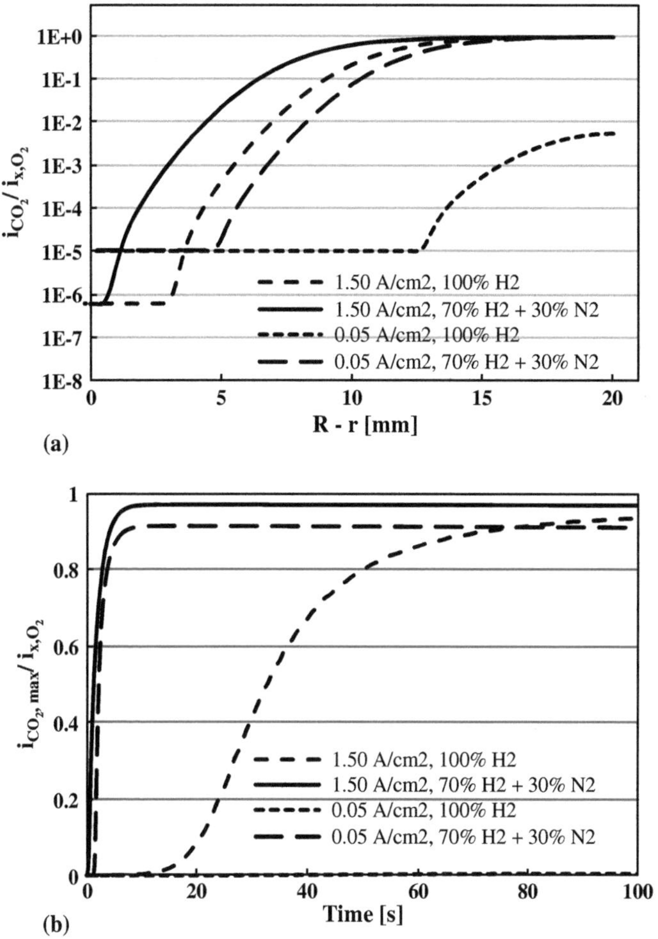

Figure 13. Effect of N_2 dilution in the anode on length scale and time constant for carbon corrosion in a local H_2-starved region of 20 mm in radius. The cell operates on neat (100%) and N_2–diluted (70%) H_2/air (80 °C, 150 kPa_{abs}, 100% RH_{in}).

are significantly reduced to less than 2 s for both 1.50 and 0.05 A/cm^2 cases. This shows the importance of keeping the concentration of N_2 in the anode at a low level to mitigate cathode carbon corrosion.

3.3. Start–Stop Simulation

The two-dimensional, coupled kinetic and transport model can also be used to simulate start–stop processes. Figure 14 plots cathode potential and carbon corrosion current distribution at three instants when the H_2/O_2 front passes through 10, 50, and 90% of anode flow path during the start process. As H_2 displaces air in the anode flow-field, the size of the power source increases and the load size decreases accordingly. The balanced current density becomes larger, causing higher carbon corrosion current density.

In a controlled H_2/O_2-front, start–stop test, the stop process essentially mirrors the start process as H_2 and air switch periodically at the anode inlet. As a result, the predicted accumulated carbon loss distribution down the anode channel, as shown in Fig. 15, is nearly symmetric in regions opposing the anode inlet and outlet. Figure 16 (a) shows real time CO_2 measurements at the cathode exit during the controlled start–stop test.[15] It shows more carbon corrosion occurring during start than during stop, suggesting that the accumulated

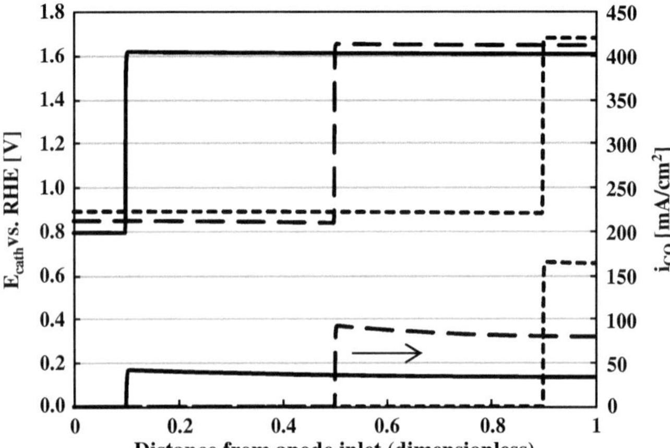

Figure 14. Calculated cathode potential and carbon corrosion current distributions when H_2/O_2 front passes through 10%, 50%, and 90% of anode flow-field during a cell start from air/air state (80 °C, 101 kPa_{abs}, 66% RH_{in}). The cell has a catalyst loading of 0.4 mg_{Pt}/cm^2 using a 50%wt Pt/Vulcan catalyst in both anode and cathode electrodes.

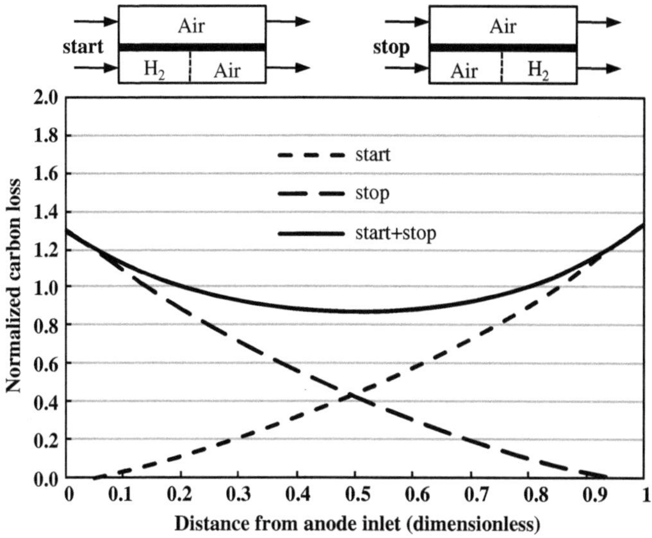

Figure 15. Predicted accumulated carbon loss distribution along anode flow-field over a complete start–stop cycle for a controlled start–stop experiment as shown above the plot at 80 °C, 101 kPa_{abs}, 66% RH_{in}, and residence time of 1.5 s based on anode void volume (including flow-field and diffusion medium). The model predicts nearly symmetric carbon loss at anode inlet and outlet because the stop process essentially mirrors the start process by switching H_2 and air periodically at anode inlet.

carbon loss distribution is asymmetrical unlike what the model predicts. Figure 16(b) plots the ratio of the integrated carbon loss (equal to the ratio of the integral of the measured CO_2 concentration over time for stop over start) with respect to the residence time of the H_2/O_2 front passing through the anode flow-field. It can be seen that the stop process causes approximately 30% less carbon corrosion than the start process for the case shown in Fig. 16(a). This asymmetry becomes more significant with decreasing residence time and gradually diminishes as the residence time increases.

Figure 17 shows the measured spatially resolved mass activity map for the aged MEA from the controlled start–stop test in comparison with a new one.[55] It also indicates more carbon loss near the anode inlet, because one can easily imagine that when carbon support particles collapse, the platinum particles can come in contact with one another, which leads to coalescence of the platinum

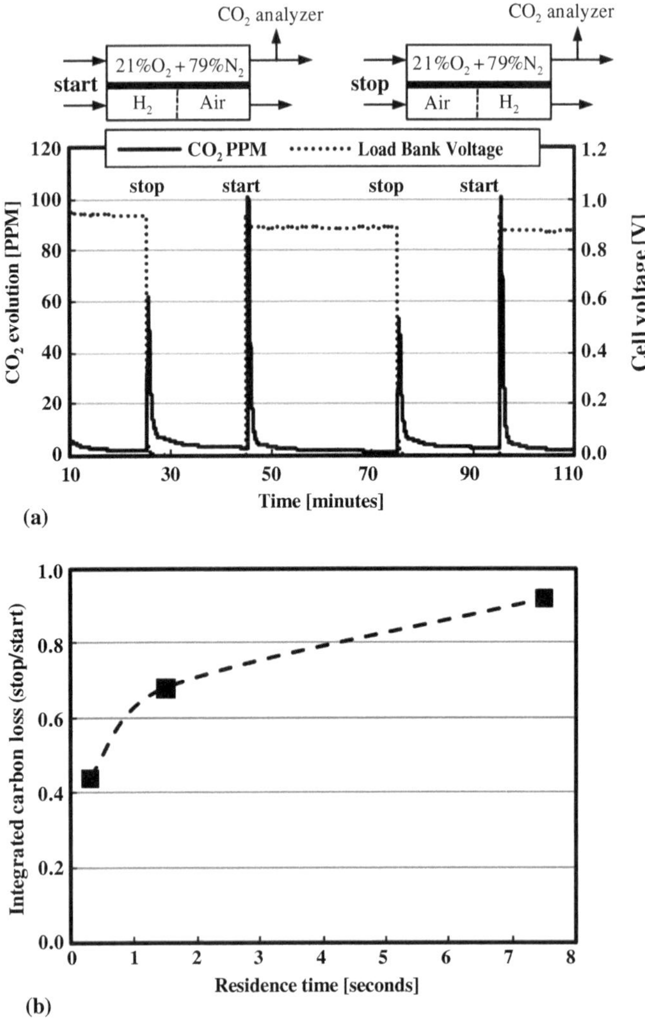

Figure 16. (a) Real time CO_2 measurement at the exit of the cathode during H_2/air-front start–stop events with a residence time of 1.5 s based on anode void volume (including flow-field and diffusion medium). More carbon corrosion occurs during start than during stop. (b) Ratio of the integrated carbon loss (proportional to the integral of CO_2 concentration over time) for stop over start versus residence time. The ratio approaches unity as residence time increases.

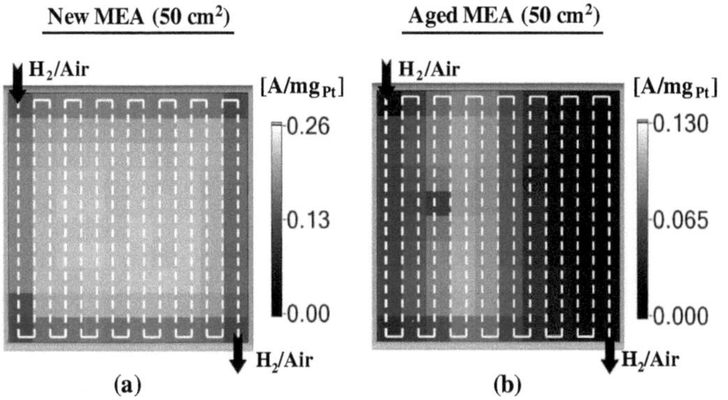

Figure 17. Mass activity distribution maps for both new and aged MEAs showing impact of 200 unmitigated start–stop cycles. Test conditions: $T_{cell} = 80\ °C$, $T_{dewpt} = 85\ °C$ (anode and cathode), $P = 250/270\ kPa_{abs}$ (anode/cathode), high stoichiometry flow for anode (H_2) and cathode (O_2).

particles. Additionally, as the carbon support corrodes, some of the platinum particles can become isolated from electronic conduction. In other words, more carbon loss leads to lower mass activity. The pattern of mass activity loss agrees with the CO_2 measurements discussed above, in that the outlet region (mostly damaged during start) shows a higher degree of degradation than the inlet region (mostly damaged during stop).

The asymmetry in the accumulated carbon loss (reflected by the CO_2 release rate) associated with start–stop cycling can be attributed to the difference in the state of the electrodes before a start or stop process occurs as well as the difference in the way that the electrode potentials change after a start or stop process takes place, as detailed in the following section. Asymmetry in CO_2 release rate was observed in potential cycling tests.[56] It appears that carbon corrosion occurs via a different reaction path in potential cycling compared to at a constant potential.[56,57] However, diminishing asymmetry with increasing residence time as shown in Fig. 16 (b) implies that the effect of potential change should be secondary. The fact that the model neglects the pseudo-capacitive effect explains why it fails to predict the asymmetrical distribution in accumulated carbon loss observed in the experiments. The pseudo-capacitive effect can be rigorously accounted for by incorporating platinum reactions that supply or consume protons in transient start–stop events. These reactions are normally neglected in a carbon corrosion model.[58] Next, we will describe a model that incorporates the pseudo-capacitance obtained from an AC-impedance measurement.

4. PSEUDO-CAPACITANCE MODEL

4.1. Mechanism Description

As shown in Fig. 14, the cathode potential changes abruptly across the H_2/air-front. This fact warrants the inclusion of the pseudo-capacitance into the previous steady-state kinetic model.[12] It is clear that the electrode's pseudo-capacitance can supply protons in transient events and thereby reduce the cathode carbon-support corrosion rate in the case of fast moving H_2/air-fronts. Figure 18

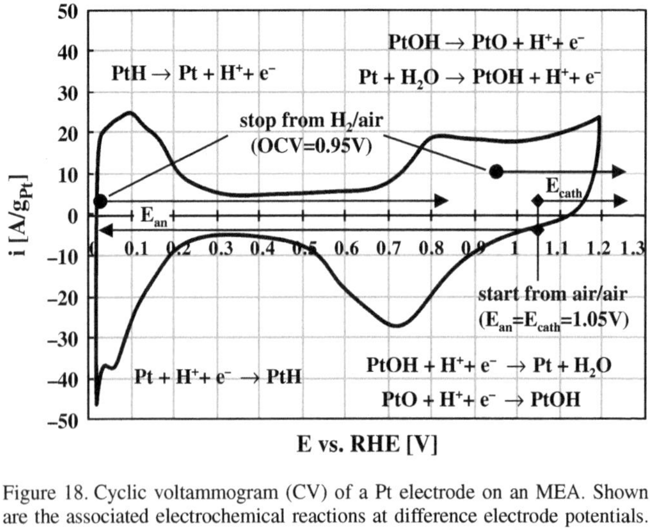

Figure 18. Cyclic voltammogram (CV) of a Pt electrode on an MEA. Shown are the associated electrochemical reactions at difference electrode potentials. Notice how the electrode potential changes as the cell starts from air/air conditions (diamond-tailed arrows) and stops from H_2/air conditions (dot-tailed arrows), respectively. Experimental conditions: 25 °C, 20 mV/s, 0.4 mg_{Pt}/cm^2 loading.

illustrates how the electrode potential changes as the cell either starts from the air/air state or stops from the H_2/air state; it also lists the appropriate pseudo-capacitive reactions. To better understand the effect of pseudo-capacitive charging and discharging, we will examine each of the processes in turn.

Start process (H_2 entering an air-filled anode): Originating from the air/air condition, with an approximate electrode rest potential of 1.05 V[11] on both anode and cathode, introduction of H_2 on the anode side creates a power source driving a load (see Fig. 1) and facile HOR forces the anode potential to near zero volts. In the portion of the anode that still has air, ORR takes place and the oxidized Pt surface is reduced according to $PtOH + H^+ + e^- \rightarrow Pt + H_2O$ plus $Pt + H^+ + e^- \rightarrow PtH$ in the potential down-transient. The protons required for these reactions are supplied from the cathode, causing the cathode potential to rise above its initial rest potential. Protons are generated at the cathode, first via further oxidation of the platinum hydroxide ($PtOH \rightarrow PtO + H^+ + e^-$), the contribution of which is determined by the pseudo-capacitance of the cathode in the potential region between $\approx$1.05 V and that where carbon corrosion sets in (ca. >1.4 V for conventional carbon-supports), and then via on-setting carbon-support oxidation and O_2 evolution when the cathode's pseudo-capacitance is expended. Thus, the cathode's pseudo-capacitance acts as a buffer against the carbon corrosion reaction during the start process. In this case, the capacitive buffer of the cathode is offset by the capacitive consumption of protons at the anode experiencing a downward potential change. If the total pseudo-capacitance of the anode between $\approx$1.05 V and that determined primarily by the ORR therein is significantly lower than the total pseudo-capacitance of the cathode, the pseudo-capacitive processes will lead to a substantial reduction in the overall H_2/air-front start damage. An example of where this could be relevant is where low anode Pt loadings are used.

Stop process (air entering an H_2-filled anode): Originating from the H_2/air condition, with an anode rest potential of $\approx$0 V, introduction of air creates a load driven by a power source (see Fig. 1) and slow ORR forces the anode potential to rise significantly in the portion of the anode that air has replaced H_2. In the potential up-transient, the PtH-terminated anode surface is oxidized according to $PtH \rightarrow Pt + H^+ + e^-$ plus $Pt + H_2O \rightarrow PtOH + H^+ + e^-$, producing protons. In the immediate vicinity of the H_2/air-front, the protons required for oxygen reduction can also be supplied from the anode oxidation reactions in the portion of the anode that still has H_2, but at distances more than a few membrane-thicknesses away, they are preferentially supplied from the cathode because the in-plane transport resistance of protons makes it more favorable for them to be supplied from the cathode. Similarly to the start process, proton generation at the cathode will initially be via further oxidation of the platinum hydroxide ($PtOH \rightarrow PtO + H^+ + e^-$) and then proceed to higher potential reactions (carbon corrosion and oxygen evolution). Thus, the pseudo-capacitive reactions on *both* the anode and cathode combine to maximize the amount of protons that can be supplied in support of the O_2 reduction reaction on the anode, so that the cathode carbon-support oxidation reaction will occur to a lesser extent than in the start process, where the anode's pseudo-capacitance is subtracted from that of the cathode.

Therefore, for equal H_2/air-front residence times, the pseudo-capacitive model would suggest lower rates of carbon-support oxidation, i.e., lower rates of CO_2 formation for the stop process if compared to the start process, which is consistent with on-line CO_2 measurements of the air exiting the cathode flow-field during H_2/air-front start–stop events, as shown in Fig. 16.

4.2. Model Description

The pseudo-capacitive effect can be incorporated in the coupled kinetic and transport model through Eqs. (19) and (20). Here we choose to illustrate the effect through the kinetic model for simplicity. With considering the pseudo-capacitive current density, the kinetic model becomes

$$i_{\text{Power}} = [i_{\text{HOR,an}} + i_{\phi,\text{an}} = i_{\text{ORR,cath}} + i_{\phi,\text{cath}}]_{\text{H}_2\text{-rich gas/Air}} \tag{34}$$

for the power source,

$$i_{\text{Load}} = [i_{\text{ORR,an}} + i_{\phi,\text{an}} = i_{\text{COR,cath}} + i_{\text{OER,cath}} + i_{\phi,\text{cath}}]_{\text{H}_2\text{-free gas/Air}} \tag{35}$$

for the load, and again

$$i_{\text{Power}} x = i_{\text{Load}}(1 - x) \tag{13}$$

as the equal amount of current flows through both H_2/air and air/air cells. The capacitive current density appearing in Eqs. (34) and (35) is given by[59]

$$i_\phi = -C_\phi \frac{dE_{\text{an or cath}}}{dt} = -C_\phi \frac{d(\phi_s - \phi_e)_{\text{an or cath}}}{dt} \tag{36}$$

and the charge capacity of the respective pseudo-capacitor is calculated by

$$q = \int_0^t i_\phi dt \leq q_{\max} \tag{37}$$

Here i is the current density with the subscript representing a specific electrode reaction, capacitive current density at an electrode, or current density for the power source or the load. The surface overpotential (defined as the difference between the solid and electrolyte phase potentials) drives the electrochemical reactions and determines the capacitive current. Therefore, the three Eqs. (34), (35), and (3) can be solved for the three unknowns: the electrolyte phase potential in the H_2/air cell ($\phi_{e,\text{Power}}$), electrolyte phase potential in the air/air cell ($\phi_{e,\text{Load}}$), and cathode solid phase potential ($\phi_{s,\text{cath}}$), with anode solid phase potential ($\phi_{s,\text{an}}$) being set to be zero as a reference. The carbon corrosion current is then determined using the calculated phase potential difference across the cathode/membrane interface in the air/air cell. The model couples carbon corrosion with the oxygen evolution reaction, other normal electrode reactions (HOR and ORR), and the capacitive current in the fuel cell during start–stop.

The capacitive current is calculated based on the pseudo-capacitance value obtained in an AC-impedance measurement.[42] Equation 37 sets the maximum stored charge, use of which can be made. In this study, $C_\phi = 0.2$ F/cm^2 and $q_{\max} = 0.06$ C/cm^2 correspond to the MEA with a Pt loading of 0.4 mg$_{\text{Pt}}$/cm^2 on both anode and cathode electrodes (see Fig. 6 in the reference[42]). Considering that the electrode's potential changes rapidly as the H_2/air-front propagates and the model caps the charge capacity, a constant value of C_ϕ suffices to capture accurately the pseudo-capacitive effects during start–stop.

4.3. The Pseudo-capacitive Effect

Figure 19 plots the normalized carbon loss (relative to the average value) for start and start–stop cycles in comparison with the loss of mass activity normalized to the new MEA value in one-dimensional fashion, showing good agreement. Incorporation of the pseudo-capacitance yields more carbon loss during start than during stop, as expected in Section 4.1. The ratio of stop carbon loss over start carbon loss (ca. 0.6) agrees reasonably well with the ratio of the integrated carbon loss for stop/start of approximately 0.7 from the real-time CO_2 measurement, as shown in Fig. 16(b). Clearly, it is essential to include the electrode's pseudo-capacitance in a start–stop model to understand the observed non-uniform mass activity loss from start–stop. This agreement between the model and the experiment again suggests the significance of the electrode pseudo-capacitance.

Figure 20 shows the impact of capacitive effects on the predicted amount of local carbon corrosion as a function of the

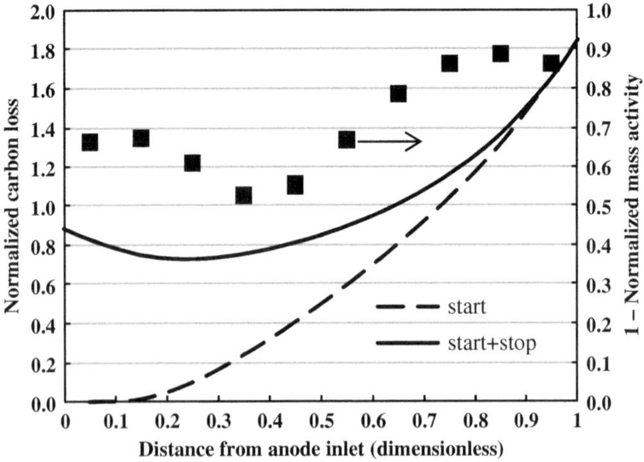

Figure 19. Predicted carbon loss distribution along anode flow-field channel over a complete H_2/air-front start–stop cycle using the pseudo-capacitance model in comparison with one-dimensional, normalized mass activity from Fig. 17. The pseudo-capacitance value used in the model is obtained from AC-impedance measurements as described in references (42, 43).

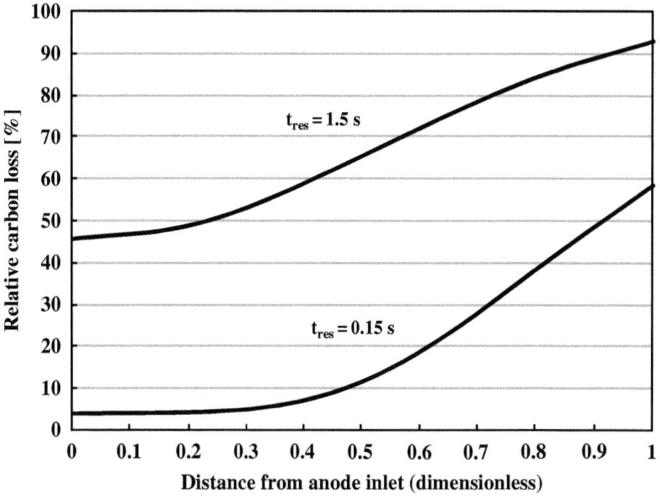

Figure 20. The impact of a faster H_2/air-front moving through the anode flow-field. 100% relative carbon loss is defined as the localized damage prediction when no pseudo-capacitance is considered in the model. Significantly less carbon corrosion is expected at the anode inlet region as the speed of H_2/air-front increases but much less benefit at the anode outlet region.

H_2/air-front residence time. 100% here refers to the extent of local carbon-support corrosion predicted without considering pseudo-capacitive effects. Apparently, pseudo-capacitive effects are expected to play a minor role, if the H_2/air-front residence time is long due to the limited capacitance. Therefore, at long residence times, the difference made by considering the pseudo-capacitive effect is quite small. This explains why simple steady-state kinetic models without including pseudo-capacitive effects work well when predicting carbon-support corrosion in accelerated H_2/air-front experiments which use long residence times to accelerate the damage.[12] On the other hand, at short H_2/air-front residence times, characteristic of what one would observe in a well-designed PEMFC system, the pseudo-capacitive effect becomes rather significant. Therefore, inclusion of pseudo-capacitive effects is critical when modeling real-world applications.

Since the time scales for establishing local H_2 starvation events are on the order of seconds or 10's of seconds,[11,14] pseudo-capacitive effects will not be important.

5. SUMMARY AND OUTLOOK

The carbon-support corrosion models developed here are based on measured electrode kinetics and the concept that, in a situation when a fuel cell has air on the cathode and a combination of H_2 and air regions on the anode, the former act as a power source (H_2/O_2 fuel cell) that drives corrosion in the latter (air/air-corrosion cell). The kinetic model has yielded much insight into the origin and impact of carbon-support corrosion. It can serve as an efficient tool to guide new material development and MEA design. For example, since corrosion can effectively be mitigated if the OER kinetics are fast enough to compete with COR, electrode materials that favor the OER reaction can serve to make electrodes much more resistant to corrosion. Additionally, given that the rate of COR is limited by oxygen crossover in the instance of localized anode H_2 starvation, incorporation of membrane materials that minimize oxygen permeation will serve as an effective mitigation strategy.

The coupled kinetic and transport model is needed to predict when local H_2 starvation occurs and how it affects the carbon corrosion rate. The ability to predict length scales of H_2 depletion is helpful in designing advanced flow-fields that prevent liquid water from accumulating to form an H_2-starved region of sufficient size for corrosion to be possible. Similarly, predicting time scales for these starved regions to develop is valuable for developing system mitigation strategies that help to remove accumulated liquid water and N_2 before a local H_2-starved region becomes fully developed. The coupled kinetic and transport model is also necessary for simulating an extended stop process in which the location and longevity of an H_2/O_2 front are determined by coupled kinetic and transport processes.

The pseudo-capacitance model is essential in predicting the observed asymmetrical carbon loss distribution along the flow channel that develops after a number of start–stop cycles. The pseudo-capacitive effect must be taken into account in simulating start–stop processes (i.e., transient H_2/air-front events). It provides useful information on the strategy of using a fast purge to mitigate cathode carbon-support corrosion, that is, how fast the H_2/O_2 front needs to be to make full use of the stored charge capacity. Because it takes longer for local H_2 starvation to fully develop, the capacitive effect is expected to be less significant in the case of local H_2 starvation.

Recent kinetic studies indicate that carbon corrosion can be significant under normal transient operation.[56,57,60–62] The rate of voltage change, common in the automotive application, enhances cathode carbon-support corrosion.[16] Hence, further model improvement shall be focused on finding the carbon corrosion kinetics associated with voltage cycling. Currently, the relationship between fuel cell performance decay and accumulated carbon-support loss is only empirical.[22] More effort has to be made to incorporate mechanisms that can accurately quantify voltage decay with carbon-support loss.[31,32]

ACKNOWLEDGEMENTS

The authors thank Dr. Frederick T. Wagner for useful discussions.

LIST OF SYMBOLS

a	electrochemically active surface area of an ingredient in an electrode, cm^2/mg
C_ϕ	electrode pseudo-capacitance, $F/cm^2_{electrode}$
c_i	molar concentration of species i, mol/cm^3
D_i	effective diffusion coefficient of species i, cm^2/s
$D_{i,mix}$	molecular diffusion coefficient of species i in a multi-component mixture, cm^2/s
$D_{K,i}$	Knudsen diffusion coefficient of species i, cm^2/s
d	pore diameter of an electrode, cm
E^o	standard equilibrium (or reversible) potential of an electrode reaction, V
$E^{\text{dry or wet}}$	activation energy for gas transport in the dry (or wet) phase of membrane, J/mol
E^{rev}	activation energy of an electrode reaction at zero overpotential, J/mol
$E_{\text{an or cath}}$	electric potential of anode (or cathode) electrode, V
F	Faraday constant, C/equiv
i	current density, A/cm^2

i_o	exchange current density of an electrode reaction, $A/cm^2{}_{Pt\ or\ C}$
$i_{x,O2}$	O_2 crossover current density, A/cm^2
K^g	permeability of a porous medium, cm^2/s
K^m	permeability of gaseous species through membrane, mol cm/(cm^2 s kPa)
L	loading of an ingredient in an electrode, mg/cm^2
M	molecular weight of a species, g/mol
m	index for carbon weight loss dependence in the COR kinetics equation
n	number of electrons transferred in an electrode reaction
p	pressure, partial pressure of a species, kPa
$q_{\max}$	maximum stored charge, $C/cm^2{}_{electrode}$
R	universal gas constant, J/mol/K
R_{H^+}	proton transport resistance, $\Omega\ cm^2$
$r_{Pt/C}$	weight ratio of ingredient Pt to carbon support in an electrode
RH	relative humidity, %
s	stoichiometry of a species in an electrode reaction
T	temperature, °C or K
t	time, s
$\mathbf{v}$	gas velocity, cm/s
x	normalized location of H_2/O_2 front at the anode
x_i	mole fraction of species i in a gas mixture

Greek Symbols

α_a	anodic transfer coefficient of an electrode reaction
α_c	cathodic transfer coefficient of an electrode reaction
δ	thickness, cm
ε	porosity
ϕ	electric potential, V
γ	reaction order of a species in an electrode reaction
η	charge transfer overpotential of an electrode reaction, V
κ	proton conductivity, S/cm
σ	electron conductivity, S/cm
τ	tortuosity

μ_{mix}	viscosity of a gas mixture, Pa s
θ	Mass fraction of carbon support that has been lost

Subscripts

an	anode
cath	cathode
CL	catalyst layer
e	electrolyte conducting protons
mem	membrane
s	solid conducting electrons

Superscripts

ref	reference
sat	saturated

REFERENCES

[1] M.F. Mathias, R. Makharia, H.A. Gasteiger, J.J. Conley, T.J. Fuller, C.J. Gittleman, S.S. Kocha, D.P. Miller, C.K. Mittelsteadt, T. Xie, S.G. Yan, P.T. Yu, Electrochem. Soc. Interface **14**, 24 (2005)

[2] R. Borup, J. Meyer, B. Pivovar, Y.S. Kim, R. Mukundan, N. Garland, D. Myers, M. Wilson, F. Garzon, D. Wood, P. Zelenat, K. More, K. Stroh, T. Zawodzinski, J. Boncella, J.E. McGrath, T. Inaba, K. Miyake, M. Hori, K. Ohto, Z. Ogumi, S. Miyata, A. Nishikata, Z. Siroma, Y. Uchimoto, K. Yasuda, K. Kimijima, N. Iwashita, Chem. Rev. **107**, 3904 (2007)

[3] W. Schmittinger, A. Vahidi, J. Power Sources **180**, 1 (2008)

[4] J. Wu, X.Z. Yuan, J.J. Martin, H. Wanga, J. Zhang, J. Shen, S. Wu, W. Merida, J. Power Sources **184**, 104 (2008)

[5] H.A. Gasteiger, W. Gu, B. Litteer, R. Makharia, M. Budinski, E. Thompson, F.T. Wagner, S.G. Yan, P.T. Yu, in *Mini-Micro Fuel Cells*, ed. by S. Kakaç, A. Pramuanjaroenkij, L. Vasiliev (Springer, Dordrecht, 2008)

[6] J. Zhang, R.N. Carter, P.T. Yu, W. Gu, F.T. Wagner, H.A. Gasteiger, in *Encyclopedia of Electrochemical Power Sources, Volume 2*, ed. by J. Garche, C. Dyer, P. Moseley, Z. Ogumi, D. Rand, B. Scrosati. (Elsevier B.V., Amsterdam, 2009)

[7] T. Hatanaka, T. Takeshita, H. Murata, N. Hasegawa, T. Asano, M. Kawasumi, Y. Morimoto, ECS Trans. **16**(2), 1961 (2008)

[8] C.A. Reiser, L. Bregoli, T.W. Patterson, J.S. Yi, J.Y. Yang, M.L. Perry, T.D. Jarvi, Electrochem. Solid-State Lett. **8**, A273 (2005)

[9] T.W. Patterson, R.M. Darling, Electrochem. Solid-State Lett. **9**, A183 (2006)

[10] H. Tang, Z. Qi, M. Ramani, J.F. Elter, J. Power Sources **158**, 1306 (2006)

[11] R. Makharia, S. Kocha, P. Yu, M.A. Sweikart, W. Gu, F. Wagner, H.A. Gasteiger, ECS Trans. **1**(8), 3 (2006)

[12] P.T. Yu, W. Gu, R. Makharia, F.T. Wagner, H.A. Gasteiger, ECS Trans. **3**(1), 797 (2006)

[13] Z.Y. Liu, B.K. Brady, R.N. Carter, B. Litteer, M. Budinski, J.K. Hyun, D.A. Muller, J. Electrochem. Soc. **155**, B979 (2008)
[14] W. Gu, R. Makharia, P.T. Yu, H.A. Gasteiger, Preprint – Am. Chem. Soc. Div. Fuel Chem. **51**(2), 692 (2006)
[15] W. Gu, R.N. Carter, P.T. Yu, H.A. Gasteiger, ECS Trans. **11**(1), 963 (2007)
[16] H. Chizawa, Y. Ogami, H. Naka, A. Matsunaga, N. Aoki, T. Aoki, K. Tanaka, ECS Trans. **11**(1), 981 (2007)
[17] J. Kim, J. Lee, G. Lee, Y. Tak, ECS Trans. **16**(2), 961 (2008)
[18] A.B. Ofstad, J.R. Davey, S. Sunde, R.L. Borup, ECS Trans. **16**(2), 1301 (2008)
[19] W.R. Baumgartner, E. Wallnöfer, T. Schaffer, J.O. Besenhard, V. Hacker, V. Peinecke, P. Prenninger, ECS Trans. **3**(1), 811 (2006)
[20] R. Dross, B. Maynard, ECS Trans. **11**(1), 1059 (2007)
[21] Q. Shen, M. Hou, D. Liang, Z. Zhou, X. Li, Z. Shao, B. Yi, J. Power Sources **189**, 1114 (2009)
[22] P.T. Yu, W. Gu, J. Zhang, R. Makharia, F.T. Wagner, H.A. Gasteiger, in *PEFC Durability and Degradation*, ed. by F.N. Büchi, M. Inaba, T.J. Schmidt (Springer, New York, NY, 2009)
[23] P.T. Yu, W. Gu, F.T. Wagner, H.A. Gasteiger, Preprint – Am. Chem. Soc. Div. Fuel Chem. **52**(2), 386 (2007)
[24] J.P. Meyers, R.M. Darling, J. Electrochem. Soc., **153**, A1432 (2006)
[25] T.F. Fuller, G. Gray, ECS Trans. **1**(8), 345 (2006)
[26] N. Takeuchi, T.F. Fuller, ECS Trans. **11**(1), 1021 (2007)
[27] N. Takeuchi, T.F. Fuller, J. Electrochem. Soc. **155**, B770 (2008)
[28] J. Hu, P.C. Sui, S. Kumar, N. Djilali, ECS Trans. **11**(1), 1031 (2007)
[29] J. Hu, P.C. Sui, N. Djilali, S. Kumar, ECS Trans. **16**(2), 1313 (2008)
[30] A. Gidwani, K. Jain, S. Kumar, J.V. Cole, ECS Trans. **16**(2), 1323 (2008)
[31] A.A. Franco, M. Gerard, J. Electrochem. Soc. **155**, B367 (2008)
[32] A.A. Franco, M. Gerard, M. Guinard, B. Barthe, O. Lemairea, ECS Trans. **13**(15), 35 (2008)
[33] K.C. Neyerlin, W. Gu, J. Jorne, H.A. Gasteiger, J. Electrochem. Soc. **154**, B631 (2007)
[34] K.C. Neyerlin, W. Gu, J. Jorne, H.A. Gasteiger, J. Electrochem. Soc. **153**, A1955 (2006)
[35] P.T. Yu, W. Gu, H.A. Gasteiger, *Internal experimental data*, Electrochemical Energy Research Laboratory, General Motors Research and Development (2007)
[36] S.G. Bratsch, J. Phys. Chem. Ref. Data **18**, 1 (1989)
[37] K. Kinoshita, *Electrochemical Oxygen Technology* (Wiley, New York, NY, 1992)
[38] L.B. Kriksunov, L.V. Bunakova, S.E. Zabusova, L.I. Krishtalik, Electrochim. Acta **39**, 137 (1994)
[39] A.E. Bolzan, A.J. Arvia, J. Electroanal. Chem. **375**, 157 (1994)
[40] I.V. Barsukov, M.A. Gallego, J.E. Doninger, J. Power Sources **153**, 288 (2006)
[41] K. Kinoshita, *Carbon* (Wiley, New York, NY, 1988)
[42] Y. Liu, M. Murphy, D. Baker, W. Gu, C. Ji, J. Jorne, H.A. Gasteiger, ECS Trans. **11(1)**, 473 (2007)
[43] Y. Liu, M.W. Murphy, D.R. Baker, W. Gu, C. Ji, J. Jorne, H.A. Gasteiger, J. Electrochem. Soc. **156**, B970 (2009)
[44] C.K. Mittelsteadt, H. Liu, in *Handbook of Fuel Cells – Fundamentals, Technology and Applications, Volume 5: Advances in Electrocatalysis, Materials, Diagnostics and Durability*, Part 1, ed. by W. Vielstich, H. Yokokawa, H.A. Gasteiger (Wiley, New York, NY, 2009)

[45] A.E. Fischer, G.M. Swain, J. Electrochem. Soc. **152**, B369 (2005)
[46] E. Antolini, Appl. Catalysis B: Environ. **88**, 1 (2009)
[47] S.D. Knights, K.M. Colbow, J. St.-Pierre, D.P. Wilkinson, J. Power Sources **127**, 127 (2004)
[48] T.R. Ralph, S. Hudson, D.P. Wilkinson, ECS Trans. **1**(8), 67 (2006)
[49] C.R. Wilke, J. Chem. Phys. **18**, 517 (1950)
[50] C.R. Wilke, Chem. Eng. Prog. **46**, 95 (1950)
[51] D. Baker, C. Wieser, K.C. Neyerlin, M.W. Murphy, ECS Trans. **3**(1), 989 (2006)
[52] D.R. Baker, D.A. Caulk, K.C. Neyerlin, M.W. Murphy, J. Electrochem. Soc., **156**, B991 (2009)
[53] S.V. Patankar, *Numerical Heat Transfer and Fluid Flow* (Taylor & Francis, New York, NY, 1980)
[54] R.N. Carter, W. Gu, B. Brady, K. Subramanian, H.A. Gasteiger, in *Handbook of Fuel Cells – Fundamentals, Technology and Applications, Volume 6: Advances in Electrocatalysis, Materials, Diagnostics and Durability*, Part 2, ed. by W. Vielstich, H. Yokokawa, H.A. Gasteiger (Wiley, New York, NY, 2009)
[55] R.N. Carter, S.S. Kocha, F.T. Wagner, M. Fay, H.A. Gasteiger, ECS Trans. **11**(1), 403 (2007)
[56] K.G. Gallagher, D.T. Wong, T.F. Fuller, J. Electrochem. Soc. **155**, B488 (2009)
[57] S. Maass, F. Finsterwalder, G. Frank, R. Hartmann, C. Merten, J. Power Sources **176**, 444 (2008)
[58] K.G. Gallagher, R.M. Darling, T.F. Fuller, in *Handbook of Fuel Cells – Fundamentals, Technology and Applications, Volume 6: Advances in Electrocatalysis, Materials, Diagnostics and Durability*, Part 2, ed. by W. Vielstich, H. Yokokawa, H.A. Gasteiger (Wiley, New York, NY, 2009)
[59] B.E. Conway, V. Birss, J. Wojtowicz, J. Power Sources **66**, 1 (1997)
[60] L.C. Colmenares, A. Wurth, Z. Jusys, R.J. Behm, J. Power Sources **190**, 14 (2009)
[61] Y. Shao, J. Wang, R. Kou, M. Engelhard, J. Liu, Y. Wang, Y. Lin, Electrochim. Acta **54**, 3109 (2009)
[62] N. Takeuchi, T.F. Fuller, ECS Trans. **16**(2), 1563 (2008)

3

Cold Start of Polymer Electrolyte Fuel Cells

Kazuya Tajiri[1] and Chao-Yang Wang[2]

[1]*Argonne National Laboratory, 9700 S. Cass Avenue, Argonne, IL 60439, USA*
[2]*Electrochemical Engine Center (ECEC), and Department of Mechanical and Nuclear Engineering, The Pennsylvania State University, University Park, PA 16802, USA*

1. INTRODUCTION

The ability of polymer electrolyte fuel cells (PEFCs) to startup and operate under subzero temperatures has been an issue for the commercialization of the fuel cell vehicle (FCV). It is widely believed that during PEFC operation in a subzero temperature environment a portion of water produced from the oxygen reduction reaction (ORR) forms ice in the catalyst layer (CL) that hinders the oxygen transport to the reaction sites, until the PEFC eventually stops operation due to oxygen starvation. For the automotive application, successful cold start is defined as PEFC temperature increase above 0°C with self-heating before the cell shutdown due to oxygen starvation. Several automakers have already claimed capability of FCV startup from a subzero temperature environment. However, the underlying physics has only begun to emerge in the most recent literature.

This chapter describes PEFC cold-start characteristics and performance, focusing on water management and electrochemical kinetics at subzero temperatures. Other issues related to low

U. Pasaogullari, C.-Y. Wang (eds.), *Modeling and Diagnostics of Polymer Electrolyte Fuel Cells*, Modern Aspects of Electrochemistry 49,
DOI 10.1007/978-0-387-98068-3_3, © Springer Science+Business Media, LLC 2010

temperatures, such as degradation due to thermal cycling without electrochemical water production or due to cold start, are not covered in this chapter but have been presented elsewhere; see Wang et al.[1] In addition, the chapter is limited to the so-called "isothermal cold start." Temperature increase is expected in a fuel cell stack due to heat generation during cold start. However, the dynamically evolving temperature makes it difficult to study the basic phenomena occurring in fuel cells and to understand the underlying physics, because all material properties and kinetic parameters are temperature-dependent. In order to avoid this complexity the method called isothermal cold start is often employed. In this method the thermal mass of each component is made so large that the temperature increase during cold-start operation becomes negligibly small. Thus the temperature during cold start is nearly constant, making analysis relatively easy.

As mentioned earlier, cold start of the fuel cell system is considered successful when the cell temperature increases above 0°C before the cell shuts down due to oxygen starvation. Next, how long must the cell be able to run under such a condition? A simple, back-of-envelope calculation can be made. Fuel cell operation generates both heat and water. Heat generation from the fuel cell operation is calculated with the relation

$$q\left(\mathrm{J/cm^2}\right) = \int (E_\mathrm{h} - V_\mathrm{cell}) I \ \mathrm{d}t \tag{1}$$

where q is the heat generation, E_h and V_cell are the thermodynamic equilibrium potential and the cell voltage in V, respectively, and I is the current density in A/cm^2. The water production is

$$\Delta m_{\mathrm{H_2O}}\left(\mathrm{mg/cm^2}\right) = \frac{\int I\,\mathrm{d}t}{2F} M_{\mathrm{H_2O}} \tag{2}$$

where $\Delta m_{\mathrm{H_2O}}$ is the produced water, F is the Faraday constant, and $M_{\mathrm{H_2O}}$ is the molecular weight of water. Assuming that the V_cell is constant at $V_\mathrm{cell} = 0.5$ V and that the temperature is –30°C, the relation between heat and water generation is obtained as below.

$$q\left(\mathrm{J/cm^2}\right) = \frac{2F\left(E_\mathrm{h} - V_\mathrm{cell}\right)}{M_{\mathrm{H_2O}}} \Delta m_{\mathrm{H_2O}} \approx 9.43 \Delta m_{\mathrm{H_2O}} \left(\mathrm{mg/cm^2}\right) \tag{3}$$

Thus, the heat release is directly related to the amount of product water. The next consideration is the amount of heat needed to raise fuel cell temperature from, for example, −30 to 0°C ($\Delta T = 30$ K). The thermal mass of the fuel cell components comes in large part from the bipolar plates (BPPs), neglecting the end plates. With graphite bipolar plates of 1 mm thickness each, and assuming an adiabatic system, the required heat is

$$q = mc_p\Delta T \approx 9.4\ \left(\mathrm{J/cm^2}\right) \tag{4}$$

Combining Eqs. (3) and (4), we can conclude that product water of approximately 1 mg/cm^2 is a criterion for successful self-startup. As shown with Eqs. (3) and (4), the temperature increase is proportional to the product water, and therefore, the product water from the fuel cell operation, Δm_{H_2O}, is considered a quantitative index to measure the cold-start performance.

Knowing that 1 mg/cm^2 of product water is a threshold, how much water can be stored at maximum within each component of the fuel cell, and how much can be removed to the outside? For the cathode catalyst layer (CCL) with typical thickness of 10 μm and 50% pore volume fraction, the CCL water storage capacity is approximately 0.5 mg/cm^2. A 30-μm-thick membrane can store ~1.5 mg/cm^2 of water, but its actual water storage capacity depends on the initial water content, λ_i, and therefore is proportional to (λ_{sat}–λ_i), where λ_{sat} denotes the water content of a fully hydrated membrane. The escape of water into the GDL is unlikely due to the very low vapor pressure at cold-start temperatures ($P_{v,sat} = 40$ Pa at −30°C). For reference, the GDL with 300 μm thickness and 50% porosity would store about 15 mg/cm^2 of water, if it could be fully utilized. This capacity is too large to be used for cold start. From this simple estimation we can conclude that the CCL water storage capacity alone is not sufficient for successful cold start and that a successful strategy is to store water in the membrane.

Only a few years ago very little was known about the physical nature of PEFC cold start, and despite its importance to fuel cell deployment, there was limited active research in this area (albeit the existence of a large number of system-based patents that do not necessarily address fundamental questions). Recently, however, an increasing number of basic studies have been reported.

Kagami et al.[2] and Hishinuma et al.[3] explored the cold-start performance of a 104-cm^2 single cell both experimentally and numerically. In their cold-start experiment, they varied several operating parameters, such as current density, startup temperature, and gas pressure, and recorded the voltage change with time. In their numerical model, the CL and GDL were treated as an interfacial layer between the membrane and the bipolar plate with zero thickness. Furthermore, the water storage role of the membrane was neglected, which later turned out to be quite important for cold start. With these assumptions they concluded that from the steady-state water balance between the water production rate and the water vapor removal rate, self-startup was possible at −5°C or higher, but impossible at lower temperature without external heat assist.

Oszcipok et al.[4] conducted cold-start experiments and statistical analysis of the experimental results. Notable features of their experiments are as follows. First, they began cold start from very low proton conductivity initial condition (1 kHz AC impedance of ~15 Ω cm^2). Second, they used constant voltage operation that is not common in the practical FC operational mode. Because of these two features, their fuel cell could survive a very long period of cold start (~30 min). They defined the cumulative charge transfer density as a measure of cold-start capability, and fitted it statistically with the air flow rate during startup, membrane water content prior to startup, and cell performance at room temperature prior to cold start. With their fitting data it may be possible to predict the cumulative charge transfer density from the initial and operating conditions. Subsequently, Oszcipok et al.[5] experimentally studied the effects of membrane thickness, types of GDL, and gas flow rates on the cumulative charge transfer density, and developed a similar empirically based statistical relation to the previous study.

Ge and Wang also visualized the fuel cell cathode during cold start.[6] Using a silver mesh as cathode GDL, they observed that when product water was less than 0.56 mg/cm^2 water was not seen on the catalyst layer surface and that when it reached 1.12 mg/cm^2 the liquid water emerged from the catalyst layer surface. This is consistent with the roughly estimated value of cathode catalyst layer water storage capacity of ~0.5 mg/cm^2. They also estimated from their experiment that the freezing-point depression of water in the catalyst layer was at most 2°C. Then, they made the fuel cell optical configuration more realistic using carbon paper GDL punched with microholes instead of silver mesh.[6] In this paper they reported the importance of the purge duration prior to cool down, with the most useful purge duration being between 90 and 120 s, and noted that cell HFR relaxation occurs after purge because of water redistribution. Subsequently, they used a cyclic voltammetry technique to examine the ECA change due to cold start.[7] It was found that the Pt area decreases as the product water increases during cold start and that the Pt area loss substantially recovers with higher temperature operation. However, both permanent loss in the active Pt area and cell performance degradation are observed and become more severe as the cold-start temperature decreases, or as the duration of cold start increases.

Thompson et al. reported a series of studies on PEFC cold start. First, they measured the characteristics of Nafion membrane under low temperature, such as the proton conductivity or the activation energy of proton conduction.[8] Second, they investigated the ORR kinetics under subfreezing temperature.[9] They used thick membranes and short-duration testing for maximum control of the membrane water content, and found that, as observed at above-freezing temperatures, the ORR kinetics decreased with decreasing water content at subfreezing temperatures. This was attributed to the effect of reduced proton activity. The kinetic parameters, such as the exchange current density, specific activity, and Tafel slope, were found to be consistent with the kinetics observed at above-freezing temperatures, and the activation energies were found to be constant within the temperature range of −40 to 55°C, indicating no change in the fundamental reaction mechanisms. They subsequently studied the water storage characteristics of membrane and catalyst layer during cold start using galvanostatic discharge operation and cryo-scanning electron microscopy.[10] At low current densities the membrane absorbed product water to its full capacity, and the catalyst layer pores were completely filled with product water. With increasing current density, the fraction of water storage utilization in the membrane, as well as the fraction of water-filled CL pores, decreased. As such, they concluded that water storage utilization was mainly controlled by current density, and less dependent on the initial water content or electrode thickness. Most recently they reported their study on ice formation dynamics and current distribution within the electrode, along with the overpotential decomposition at the beginning of cold start with various current densities.[11]

Li et al. visualized ice formation and distribution in the catalyst layer during cold start process using a cryo-scanning electron microscope.[12] The fuel cell was operated to control the amount of product water during cold start, and then immediately cooled down to −75°C for the sample preparation. SEM images were taken at −170°C environment. They observed that even after the purge process, 75% of the cathode catalyst layer pores were filled with ice. After cold start, the porosity decreased from 65% in its dry condition to 13% after cold start with ~0.3 mg/cm^2 product water, and to 8% with ~0.6 mg/cm^2 product water, which caused cell shutdown. They also observed that after ~0.6 mg/cm^2 water production operation, ~0.25 mg/cm^2 of product water remained in ice phase in the catalyst layer and the microporous layer. The remaining product water was thought to be absorbed in the membrane.

For computational modeling of PEFC cold start, Mao and Wang[13] first developed an analytical model, and others followed. The model of Mao and Wang[13] described heat balance, ice formation, water transport, and voltage change during cold start. The ice fraction in the catalyst layer is modeled to affect cold start in two ways: one is the blockage of oxygen diffusion and the other is the surface coverage of the catalyst. With this model they identified the key parameters controlling cold-start performance as the initial membrane water content and the thermal mass of the bipolar plates. Next, Mao et al.[14] developed a multiphase, transient computational model based on commercial CFD code to describe cold-start phenomena, including mass transport, electrochemistry, and ice formation. The results were validated with experimental data and the effects of startup current density and membrane thickness were numerically explored.

Wang[15] investigated heat and mass transport and electrochemical kinetics in the cathode catalyst layer during cold start, and identified the key parameters characterizing cold-start performance. He found that the spatial variation of temperature was small under low current density cold start, and thereby developed the lumped thermal model. A dimensionless parameter, defined as the ratio of the time constant of cell warm-up to that of ice volume growth, was found to be important for successful self-cold start.

Jiang et al.[16] developed a multiphase, three-dimensional model to describe non-isothermal cold start and to study the effect of temperature rise. Due to the temperature rise during cold start, more water was transported into the membrane and less ice formation occurred in the catalyst layer. It was also found that a lumped thermal analysis significantly overestimated the overall thermal requirement for successful self-cold start. In addition, pre-startup conditions such as gas purge had significant impact on cold start that implied the importance of the shutdown process.

Recently Meng[17] developed a transient, multiphase, multidimensional PEFC model to elucidate the fundamental physics of cold start. The results showed the importance of water vapor concentration in the gas channels, which implies that large gas flow rates benefit cold-start performance. They also found that ice growth in the cathode catalyst layer during cold start was faster under the land than under the gas channels, and accumulated more at the interface between the cathode catalyst layer and GDL.

The remainder of this chapter consists of two major topics followed by concluding remarks. Section 2 discusses startup under subzero temperatures, while Section 3 examines shutdown condition prior to cool down. In Section 2, the concept of "isothermal cold start" is introduced, and the cold-start capability and characteristics of the PEFC are discussed with idealized initial condition. Since cold start is inherently transient phenomena with the time scales in the order of minutes at maximum, the initial condition plays a dominant role. In order to examine the cold-start intrinsic capability, the initial condition needs to be precisely controlled, after which a more practical procedure for fuel cell shutdown is used to simulate FCV operation. It will be shown that any practical shutdown method can utilize only a portion of the MEA cold-start capability, demonstrating the importance of the procedure used for FC system shutdown. Subsequently, Section 3 studies FCV shutdown procedure, discussing water removal from the PEFC during the gas purge process. For vehicular application, the gas purge should be of short duration, at most 1 min, but must still remove the water effectively. Hence, the key factors controlling purge effectiveness are reviewed and the method to predict purge effectiveness is identified.

2. EQUILIBRIUM PURGE COLD START

2.1. Equilibrium Purge

Gas purge is a common and integral part of PEFC cold start in the industry. The primary purpose of gas purge is to remove residual water from a PEFC prior to cell shutdown and cool down. The gas purge process thus defines the initial condition of water distribution throughout a cell and plays a crucial role in cold start. While gas purge is commonly conducted using dry nitrogen/air for durations ranging from tens of seconds to a few minutes, here we introduce a method using partially humidified gas with well-controlled relative humidity (RH) to purge for extended periods of time (in the order of several hours). After this purge, all liquid water should be evaporated away from all components of the cell, and the membrane water content reaches thermodynamic equilibrium with the RH of purge gas (thus termed equilibrium purge).[18] Equilibrium purge permits the initial water content in the membrane to be precisely defined and controlled (through the RH control in purge gas), and hence, the important effect of initial water distribution on cold-start performance can be investigated, as will be illustrated in this section.

In equilibrium purge, water content of the MEA can be easily calculated from the purge gas RH via the membrane water uptake curve such as[19]

$$\lambda = 0.043 + 17.81a - 39.85a^2 + 36.0a^3 \tag{5}$$

where λ is water content in the membrane and a is the relative humidity of purge gas. Typically the equilibrium purge takes 2–3 h to complete.

Once equilibrium purge is concluded, the inlet and outlet gas lines are closed for both anode and cathode in order to keep the cell's internal condition unchanged, and the cell is then cooled down in an environmental chamber. When the cell temperature reaches a specified startup temperature, the cell is left at that temperature for more than 1 h before proceeding to the next step in order to ensure that the entire cell becomes isothermal at the ambient temperature.

During cool down, water content in the membrane can be safely assumed to remain constant or "locked" in place. This assumption can be justified by the fact that the cell compartment on each side is about 2 cm^3 for typical laboratory-scale cell with 25 cm^2 active area, and its filling gas at the dew point of 30°C after equilibrium purge contains ~0.06 mg of water, equivalent to an increase in membrane water content of $\Delta\lambda = 0.02$. Thus, the change in membrane water content during cool down is negligible, making it possible to predetermine the membrane water content at the beginning of cold start.

2.2. Isothermal Cold Start

In practice, cold start of a PEFC stack is inherently non-isothermal, as power generation produces waste heat that causes the cell temperature to rise. However, non-isothermal cold-start data vary greatly from one system to another and strongly depend on the thermal mass of the test cell/stack. In order to determine key MEA characteristics to ensure self-startup of an automotive PEFC from, for example, −30°C, and ultimately to relate these characteristics to materials and control requirements, isothermal cold start as proposed here is uniquely useful. Isothermal cold start can be easily realized in laboratories by using a single cell fixture with sufficiently large thermal mass. While isothermal cells are bound to shut down (the cell temperature is always kept constant at the ambient temperature of startup), the product water so measured before cell shutdown is a quantitative measure of the intrinsic cold-start capability of a MEA that is completely independent of the thermal mass of a specific cell. Such experimental data are therefore of benchmark quality, highly reproducible in laboratories (independent of cell fixtures) and practically useful for stack design as the amount of product water also roughly corresponds to the amount of waste heat able to be generated. Lastly, isothermal cold start provides the most conservative scenario by which the impact of MEA material and design on performance in PEFC cold start is examined.

2.3. Proton Conductivity at Low Temperature

There are a substantial number of studies reported on the characteristics of the membrane at low temperatures. This may be because Nafion membrane has long been used for many industrial and research applications, for instance in electrolysis and sensors.[20] The techniques used for these studies include nuclear magnetic resonance (NMR) and differential scanning calorimetry (DSC). Most of these studies are, however, ex-situ measurements, in conditions not directly applicable for PEFC cold-start study.

One study specifically designed for PEFC was reported by Thompson et al.[8] They used a direct current to measure the proton conductivity at low temperature. In conjunction with the DSC data, they found the dependency of crossover temperature (temperature where the activation energy changes) on water content and hysteresis between freezing and melting.

Most recently, Gallagher et al.[21] measured the water uptake of Nafion membrane under subfreezing temperatures, which showed a significant reduction in the maximum water content corresponding to membrane full hydration. The Nafion membrane with 1,100 equivalent weight, for example, uptakes $\lambda \sim 8$ of water at $-25°C$ when it equilibrates with vapor over ice because of the low vapor pressure of ice compared to supercooled liquid water. They also found the electro-osmotic drag coefficient to be ~ 1 for Nafion membrane under subfreezing temperatures.

The cool-down process of the cold-start experiment also provides an opportunity to obtain the membrane proton conductivity as a function of temperature at a known water content. Note that the temperature dependence of proton conductivity with low membrane water content is of particular interest here as PEFC cold start rarely involves fully hydrated membranes after gas purge. In addition, unlike PEFCs operated under normal temperatures, the membrane resistance under low water content and low temperature typical of cold start conditions is much greater than the contact resistance, making in-situ measurements of the membrane proton conductivity in a PEFC a simple but accurate method.

Figure 1 shows proton conductivity variation with temperature for varying membrane water content.[18] In the Arrhenius form the lines displayed in Fig. 1 are nearly straight and parallel to each other, indicating no noticeable phase transition of water occurring in the membrane in this temperature range between room temperature and $-30°C$ and water content range from 2.4 to 14. This finding is in

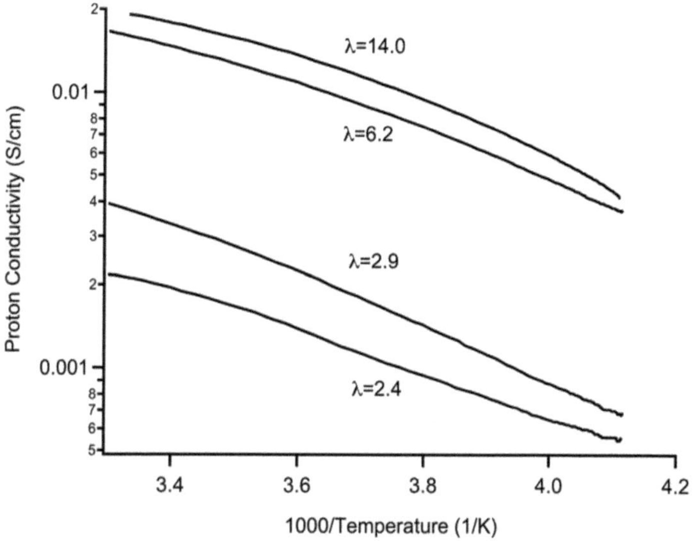

Figure 1. Proton conductivity as a function of temperature for various membrane water content λ between room temperature and $-30°C$. (reproduced with permission from Tajiri et al.[18])

good agreement with observations of McDonald et al.[22] for partially hydrated Nafion 112-based MEAs. For $\lambda_i = 14$ case, the conductivity curve shown in Fig. 1 is slightly convex. This may be because the 100% RH purge gas used in equilibrium purge could not remove all liquid water by evaporation from the interfaces between GDL and current-collecting lands, or between CL and GDL, resulting in water freezing at these interfaces during cool down and hence a change in contact resistance.

The conductivity data measured in situ in Fig. 1 are within the temperature range from room temperature (27°C) to $-30°C$. In contrast, Cleghorn et al. reported the proton conductivity for Gore-Select membranes in the temperature range of 40–100°C.[23] Extrapolating the correlation of Cleghorn et al. to 27°C and at 100% relative humidity, the membrane conductivity is calculated to be 0.027 S/cm, which is in reasonable agreement with our in-situ measurement of 0.021 S/cm.

2.4. Effects of Key Parameters

2.4.1. Initial Membrane Water Content

The voltage curve during cold start is shown in Fig. 2 for various initial membrane water contents prior to cold start.[18] In general, the voltage curve can be divided into three stages. It should be noted that the applied current initially increases in the first 80 s, and then is kept at a constant value (Fig. 2). In the first stage the cell voltage decreases as the applied current density increases. The voltage slightly increases in the second stage, primarily because of membrane rehydration from the product water and partly because of the reaction kinetics improvement due to increased water activity. The rapid voltage drop in the final stage is apparently due to hindrance of oxygen transport into an ice-filled CL.

During the initial current ramp period the cell voltage is higher as the initial membrane water content increases, as expected. In the

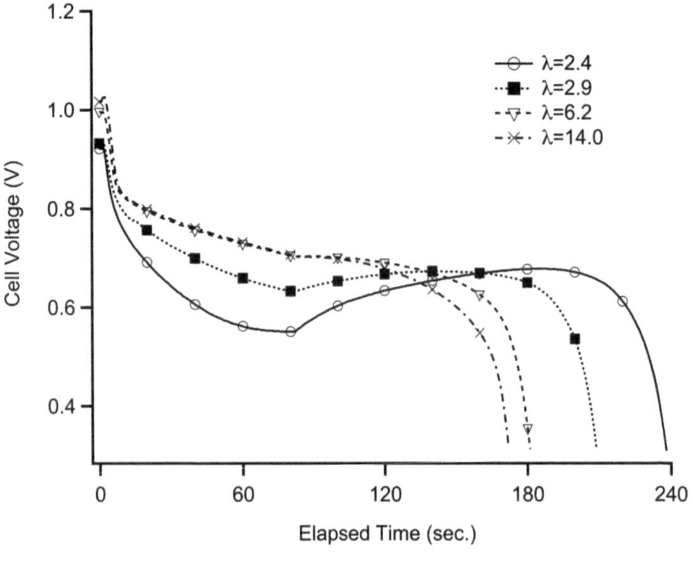

Figure 2. Effect of initial water content λ in the membrane on cold-start performance. The environmental temperature is −30°C. (reproduced with permission from Tajiri et al.[18])

voltage recovery period two characteristics can be noticed. First, the voltage recovery due to membrane rehydration decreases as the initial membrane water content increases. Secondly, the duration of this period is greater for lower initial water content. For $\lambda_i = 14$ case the voltage recovery is hardly seen and the profile moves directly from the initial period to the final voltage drop-down stage. This is because the membrane is already nearly saturated at the beginning of cold start and hence no room remains to accommodate product water diffusion. Consequently, the total operational time of cold-start discharge increases as the initial membrane water content decreases, demonstrating the profound significance of gas purge.

In light of the importance of water storage in partially dry membranes, it is instructive to correlate the cold-start performance quantified by product water (m_{H_2O}, mg/cm^2) with the membrane water uptake potential ($\Delta\lambda$) defined as:

$$\Delta\lambda = \lambda_{sat} - \lambda_i \tag{6}$$

where λ_{sat} denotes the water content of a fully hydrated membrane (taken to be 14 in this study but the analysis remains the same for other values of λ_{sat}) and λ_i the initial membrane water content prior to cold start. If a membrane is initially fully hydrated prior to cold start, i.e., $\lambda_i = \lambda_{sat}$, then the water uptake potential in the membrane is zero, and all product water must accumulate in the CL at the startup temperature of $-30°$C. In the other extreme, if gas purge is highly effective and the membrane is dry initially, then the water uptake potential reaches the maximum, thus resulting in more product water. Figure 3 displays such an experimental relationship between product water and membrane water uptake potential.[18] It can be said that in the case of $\Delta\lambda = 0$ (i.e., $\lambda_i = 14$) the majority of product water resides in the cathode CL. Thus, the measured product water (i.e., $\sim$0.5 mg/cm^2) in this condition is indicative of the water storage capacity in the CL. One can quickly estimate the latter by the following equation:

$$m_{CL} = \delta_{CL}\left(\varepsilon\rho_{ice} + \varepsilon_e c_{f,dry}\Delta\lambda M_{H_2O}\right) \tag{7}$$

where the first term on the right hand side accounts for ice storage in open pores of the CL and the second term for water storage in the electrolyte phase. Assuming typical CL properties (δ_{CL}=12 μm,

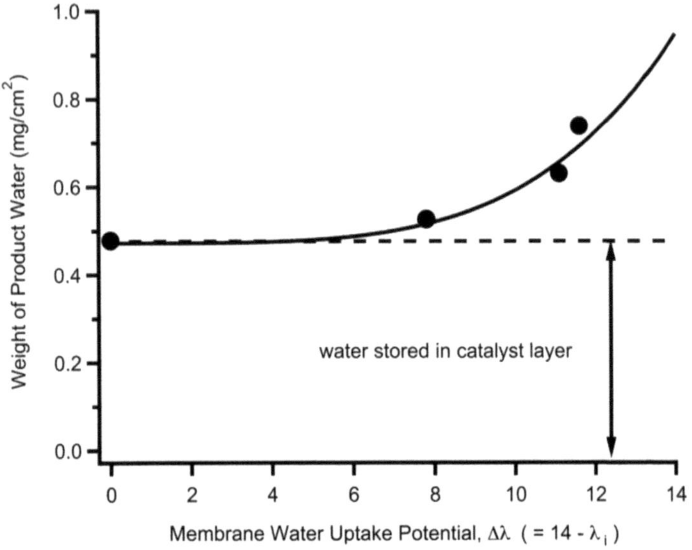

Figure 3. Relation between product water and membrane water uptake potential for the startup from −30°C. The dashed line indicates the amount of water stored in the catalyst layer and the extra portion above it denotes the amount of water diffused into the membrane. (reproduced with permission from Tajiri et al.[18])

$\varepsilon = 0.5$, $\rho_{ice} = 0.9$ g/cm^3, $\varepsilon_e = 0.2$, and $c_{f,dry} = 1.8\times10^{-3}$mol/cm^3), the CL water storage capacity is estimated to be 0.54 mg/cm^2 for $\Delta\lambda = 0$, which is in good agreement with the measured product water. This analysis further suggests that the vapor phase transport of product water out of CL into GDL is indeed negligible at the startup temperature of −30°C.

Figure 3 also shows that the total product water can be separated into the CL and membrane contributions for each water uptake potential in the membrane. While the CL contribution depends mainly on the CL pore volume and weakly on the water uptake potential $\Delta\lambda$ as indicated in Eq. (7), the membrane contribution varies nonlinearly with $\Delta\lambda$. This is because longer CL operation before ice filling and shutdown allows more membrane water uptake.

A fundamental map like Fig. 3 is useful for developing novel MEA materials and designs in order to enhance the intrinsic capability of PEFC cold start. For example, in order to earn more time for product water diffusion into the membrane, low water production rate or current density is preferred for isothermal cold start from startup temperatures as low as −20°C or −30°C. This conclusion implied from Fig. 3 can be experimentally verified as discussed below.

2.4.2. *Startup Current Density*

To see the effect of the startup current density, the current density profile applied to the cell is doubled, as shown in Fig. 4.[18] The initial membrane water content is $\lambda_i = 6.2$ for both cases. For the higher current density case the cell voltage drops gradually in the first 1 min and then falls into the voltage drop-down period quickly. Soon after the current density reaches 80 mA/cm^2, the voltage drops below 0.3 V and the cold-start operation stops. Thus, only two stages of voltage evolution are evident at this higher current density, as compared to the three stages experienced in the lower current density case shown in Fig. 4.

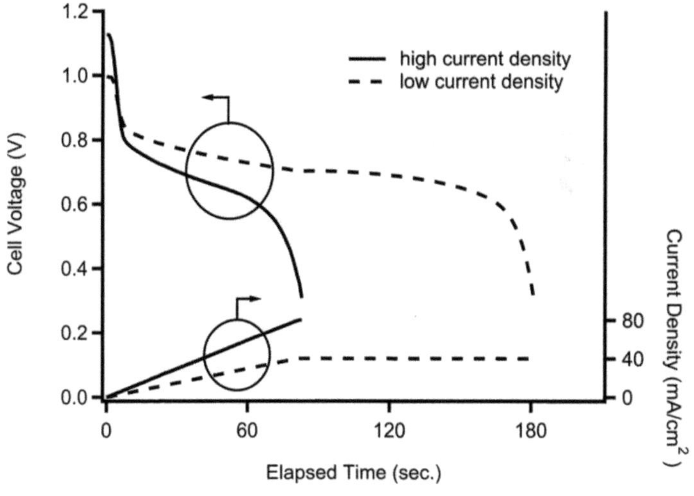

Figure 4. Effect of current density on voltage evolution during cold start from −30°C with initial membrane water content of $\lambda_i = 6.2$. (reproduced with permission from Tajiri et al.[18])

The cumulative product water at cell shutdown is 0.32 mg/cm^2 for the high current density, significantly smaller than the catalyst layer storage capacity. Mass transport limitation in the CL with lower product water at higher current density can be explained by the water distribution in the CL as schematically shown in Fig. 5. For the low current density case (Fig. 5a), the reaction current distribution is more uniform and product water has sufficient time to redistribute uniformly throughout the pores in the CL. Thus, the CL water storage capacity can be fully utilized. On the other hand, in the high current density case (Fig. 5b), the reaction current is non-uniform and is determined by a delicate balance between proton conduction through ionomers and O_2 transport through pores of the catalyst layer. Further, the dominance between the two resistances changes with time during cold start; the proton resistance continues to decrease with more water production, but ice formation exacerbates the O_2 transport. As a result, the peak reaction current tends to shift toward the CL/GDL interface in favor of O_2 transport, which leads to even more ice formation and aggravated mass transport limitation in that region. In addition, product water formed in the CL near the CL/membrane interface is likely to be absorbed in the membrane, leaving less water to precipitate as ice in CL pores. That is, product water may concentrate in the front portion of the electrode interfacing with GDL and form an ice sheet there to block oxygen

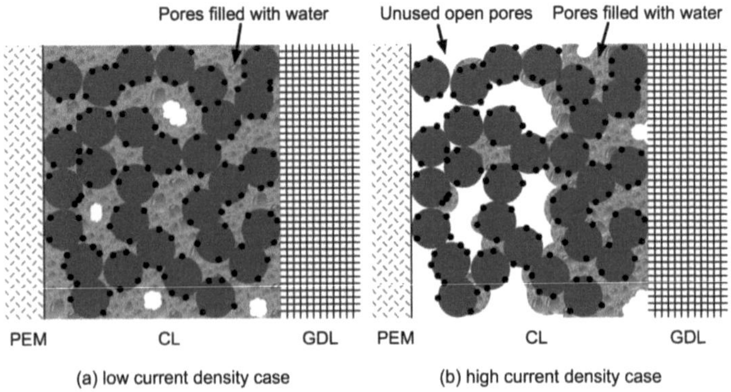

Figure 5. Schematic of water distribution at the end of cold start for (a) low current density and (b) high current density. (reproduced with permission from Tajiri et al.[18])

transport into the CL before all pores are occupied by product water. Thus the CL water storage capacity cannot be fully utilized and the CL is prematurely shut down under the high current density. The differing utilization in the CL water storage capacity is responsible for ~60% reduction in cold-start product water between the high and low current densities.

Thompson et al.[10] investigated the effect of a wide range of current density values as seen in Fig. 6. In this figure the vertical axis shows the charge storage utilization defined as

$$\frac{Q_{\text{measured}}}{Q_{\max}} = \frac{Q_{\text{measured}}}{Q_{\text{membrane}} + Q_{\text{electrode}}} \tag{8}$$

where Q_{membrane} and $Q_{\text{electrode}}$ denote the maximum charge storage in the membrane and electrode, respectively, and $Q_{\max}$ is the sum of these two. For effective cold start this parameter is desired to

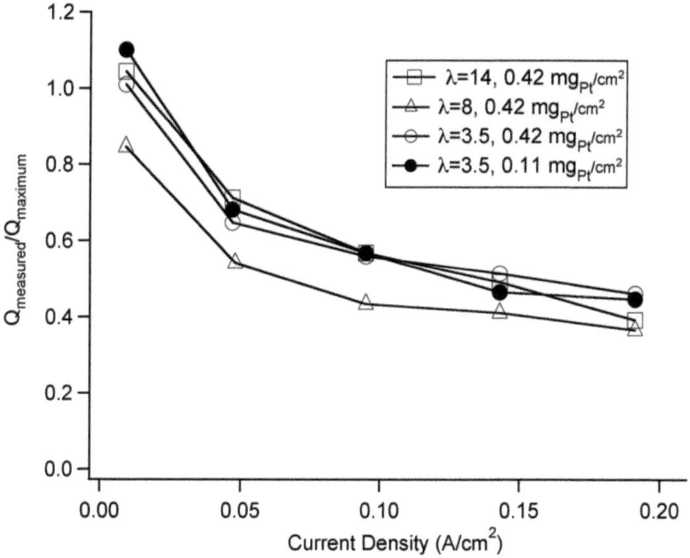

Figure 6. Charge storage utilization versus current density at −20°C for various initial water contents and electrode thickness. (reproduced with permission from Thompson et al.[10])

approach unity. This result shows the same trend observed in Fig. 4, that is, the current density increase leads to water storage utilization decrease. Furthermore, it can be observed that the charge storage utilization is relatively insensitive to the initial membrane water content λ while it strongly depends on the current density.

2.4.3. *Startup Temperature*

The effect of startup temperature was studied by starting the cell at ambient temperatures of −3, −10, −15, −20, and −30°C. Figure 7 depicts a strongly nonlinear relationship between the product water and the startup temperature. Primary reasons for large product water at high temperatures include high proton conductivity, large water diffusivity in the membrane, facile ORR kinetics, and high saturation vapor pressure.[24]

As implied in Fig. 7, at −3°C the cell can operate indefinitely at steady state. How, then, is the water production rate of 0.0037 mg/cm^2/s (40 mA/cm^2 current density) balanced by the water removal rate from the cathode CL? Assuming the exhaust gases from the cell are fully saturated with water vapor at cell

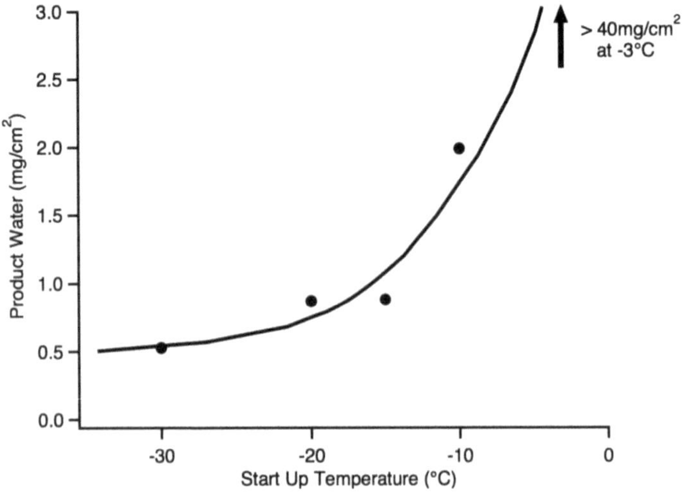

Figure 7. Relation between product water and startup temperature under equilibrium purge ($\lambda_i = 6.2$) and current density of 40 mA/cm^2. (reproduced with permission from Tajiri et al.[24])

temperature of −3°C, 0.0012 mg/cm²/sec of water vapor can be removed from the cell, which is only 30% of the production rate. Therefore, to achieve the steady-state operation at −3°C, 0.0025 mg/cm²/s of water must be removed from the cathode CL either in liquid or vapor. Indeed, Ge and Wang[6] observed that liquid water emerges from the CL surface at the cell plate temperature of −3°C, implying that the CL temperature is slightly higher than the ambient temperature because of the thermal gradient in the GDL resulting from heat generation. Thus, it is likely that the water produced in the cathode CL is transported to the GDL in liquid phase by capillary forces and may freeze in the GDL pores or be further transported to the gas channels.

2.5. ORR Kinetics at Low Temperatures

Detailed investigation of ORR kinetics at subzero temperature was first reported by Thompson et al.[9] Fig. 8 shows the Tafel plots measured at −20°C with various membrane water content. At high

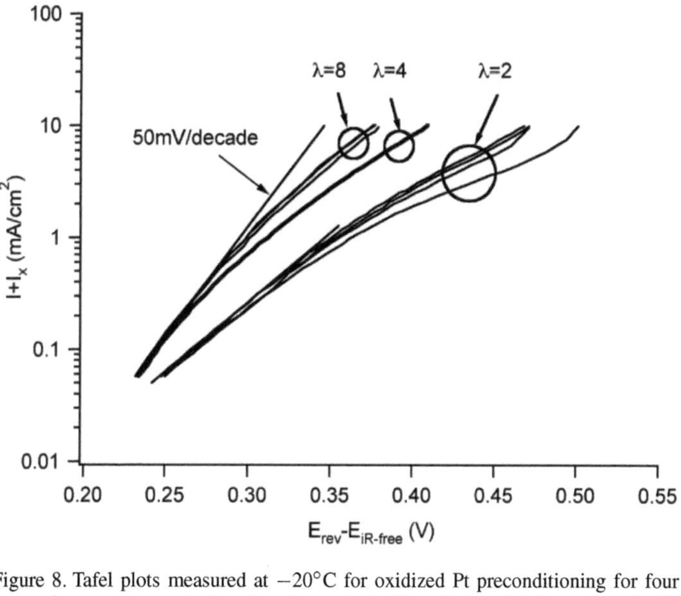

Figure 8. Tafel plots measured at −20°C for oxidized Pt preconditioning for four successive measurements at each water content of λ = 2, 4, and 8. (reproduced with permission from Thompson et al.[9])

water content ($\lambda = 4$ and $\lambda = 8$), the four consecutive experiments are reproducible and the Tafel slope is 50 mV/dec at the low current density. On the other hand, the low water content case ($\lambda = 2$) shows a much higher Tafel slope. Figure 9 is the Arrhenius plots of current density for various overpotentials between +55°C and −40°C. Following the Arrhenius-type behavior, the logarithm of the crossover-corrected current density ($\ln[i+i_x]$) is inversely proportional to the temperature. The dependency of the cathode activation energy $E_c{}^{\eta}$ on the cathode overpotential η is expressed as

$$E_c^{\eta} = E_c^{rev} - \alpha_c F \eta \tag{9}$$

where $E_c{}^{rev}$ is the cathode activation energy at zero overpotential and $\alpha_c = 1$. The activation energy shown in Fig. 9 decreases with increasing overpotential as expressed in Eq. (9). This indicates that the ORR kinetics do not change over the wide temperature range.

Another interesting observation was made by Ge and Wang.[25] They conducted cyclic voltammetry at subzero temperatures on

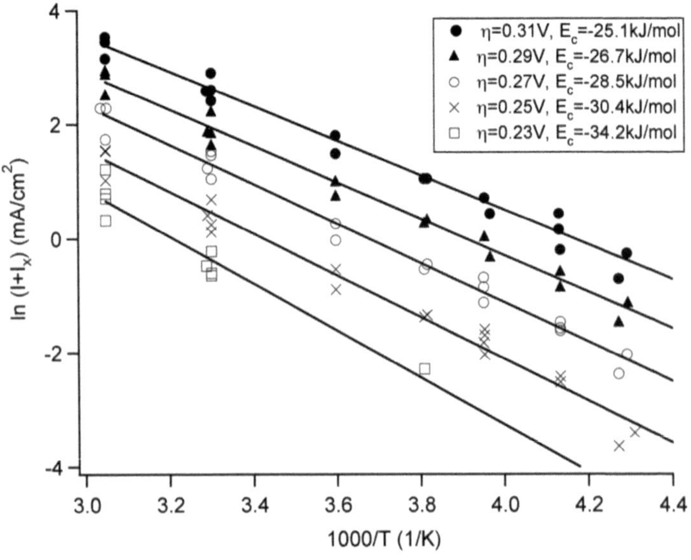

Figure 9. Arrhenius plots of current density for various overpotentials measured from +55°C down to −40°C. Activation energies for each overpotential are indicated in the legend. (reproduced with permission from Thompson et al.[9])

partially ice-filled electrode and correlated the ECA loss with the product water during cold start and with startup temperature (Fig. 10). They hypothesized that the ECA loss is attributed to ice

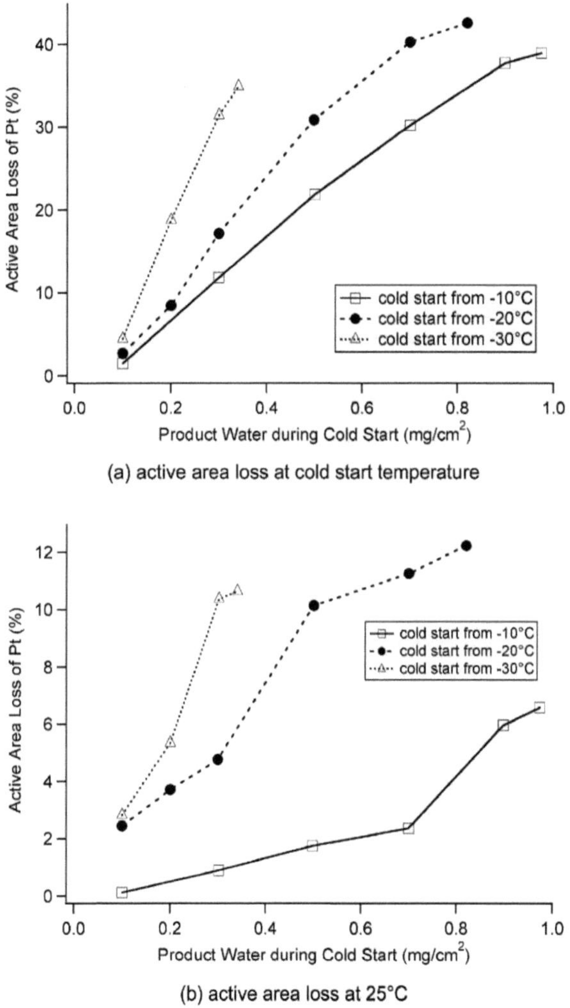

Figure 10. Pt active area loss at (a) startup temperature after cold start and (b) 25°C. (reproduced with permission from Ge and Wang[25])

sheet formation between the Pt particles and the ionomers in the cathode catalyst layer. When the cell was warmed up to 25°C after cold start, the Pt area loss still existed, but the extent of the loss was smaller than observed immediately after cold start. It is believed that this Pt area loss at 25°C was caused when, even after the ice sheet melted, a thin film of liquid water was trapped between the Pt particles and ionomers. They also found that the Pt area loss is mostly recovered once the cell was operated at 70°C.

2.6. Short-Purge Cold Start

The second purge method "short purge" was designed to simulate realistic operation in a FCV. When a FCV is stopped and turned off, the fuel cell is purged to remove water from the open pores of the cell and from the membrane to prepare for the next startup, which may be at subzero temperature. Unlike the equilibrium purge, which was developed solely for research purposes, the purge conducted in the FCV must be of short duration to reduce the energy required, and must also be sufficiently effective to make the upcoming startup successful. Therefore, it is important to make the product water in short-purge cold start as close as possible to that of equilibrium-purge cold-start performance, since the latter indicates the intrinsic cold-start capability of the MEA at given conditions. Three factors in this short-purge condition differ from the equilibrium-purge cold start: (1) the purge duration is at most 2 min, (2) the purge gas is dry, and (3) the applied current density during cold start is 100 mA/cm^2.

Again unlike the equilibrium purge where membrane water content is controlled by the purge gas RH, in short purge the membrane water content cannot be predicted prior to purge. Instead, the membrane water content λ is inversely calculated from the membrane proton conductivity κ_e (in S/cm) at a purge temperature using the correlation proposed by Springer et al.[19]

$$\kappa_e = \exp\left[1268\left(\frac{1}{303} - \frac{1}{T}\right)\right](0.005139\lambda - 0.00326) \qquad (10)$$

The cold-start performance for two different purge modes are directly compared in Fig. 11 for −20°C startup with various initial

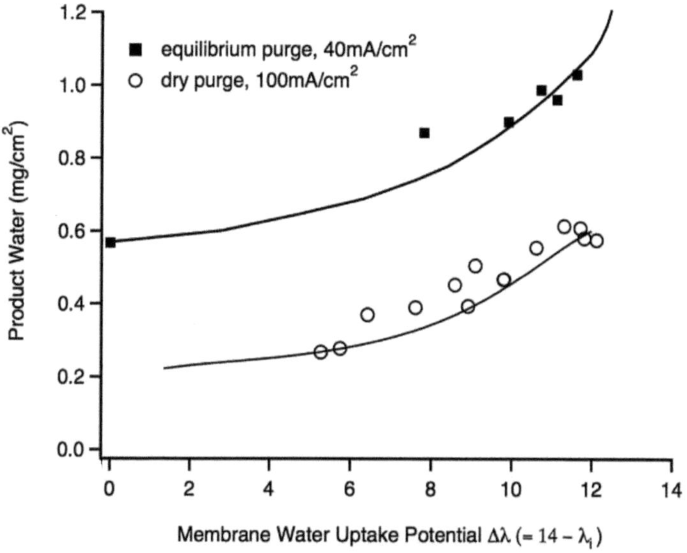

Figure 11. Relation between product water and membrane water uptake potential for the startup from −20°C for different purge methods. (reproduced with permission from Tajiri et al.[24])

membrane water content.[24] It should be noted that two major factors differ between these two experimental sets: the purge method as well as the startup current density (40 mA/cm^2 versus 100 mA/cm^2). Therefore, the difference in product water shown in Fig. 11 is contributed by both factors.

Generally, the product water is seen to increase with the membrane water uptake potential $\Delta\lambda$. Furthermore, the equilibrium purge startup has roughly 0.5 mg/cm^2 higher product water than that of the short-purge startup at −20°C. The reasons for lower product water in short-purge startup are thought to be the higher current density and the presence of residual water in the CL pores.

Thus, it seems the short-purge cold start does not fully utilize the cold-start capability of MEA. Deeper understanding of the purge mechanisms will help to greatly improve short-purge cold-start performance. This is the motivation of the next section.

3. WATER REMOVAL DURING GAS PURGE

3.1. Introduction

Gas purge is an integral part of the frequent shutdown process of PEFC, as a FCV typically goes through a large number of start and stop cycles. In some situations, gas purge serves to avoid or minimize various types of material degradation in a fuel cell stack, some of which are permanent and catastrophic. In other situations, gas purge is intended to remove water from the fuel cell to avoid ice formation and blockage when exposed to subfreezing temperatures.

Much research has reported possible carbon corrosion on the air electrode due to residual hydrogen in the anode during idle state. This degradation is primarily caused by the presence of a H_2/O_2 front in the anode along with high cathode voltage (open circuit voltage), and can be prevented by fully removing the hydrogen from the anode compartment. Lee et al. studied the effect of the residual hydrogen in the anode compartment during the idling using polarization measurement, cyclic voltammetry, and electron microscopy, and found that hydrogen removal from the anode channels by gas purge helps prevent fuel cell degradation.[26]

On the cathode side, on the other hand, the primary purpose of gas purge is to remove water from the cathode compartment, particularly in preparation for cold start from subzero temperatures. As gas purge defines the initial condition of water distribution in a cell, it is a crucial step in PEFC cold start. Recent experimental studies have amply shown that not only performance but also material durability of PEFC hinges strongly upon the gas purge process prior to cool down and cold start. This is because an effective gas purge can remove water from the catalyst layer and membrane, thereby creating space for water produced in cold start to be stored.

The method of equilibrium purge was introduced in the preceding section. The distribution of water in the cell is well controlled in equilibrium purge, thus providing an excellent experimental technique for fundamental research of PEFC cold start and other problems requiring highly reproducible gas purge practice. Whereas the equilibrium purge is useful for laboratory experiments, practical gas purge for FCVs requires much shorter duration, preferably less than 60 s, and high efficiency. Therefore, a fundamental understanding of water removal during practical gas purge is necessary. Unfortunately, very little is found in the literature on purge mechanisms for automotive application. St-Pierre et al.[27] showed that cool purge at 20°C is better than hot purge at 85°C. Performance losses were not observed in a cell purged with dry gas at 20°C after the freeze/thaw cycling. Bradean et al.[28] touched upon purge effectiveness, using one section of their paper to briefly present a one-dimensional model and experimental results, and concluded that the purge cell temperature is the most critical parameter for an effective purge. However, they did not seek a fundamental understanding of purge mechanisms.

Ge and Wang[7] measured the cell high-frequency resistance (HFR) during gas purge with various purge durations and demonstrated that the cell HFR directly impacts the amount of product water generated or cell operational time in isothermal cold start. Sinha et al.[29] used X-ray microtomography to dynamically visualize the liquid water in the gas diffusion layer (GDL) and calculated the variation of liquid saturation with time during room-temperature purge. St-Pierre et al.[30] applied a residence time distribution method and demonstrated its potential to detect the amount of liquid water in the gas channels and the gas diffusion electrodes. Lee et al.[31] studied water removal characteristics during gas purge, where the residual water in a fuel cell after gas purge was forced to evaporate under low pressure and the gaseous pressure was measured to estimate the amount of residual water. They reported the general trend of the water removal with various parameters such as flow rate and purge temperature, but without much discussion of water removal mechanisms. Most recently, Sinha and Wang[32] presented a comprehensive theoretical description of water removal phenomena during gas purge, classifying the process into four stages: through-plane drying, in-plane drying, vapor transport, and membrane equilibrium stages. The theoretical predictions further concluded that high cell temperature with low RH and high flow rate of purge gas facilitates water removal.

This section concerns experimental characterization of gas purge typical of automotive applications. Much attention has been paid to finding a procedure to achieve reproducible purge experiments. A purge curve is defined and stages of gas purge are elaborated. Two characteristic parameters are then introduced to describe purge performance. Finally, a comprehensive set of experimental results characterizing automotive gas purge is reviewed.

3.2. Purge Curve

It should be noted that the purpose of gas purge is not to address how much water is removed from the cell or how thoroughly the GDL is dried. Instead, gas purge is ultimately intended to remove water from the catalyst layer and membrane, thereby creating storage space for water produced in cold start. Thus, it is proposed to use the membrane HFR (or more conveniently cell HFR) as the indicator to measure effectiveness of gas purge. A purge curve is thus defined as the cell HFR versus purge time. An effective purge protocol is one that reaches a certain HFR within the shortest purge time with the least energy consumption. Furthermore, because the cell HFR or membrane proton conductivity is also a function of temperature, the raw HFR data at different purge temperatures are converted to the single reference temperature of 30°C via the following temperature-dependent correlation obtained empirically[33]:

$$\text{T-compensated HFR} = \exp\left[1455\left(\frac{1}{303} - \frac{1}{T}\right)\right]\text{HFR}_{\text{purge}} \quad (11)$$

The temperature-compensated HFR is then essentially indicative of membrane water content only, and the temperature dependence is removed from actually measured HFR data.

The practical, short-duration gas purge of a fuel cell is inherently transient and extremely sensitive to the initial water distribution inside the cell. Therefore, controlling the initial conditions prior to gas purge is of paramount importance to ensure reproducibility and consistency of purge results. Extreme care must be taken to establish a controllable pre-purge condition and demonstrate the repeatability of purge data. Tajiri et al.[33] proposed a reliable procedure to carry out gas purge experiments. In their work, the experimental repeatability is verified through multiple purge experiments of various duration performed under identical purge conditions. Figure 12 shows the time evolution of cell HFR during gas purge for four experiments with different purge duration. The purge cell temperature is 55°C and the purge gas is dry N_2 with flow rate of 4.5 L/min. The dry purge gas was used in this set of experiments, unlike all other experiments of this study, in order to provide the most severe condition to verify experimental repeatability. The excellent repeatability is evident by the fact that purge

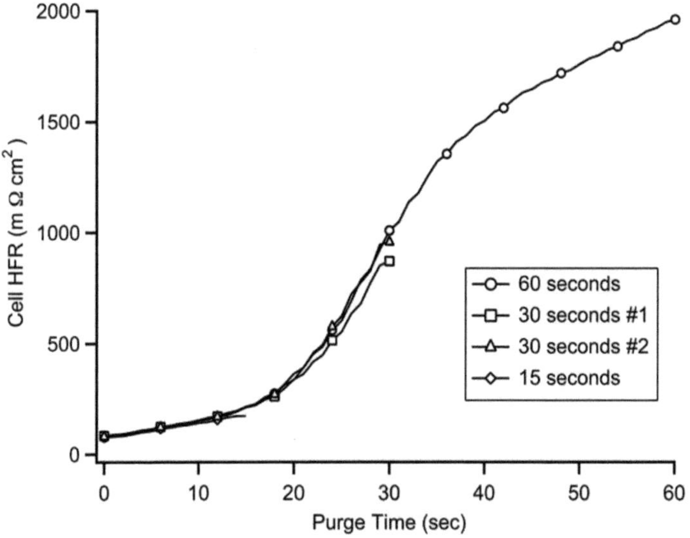

Figure 12. Time evolution of cell HFR for four experiments with different purge durations. (reproduced with permission from Tajiri et al.[33])

curves of longer duration closely overlap those of shorter durations, under otherwise identical purge conditions. We recommend that the type of repeatability tests shown in Fig. 12 be carried out as a standard procedure in all purge experiments before reporting meaningful results.

3.3. Two Characteristic Parameters for Water Removal

To better analyze and understand gas purge, two characteristic parameters are proposed based on a simplified schematic of purge physics, as shown in Fig. 13. Although this figure shows only the cathode side of the fuel cell from the polymer electrolyte membrane to the gas channel, the same conceptual depiction is applicable to the anode side. Prior to gas purge, CL and GDL are partially saturated with liquid water. It is assumed that water removal during gas purge is primarily a vapor-phase transport process, because liquid water residing in gas channels is swept away typically within the very first seconds of purge operation and the liquid water motion inside

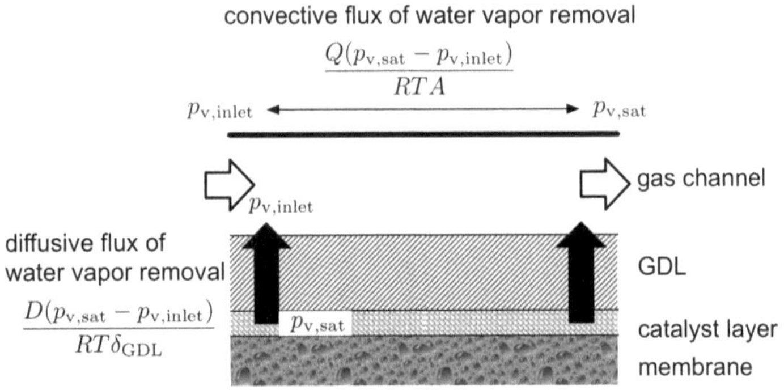

Figure 13. Schematic diagram of two limiting stages of water removal during gas purge. (reproduced with permission from Tajiri et al.[33])

CL and GDL is sufficiently slow to be considered quasi-stationary in the time scale of a gas purge. Thus, the water is removed from the liquid surfaces residing within CL/GDL to the gas channel by through-plane vapor diffusion. Subsequently, the water is removed by down-the-channel convection by purge gas. Mathematically, the first process of water vapor removal can be characterized by the through-plane diffusive flux between a liquid surface in CL/GDL to the gas channel, i.e., $D(p_{v,sat}-p_{v,inlet})/(RT\delta_{GDL})$. Here D stands for the diffusion coefficient of water in purge gas, and $p_{v,sat}$ and $p_{v,inlet,}$ the saturation vapor pressure at the purge temperature and the water vapor partial pressure in the purge gas at the inlet, respectively. R is the universal gas constant, T the purge cell temperature, and δ_{GDL} an effective diffusion length for water vapor from the membrane surface to the gas channel. This should scale mainly with the GDL thickness but could also contain effects of CL and MPL thickness and porosity, as well as the effects of the land-to-channel width ratio to account for in-plane transport effects.

The second process of water vapor removal down the channel can be described by the convective flux, $Q(p_{v,sat}-p_{v,inlet})/(RTA)$, representing the maximum amount of water vapor removed with the purge gas when the exit purge gas is fully saturated with vapor. In the above definition Q is the purge gas volumetric flow rate and A the active area of the fuel cell. Both parameters defined above have the unit of mol/s per unit of the fuel cell active area. It follows that the net rate of water removal during gas purge is determined by the two characteristic parameters and their relative magnitudes.

3.4. Stages of Purge

Figure 12 also reveals general characteristics of a purge curve, consistent with the model prediction of Sinha and Wang.[32] Several stages of gas purge are evident from the 60-s purge case shown in Fig. 12. The first stage (between 0 and ~2) can be called the slow rise period (SRP) where the membrane HFR does not rise substantially. This stage can be subdivided into SRP1 (0 to ~1) and SRP2 (10 to ~2). Physically, SRP1 coincides with the through-plane drying where liquid water is evaporated from CL/GDL underneath the channel area. This is then followed by SRP2 where the drying front propagates along the in-plane direction into the land area. During SRP1 and SRP2, the CL/GDL remains saturated with water vapor and hence the membrane remains nearly fully hydrated. The end of SRP is defined as the critical point, at which contiguous clusters of liquid water have completely disappeared in CL/GDL. While the SRP seems ineffective to increase the membrane HFR, it is a required step of gas purge to clear liquid water from CL/GDL, and every gas purge must go beyond the critical point to be effective. Thus, the critical point also defines the minimum purge duration. A residence time distribution method may be applicable to detect the liquid water saturation in the GDLs during SRP, although several issues have to be solved such as the treatment of dynamic saturation change during the measurement.[30]

After the critical point, purge enters the fast rise period (FRP) (between 20 and ~45 s in Fig. 12), where the membrane HFR is seen to rise substantially. Physically, the FRP coincides with the vapor diffusion process with the entire membrane subject to water desorption by dry gas. The FRP is finally followed by a membrane equilibration period (MEP) where water content in the membrane gradually reaches equilibrium with the relative humidity of purge gas. As a consequence, HFR asymptotically approaches the theoretical value corresponding to the equilibrium water content. In Fig. 12, due to the short duration of purge, the MEP is not clearly visible. A schematic illustration of all stages of gas purge is provided in Fig. 14. Evidently, the FRP is the most efficient period of gas purge.

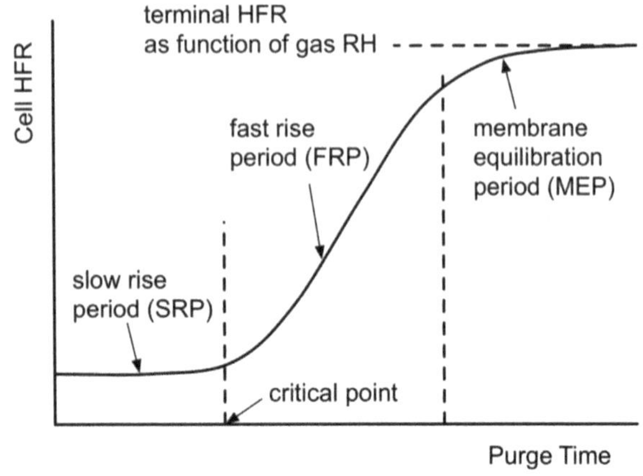

Figure 14. Stages of purge curve. (reproduced with permission from Tajiri et al.[33])

Hence, an optimized gas purge should pass beyond the critical point and take full advantage of the FRP, but not dwell on the MEP.

3.5. Effect of Key Parameters

3.5.1. Purge Cell Temperature

Based on the aforementioned hypothesis that the primary mechanism of water removal during gas purge is by the vapor phase, the purge performance is expected to strongly depend on the saturation vapor pressure or equivalently, the purge cell temperature. Figures 15, 16, 17 show the purge curves and the final HFR after 60-s purge for various purge cell temperatures. The purge curves plotted in Fig. 15 correspond to three cell temperatures and two purge gases, respectively, with a constant flow rate of 4.5 L/min for each side. Two observations can be made. First, as the purge cell temperature increases, the HFR rise becomes faster and the final value at 6 is higher. Second, at all temperatures, the He purge is more effective than N_2. The final HFR value depends primarily on the slope of the FRP (between 5 and 1 in He, 75°C purge case, for instance), and this slope is related to the through-plane diffusive flux of water vapor. Furthermore, at 75°C with He, the HFR approaches

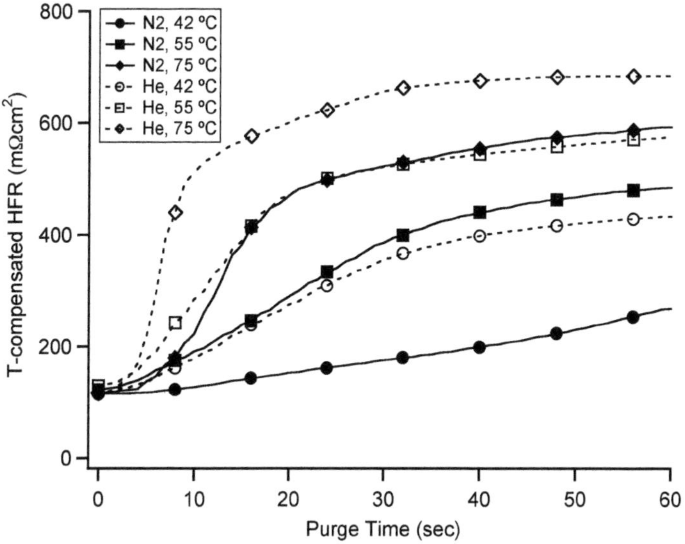

Figure 15. Time evolution of cell HFR at different cell temperatures and for various purge gases. The purge gas flow rate was 4.5 L/min. (reproduced with permission from Tajiri et al.[33])

a constant asymptote of ~650 mΩ cm^2 already in less than 60 s, signifying the attainment of MEP of gas purge. From this result it can be expected that if all purge experiments continue for a sufficiently lengthy period, the final temperature-compensated HFR values all approach ~650 mΩ cm^2.

In Fig. 16 the T-compensated HFR after 60 s purge is plotted as a function of temperature for two purge gases with two flow rates. At every temperature or flow rate, the He gas purge (open symbols) yields higher HFR than N_2 (solid symbols), due to its high water vapor diffusivity. However, with 1 L/min flow rate the advantage of He almost vanishes at 75°C, as the water removal in this case is limited by the convective flux down the channel and the enhanced diffusion does not help.

Figure 17 displays the T-compensated HFR as a function of the through-plane vapor diffusion flux. If there is sufficiently large flow rate, the resulting HFR should no longer be affected by the convective flux of water vapor down the channel, and hence the HFR would

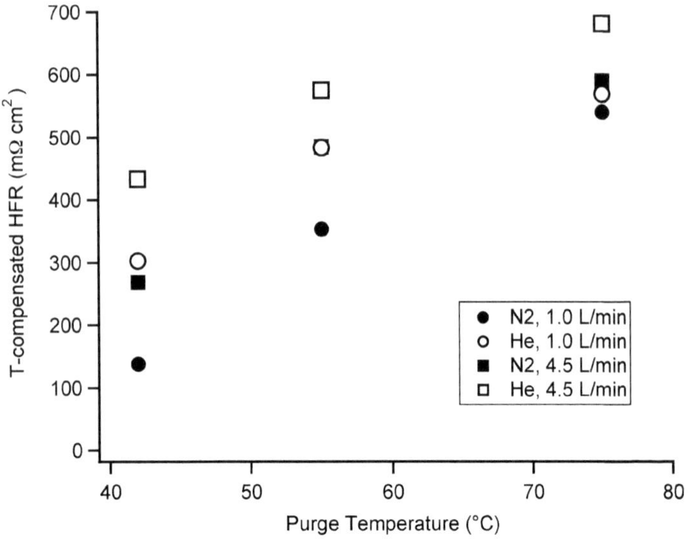

Figure 16. Cell HFR after 60-s purge as a function of purge cell temperature. (reproduced with permission from Tajiri et al.[33])

be determined only by the diffusive flux. In Fig. 17 it is seen that the purge curves for N_2 (1.0 L/min) and He (1.0 L/min) deviate from each other even at the same diffusive flux (e.g., 1×10^{-4}mol/cm^2/s). This is because the flow rate in this case is not large enough to eliminate the effect of convective flux. Similarly, the purge curves for N_2 (4.5 L/min) and He (4.5 L/min) differ, although the difference becomes smaller than the 1.0 L/min case, indicating that the contribution of convective flux becomes weaker at the higher flow rate. It is expected that if the flow rate is further increased, the purge curves for N_2 and He will collapse together over an entire range of the diffusive flux, indicative of a diffusion-controlled regime free of the convection effect down the channel.

It should be noted that the advantage of He gas purge diminishes with increasing temperature, as can be seen from the purge curves displayed in Fig. 15. At high temperature such as 75°C, the HFR evolution nearly reaches the equilibrium stage, in other words, the membrane is nearly equilibrated with the 40% RH purge gas after 60 s purge, both in He and N_2 cases. However, the transition

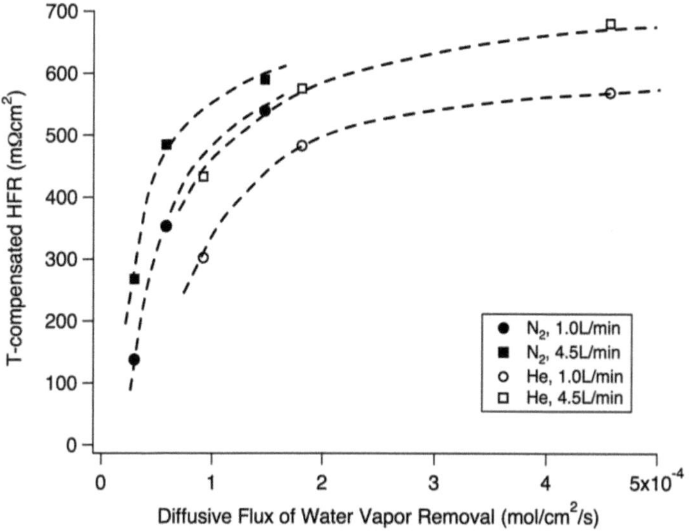

Figure 17. Cell HFR after 60-s purge as a function of diffusive flux of water vapor. (reproduced with permission from Tajiri et al.[33])

from FRP to MEP occurs much earlier for He gas purge (about 10 s at 75°C) than for N_2 (about 20 s at 75°C). Therefore, when the system requires the shorter purge duration such as 15 s, the advantage of He gas purge is still significant.

3.5.2. Purge Gas Flow Rate

The flow rate, or equivalently gas velocity, affects not only the vapor capacity rate of the channel flow but also the residence time of purge gas through the channels. The higher flow rate gas stays for a shorter time in the channel, allowing less vapor diffusion into the channel and reaching the exit with lower relative humidity, and vice versa. Therefore, even in the diffusion-dominated regime the purge curve is partly controlled by the purge gas flow rate.

The T-compensated HFR is plotted in Fig. 18 as a function of the convective flux of water vapor down the channel, where the data points for the same diffusive flux are connected by a dashed line. It can be seen that all dashed lines corresponding to various diffusive

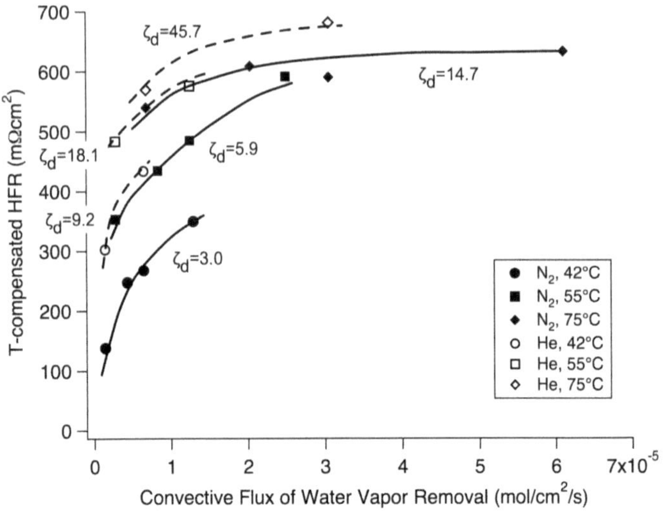

Figure 18. Cell HFR after 60-s purge as a function of convective flux of water vapor. ζ_d denotes the diffusive flux. (reproduced with permission from Tajiri et al.[33])

fluxes do not cross each other and that at a certain convective flux the resulting T-compensated HFR value after 60-s purge monotonically increases with the diffusive flux. For instance, at the convective flux of 1×10^{-5}mol/cm^2/s, the T-compensated HFR increases from 300 mΩ cm^2 at $\zeta_d = 3.0$ to 620 mΩ cm^2 at $\zeta_d = 45.7$. Thus, it can be concluded that the T-compensated HFR after 60-s purge may be expressed as a unique function of these two parameters only, namely:

$$\text{T-compensated HFR} = f\left(\frac{D\left(p_{v,sat} - p_{v,inlet}\right)}{RT\delta_{GDL}}, \frac{Q\left(p_{v,sat} - p_{v,inlet}\right)}{RTA}\right) \tag{12}$$

Equation (12) is significant, in that the HFR after purge becomes predictable based on calculations of the through-plane diffusive flux and convective flux down the channel.

3.5.3. *Matching Two Parameters*

To further test the hypothesis that the HFR after purge is uniquely determined by the two characteristic parameters defined above, two additional purge experiments were conducted. In these experiments, both diffusive and convective fluxes are matched while the purge cell temperature, purge gas species, and flow rates are varied as shown in Fig. 19. The purge curves for these four cases with the diffusive and convective fluxes closely matched are displayed in Fig. 19. It is interesting to note that if the two characteristic parameters are matched, the HFR evolution indeed exhibits a similar pattern, and the resulting T-compensated HFR after 60-s purge is nearly equal. Figure 19 strongly suggests that a purge curve can be adequately described by the diffusive and convective fluxes defined earlier.

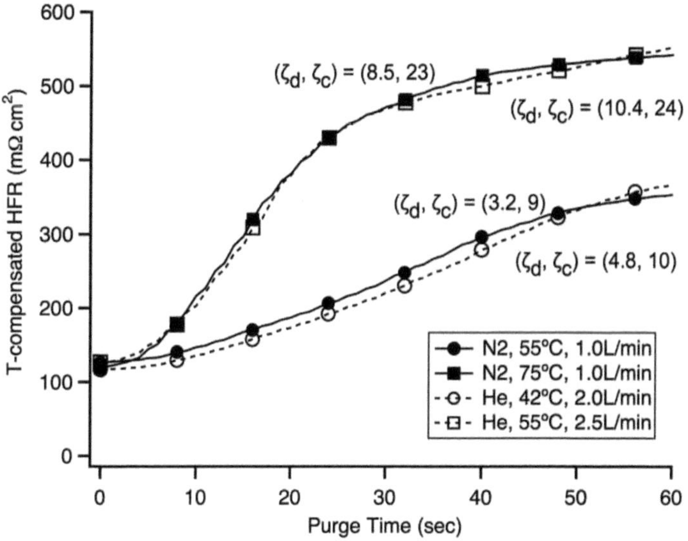

Figure 19. Time evolution of cell HFR with matched parameters. ζ_d and ζ_c are the diffusive flux and convective flux, respectively. (reproduced with permission from Tajiri et al.[33]

3.6. HFR Relaxation

Another interesting phenomenon observed so far is that after purge the cell HFR gradually decreases in a time scale of hours, which is called HFR relaxation after purge. Typical results of HFR relaxation are shown in Fig. 20. When the 60-s purge is completed, the valves at the inlet and outlet of the cell for both anode and cathode gas lines are closed and the cell temperatures are maintained constant at the purge cell temperature during the whole relaxation process.

Fundamental mechanisms of this relaxation phenomenon remain unknown and need future investigation. However, correlating the T-compensated HFR after relaxation with that after purge is of practical interest, because the initial membrane water content critically important for PEFC cold-start performance corresponds to the HFR after relaxation (during cool down), not the HFR immediately after purge. For this reason, an empirical correlation is attempted between the HFR after purge and that after relaxation, as shown in Fig. 21. It can be seen that a reasonable correlation exists over a

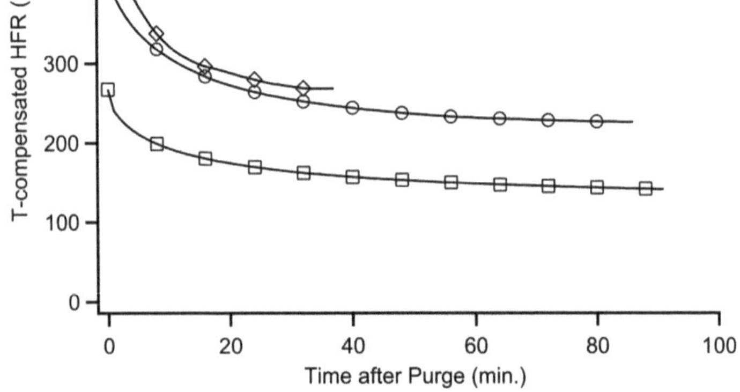

Figure 20. HFR relaxation after purge. (reproduced with permission from Tajiri et al.[33])

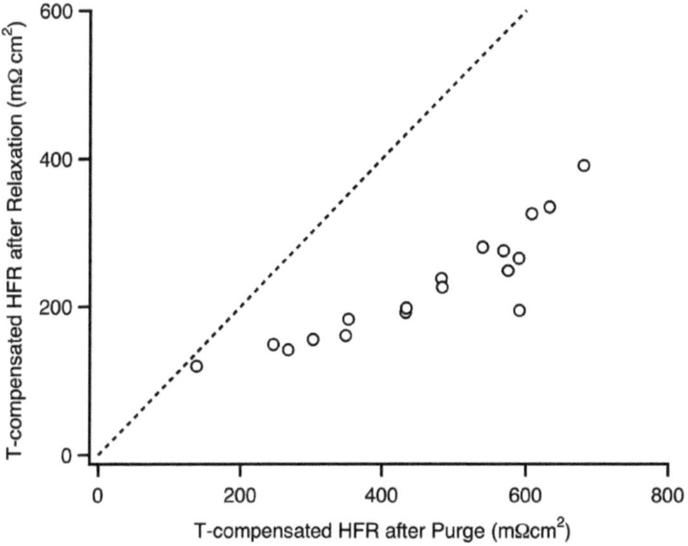

Figure 21. Relation between HFR after relaxation and HFR after 60-s purge with purge gas RH greater than 40%. (reproduced with permission from Tajiri et al.[33])

range of HFR, or equivalently membrane water content, that ranges from full humidification to that corresponding to 40% RH. It is clear from Fig. 21 that the extent of HFR relaxation after purge increases with lowering membrane water content. Any data beyond the HFR corresponding to 40% RH are not included in Fig. 21 as they become widely scattered and somewhat chaotic. It is noted that HFR relaxation after purge becomes stochastically random as soon as the membrane becomes drier than that corresponding to 40% RH.

The correlation shown in Fig. 21 provides a practical means to estimate the HFR or membrane water content as the important input to evaluate cold-start performance. That is, one can estimate the HFR after purge from Eq. (10) based on the purge conditions, and subsequently correct for HFR relaxation using Fig. 21. Based on the HFR value after relaxation or prior to cold start, one can use the analytical models and performance data developed in previous work to estimate the cold-start performance.

4. CONCLUDING REMARKS

Cold-start capability and survivability of polymer electrolyte fuel cells in a subzero temperature environment are of paramount importance for automotive applications. Fundamental mechanisms governing cold start have been reviewed, and it is recognized that product water becomes ice or frost upon startup when PEFC internal temperature is below the freezing point of water. If the local pore volume of the cathode catalyst layer is insufficient to contain all of the accumulated water before cell operating temperature rises above freezing, ice may plug the catalyst layer and stop the electrochemical reaction by starving it of reactant gases.

Simple calculations of heat and water balance indicate that for successful self-cold start from −30°C, heat release corresponding to approximately 1 mg/cm^2 of product water from ORR reaction is required. Therefore, the key challenge is the removal or storage of 1 mg/cm^2 water. The cathode catalyst layer of common design can store ~0.5 mg/cm^2 water, at most. The vapor phase removal of water from the catalyst layer to the gas channel is negligibly small at this temperature. Hence, it is necessary to store some water in the electrolyte membrane.

The method of equilibrium purge was reviewed and found capable of precisely controlling membrane water content prior to cold start thereby providing well-defined initial conditions of required water distribution. In general, a drier membrane can absorb more water during cold start and consequently more product water can be obtained. The startup current density has an impact on cold-start performance, since higher current density means higher water production rate and, in turn, higher water accumulation rate in the catalyst layer. Furthermore, at higher current densities the catalyst layer pores are not fully utilized upon shutdown, and water storage utilization decreases with current density. The startup temperature also strongly affects cold-start performance.

Detailed investigation of ORR kinetics under subzero temperature showed that the fundamental ORR kinetic mechanism remains unchanged over a wide temperature range down to −40°C. The low-temperature CV detected the loss of cathode catalyst electrochemical active area during cold start, part of which was recovered at room temperature, while others were not recovered until higher temperature (~70°C) operation.

With practical short-duration (~1 min) purge, the cold-start performance is strongly affected by the initial condition of water distribution, thereby indicative of the importance of gas purge upon FCV shutdown. A purge curve, defined as the membrane HFR varying with time, can be divided into three stages. In the slow rise period (SRP) the liquid water in CL and GDL is removed but the water in the membrane still remains, and therefore the HFR does not change significantly. In the fast rise period (FRP) the water is directly removed from the membrane, leading to fast HFR increase, followed by membrane equilibration period (MEP) where HFR increase approaches a final value. Thus, the effective purge should go beyond SRP and take full advantage of FRP, but not dwell on MEP. Under typical purge conditions, the final HFR can be predicted with two characteristic parameters: diffusive flux and convective flux of water vapor. It should be noted that while the understanding of water removal mechanisms is showing progress, the implementation of optimal gas purge practices in FCVs remains a challenge. For example, when the temperature of a fuel cell stack is low, the gas purge requires a longer duration which is not acceptable for the FCV. The large scale of automotive fuel cells also poses a problem, as the gas purge produces a significant variation in water content along the channel. Invention of practical and effective purge methods continues to be of great interest for FCV commercialization.

REFERENCES

[1] C.Y. Wang, X.G. Yang, Y. Tabuchi, F. Kagami, *Handbook of fuel cells: fundamentals, technology, and application*, vol. 6 (Wiley, New York, NY, 2009) p. 880

[2] F. Kagami, T. Ogawa, Y. Hishinuma, T. Chikahisa, in *Fuel cell seminar* (Palm Springs, CA, 2002)

[3] Y. Hishinuma, T. Chikahisa, F. Kagami, T. Ogawa, *JSME Internati. J. Series B* **47**, 235 (2004)

[4] M. Oszcipok, D. Riemann, U. Kronenwett, M. Kreideweis, M. Zedda, *J. Power Sources* **145**, 407 (2005)

[5] M. Oszcipok, M. Zedda, D. Riemann, D. Geckeler, *J. Power Sources* **154**, 404 (2006)

[6] S. Ge, C.-Y. Wang, *Electrochem. Solid-State Lett.* **9**, A499 (2006)

[7] S. Ge, C.-Y. Wang, *Electrochim. Acta* **52**, 4825 (2007)

[8] E.L. Thompson, T. W. Capehart, T. J. Fuller, J. Jorne, *J. Electrochem. Soc.* **153**, A2351 (2006)

[9] E.L. Thompson, J. Jorne, H.A. Gasteiger, *J. Electrochem. Soc.* **154**, B783 (2007)

[10] E.L. Thompson, J. Jorne, W. Gu, H.A. Gasteiger, *J. Electrochem. Soc.* **155**, B625 (2008)

[11] E.L. Thompson, J. Jorne, W. Gu, H.A. Gasteiger, *J. Electrochem. Soc.* **155**, B887 (2008)
[12] J. Li, S. Lee, J. Roberts, *Electrochim. Acta* **53**, 5391 (2008)
[13] L. Mao, C.-Y. Wang, *J. Electrochem. Soc.* **154**, B139 (2007)
[14] L. Mao, C.-Y. Wang, Y. Tabuchi, *J. Electrochem. Soc.* **154**, B341 (2007)
[15] Y. Wang, *J. Electrochem. Soc.* **154**, B1041 (2007)
[16] F. Jiang, W. Fang, C.-Y. Wang, *Electrochim. Acta* **53**, 610 (2007)
[17] H. Meng, *J. Power Sources* **178**, 141 (2008)
[18] K. Tajiri, Y. Tabuchi, C.-Y. Wang, *J. Electrochem. Soc.* **154**, B147 (2007)
[19] T.E. Springer, T.A. Zawodzinski, S. Gottesfeld, *J. Electrochem. Soc.* **138**, 2334 (1991)
[20] C. Heitner-Wirguin, *J. Membrane Sci.* **120**, 1 (1996)
[21] K.G. Gallagher, B.S. Pivovar, T.F. Fuller, *J. Electrochem. Soc.* **156**, B330 (2009)
[22] R.C. McDonald, C.K. Mittelsteadt, E.L. Thompson, *Fuel Cells* **4**, 208 (2004)
[23] S. Cleghorn, J. Kolde, W. Liu, in *Handbook of Fuel Cells: Fundamentals, Technology, and Applications*, ed. by W. Vielstich, A. Lamm, H.A. Gasteiger (Wiley, New York, NY, 2003) p. 566
[24] K. Tajiri, Y. Tabuchi, F. Kagami, S. Takahashi, K. Yoshizawa, C.-Y. Wang, *J. Power Sources* **165**, 279 (2007)
[25] S. Ge, C.-Y. Wang, *J. Electrochem. Soc.* **154**, B1399 (2007)
[26] S.-Y. Lee, E. Cho, J.-H. Lee, H.-J. Kim, T.-H. Lim, I.-H. Oh, J. Won, *J. Electrochem. Soc.* **154**, B194 (2007)
[27] J. St-Pierre, J. Roberts, K. Colbow, S. Campbell, A. Nelson, *J. New Mater. Electrochem. Syst.* **8**, 163 (2005)
[28] R. Bradean, H. Haas, A. Desousa, R. Rahmani, K. Fong, K. Eggen, D. Ayotte, A. Roett and A. Huang, *AIChE 2005 Annual Meeting* Cincinnati, OH, 2005.
[29] P.K. Sinha, P. Halleck, C.-Y. Wang, *Electrochem. Solid-State Lett.* **9**, A344 (2006)
[30] J. St-Pierre, A. Wong, J. Diep, D. Kiel, *J. Power Sources* **164**, 196 (2007)
[31] S.-Y. Lee, S.-U. Kim, H.-J. Kim, J.H. Jang, I.-H. Oh, E.A. Cho, S.-A. Hong, J. Ko, T.-W. Lim, K.-Y. Lee, T.-H. Lim, *J. Power Sources* **180**, 784 (2008)
[32] P.K. Sinha, C.-Y. Wang, *J. Electrochem. Soc.* **154**, B1158 (2007)
[33] K. Tajiri, C.-Y. Wang, Y. Tabuchi, *Electrochim. Acta* **53**, 6337 (2008)

4

Species, Temperature, and Current Distribution Mapping in Polymer Electrolyte Membrane Fuel Cells

Jonathan J. Martin, Jinfeng Wu, Xiao Zi Yuan, and Haijiang Wang

Institute for Fuel Cell Innovation, National Research Council Canada, Vancouver, BC, Canada V6T 1W5

1. INTRODUCTION

Due to the complexity of the electrochemical, fluid dynamics, and thermodynamic processes within the fuel cell and the interacting influences of different factors, much attention has been paid to various distribution mapping techniques for PEM fuel cell diagnosis. The review combines the work of many researchers to develop a context for the current fuel cell diagnostics climate. Brief summaries of the most important and commonly studied species and parameters of fuel cell operation are provided. To provide sufficient detail, selected data from the literature is presented in order to illustrate the type of data obtained by each method, and each diagnostic technique is analyzed while assessing its merits and weaknesses. Ultimately, the goal of all diagnostic tools is to obtain an improved understanding of the processes within a fuel cell and to apply this understanding towards improved fuel cell design. Design implications, illuminated by the diagnostic tools that have matured sufficiently to provide further insight, are then summarized.

U. Pasaogullari, C.-Y. Wang (eds.), *Modeling and Diagnostics of Polymer Electrolyte Fuel Cells*, Modern Aspects of Electrochemistry 49,
DOI 10.1007/978-0-387-98068-3_4,

2. SPECIES DISTRIBUTION MAPPING

2.1. Species and Properties of Interest

2.1.1. Hydrogen

As it is the typical fuel for PEM fuel cells, effective distribution and utilization of hydrogen is essential for obtaining high system efficiencies. In order to minimize the amount of waste hydrogen, fuel cells are often operated as close to a stoichiometric ratio of unity as possible, matching the fuel provided with what is required at the desired output. Common techniques to minimize waste hydrogen include recycling the outlet gas stream and dead-ended operation. Both of these approaches, however, can lead to increasing concentrations of diluents and contaminants.

2.1.2. Oxygen

Since nearly every practical PEM fuel cell application utilizes air rather than pure oxygen, the diluted oxidant is susceptible to mass transport limitations. It is therefore imperative to ensure adequate distribution of the gas throughout the fuel cell active area. While mass transport limitations can be addressed through operation at stoichiometries significantly higher than unity, doing so results in increased parasitic power loss and reduced system efficiency. For a fuel cell design to effectively address oxygen transport, it must also account for water management, pressure drop, and flow distribution.

2.1.3. Water

Water content affects many processes within a fuel cell and must be properly managed. Proton conductivity within the polymer electrolyte typically decreases dramatically with decreasing water content (especially for perfluorinated membranes such as Nafion®), while excessive liquid water in the catalyst layers (CLs) and gas diffusion layers (GDLs) results in flooding, which inhibits reactant access to the catalyst sites. Water management is complicated by several types of water transport, such as production of water from the cathode reaction, evaporation, and condensation at each electrode, osmotic drag of water molecules from anode to cathode by the protons, and back diffusion of water from cathode to anode. Careful control of the humidification of the reactant gases is a common water management technique, but results in decreased system efficiency and increased system complexity and cost.

2.1.4. Contaminants and Diluents

Contaminants and diluents in the reactant gases can reduce fuel cell performance by decreasing the effectiveness of various cell components, decreasing the activity of the catalysts, accelerating component degradation, or increasing mass transport resistance of the reactant gases. Contaminants can be introduced through the corrosion of cell components, such as constituents of the bipolar plates such as resins, fillers, and metal (Fe^{+3}, Ni^{+2}, Cr^{+3}, Ti^{+3}, Mn, Mo, etc.)[1] or of the gaskets (Si, Ca, Mg, etc.).[2,3] Alternatively, the reactant gas streams can introduce contaminants and diluents, which differ by whether the species interacts with the fuel cell components. Species that are typically of concern include fuel impurities and components of air such as N_2, CO, CO_2, H_2S and NH_3, NO_x, and SO_x.[4] Research on contaminants and diluents has shown that these species have a large effect on PEM fuel cell lifetime and durability. As PEM fuel cell technology continues to evolve (i.e., alloy catalysts, non-noble catalysts, composite plates, and metallic plates), it is necessary to ensure that previously used techniques are still relevant and to develop new techniques. Additionally, durability and reliability become increasingly important as fuel cells near commercialization. The most commonly studied contaminant and diluent are CO and N_2, respectively, and both will be discussed further in Section 2.2.

2.1.5. Pressure Drop

Pressure drop can be an important design parameter and diagnostic tool, especially at the cathode where the product water is produced. Results have even suggested that pressure uniformity may play a larger role in reactant transport than flow uniformity.[5] The cause of pressure drop is the friction between the reactant gases and the flow field passages and/or through the GDL (especially for interdigitated flow fields[6]). A pump or blower must provide any required increase in pressure, which results in an increase in parasitic power (load) loss. However, a higher pressure drop also results in more effective removal of excess liquid in the fuel cell. Therefore, the pressure drop must be carefully considered. In the flooding condition, too much liquid water in the flow channels increases gas flow resistance, which will impede reactant gas transport and leads to performance losses related to mass transport. Pressure drop on the cathode side increases with cell flooding, while it remains unchanged with cell drying, thus clearly distinguishing between the two phenomena.[7] An increase in pressure drop, particularly on the cathode side of PEM fuel cell, is a reliable indicator of PEM fuel cell flooding.

2.1.6. Flow Distribution

Flow distribution within fuel cells is relevant because it affects many of the previously mentioned species. Proper design of flow distribution can promote mixing and uniform concentrations of the reactant gases throughout the active area, can assist with effective water management, and can minimize the pressure drop. While modeling has traditionally been the most commonly used technique for investigating flow distribution, as will be detailed in Section 2.2, diagnostic techniques capable of experimentally visualizing flow distribution are experiencing increased development.

2.2. Methodology and Results

2.2.1. Pressure Drop Measurement

Pressure drops within fuel cells have been monitored to determine various parameters, such as flow field design (both anode and cathode), flooding, etc. For example, Barbir et al.[8] and He et al.[6] used pressure drop as a diagnostic tool for the detection of flooding in the fuel cell because of the strong dependence of the gas permeability of the porous electrodes on liquid water content. In the mentioned studies, the pressure drop was observed in a fuel cell with interdigitated flow fields under a number of operating conditions, which caused either flooding or drying inside the fuel cell. In addition, another example of using pressure drop, as a diagnostic tool, is the method and apparatus designed in General Motors for monitoring H_2/O_2 fuel cells to detect and correct flooding.[9] In this method pressure drop across a flow field (anode or cathode) is monitored and compared to predetermined thresholds of acceptability. If the pressure drop exceeds the determined threshold, corrective measures are automatically initiated, such as dehumidifying the gases, increasing the gas mass flow rate, reducing gas pressure, and/or reducing current drain.

Rodatz et al.[10] studied a large PEM fuel cell stack used in automotive applications at different operating conditions. One of the main parameters that were studied was the pressure drop within the stack and its relationship with the flow field (bends in the channels) and single- and two-phase flows. It was observed that once the current in the fuel cell stack was reduced (i.e., applying dry conditions) the pressure drop decreased slowly until it reached a new value. This was attributed to the fact that the reduction of current reduces the flux of product water in the flow channels, and thus, it reduces the total mass flow in the flow channels.

A transparent PEM fuel cell with a single straight channel was designed by Ma et al.[11] to study liquid water transport in the cathode channel (this study is also mentioned in Section 2.5). The pressure drop between the inlet and outlet of the channel on the cathode side was used as a diagnostic signal to monitor liquid water accumulation and removal. The proper gas velocities for different currents were determined according to the pressure drop curves.

Pei et al.[12] studied the hydrogen pressure drop characteristics in a PEM fuel cell in order to use the pressure drop as a diagnostic tool for prediction of liquid water flooding in fuel cell stacks before flow channels have been blocked.

2.2.2. Gas Composition Analysis

The species distribution within a PEM fuel cell is critical to fully characterize the local performance and accurately quantify the various modes of water transport. The most commonly used analytic technique for measuring the gas composition within a fuel cell is gas chromatography (GC). Mench et al.[13] demonstrated the measurement of water vapor, hydrogen, and oxygen concentration distributions at steady state. A micro gas chromatograph was utilized to measure the samples, which were extracted from eight different sampling ports at various locations along the anode and the cathode flow paths of a specially designed fuel cell. Kim et al.[14] utilized GC to study the effects of voltage, humidity, and flow rate on the gas composition at six different sampling ports along a serpentine cathode. The results suggested that the technique could also be used for flooding detection as a result of the impact on oxygen consumption. While gas chromatography (GC) provides molar percent level accuracy of the species, it still requires in the order of 5 min per data point and is consequently limited to analyzing steady state species distribution. This is the reason why Dong et al.[15] used an Agilent Technologies real-time gas analyzer that enabled the measurement of various species in near real time (about 1 s per data point). Another way to improve the accuracy and repeatability of the technique shown by Mench et al. was the one presented by Yang et al.[16] They utilized two sets of multiposition microactuators and micro GCs to measure gas concentration distributions on the anode and cathodes sides simultaneously. However, in general the weakness of utilizing GC to measure species distribution is that only discrete data at desired positions can be determined.

Investigators have also found success in combining GC measurements with current density distribution data. Araki et al.[17] and Onda et al.[18] created a model for current distribution inside a PEM fuel cell and used several experimental techniques to confirm the results of the model. One of these techniques was the measurement of hydrogen/oxygen concentration by gas chromatography (with a fixed nitrogen content for reference) along the gas supply grooves. More on these studies will be discussed in Section 4.1.2.2. Wang and co-workers[19] combined current mapping with species concentration data obtained by GC to calculate the distribution of the net water transport coefficient through the membrane within an operating fuel cell. In an extension of the previous work[20] they investigated the effects of inlet humidification levels on the net water transport coefficient distribution in a co-flow configuration of the flow fields. They found that a dry anode inlet led to a negative coefficient (back diffusion exceeded osmotic drag), while a dry inlet only had a negative coefficient at the outlet of the cell.

GC has also been employed in several studies to measure the crossover rate of hydrogen and oxygen through the membrane in PEM fuel cells.[21–23] H_2 or O_2 is supplied in one side of the fuel cell and N_2 or He is fed to the opposite side, functioning as a carrier gas. The carrier gas transports the H_2 or the O_2 permeating through the membrane to a GC detector.[21] Using this method, Broka et al.[22] investigated the crossover rates of hydrogen and oxygen through Nafion 117 membrane and recast Nafion film at different temperatures and gas relative humidities. Recently, Liu et al.[23] measured the crossover rates of oxygen through Nafion/PTFE composite membranes by means of GC.

Mass spectrometry (MS) is another analytical technique capable of determining the gas composition in an operating fuel cell. Partridge et al.[24] utilized spatially resolved capillary inlet mass spectrometry (SpaciMS) to successfully conduct measurements (with a temporal resolution of 104 m/s) at realistic humidity levels, despite the concern that liquid water could block the capillaries used for gas sampling throughout the active area of the fuel cell. They analyzed the effect of load switching on the species concentrations and observed concentration gradients and non-uniformities.

2.2.3. *Neutron Imaging*

Liquid water plays a key role in PEM fuel cells because its presence is closely linked to the functionality of the main components of the unit cell. The water balance in an operating cell has been studied through collection of the fuel cell effluent and condensation of the gas phase water vapor.[25,26] Unfortunately, these studies could not provide spatial information on water distribution throughout the working cell. To date several universities and institutes have explored the method of neutron imaging as an experimental tool to study water management in fuel cells. Due to the fact that the neutron incoherent scattering length for hydrogen is almost two orders of magnitude larger than the length for almost all other elements, the neutrons are ideal for studying hydrogen-containing compounds such as water.[27] Equipped with good quality neutrons to probe the cell and a high sensitivity scintillator/charge-coupled device (CCD) camera as the detector system to record the images, the neutron imaging technique shows the potential for discerning two-phase flows within the fuel cell in real time or steady state.

Bellows et al.[28] first used the neutron imaging technique to measure water gradient profiles within the Nafion membrane of an operating PEM fuel cell. Preliminary neutron intensity gradients showed qualitative agreement with the expected response of the membrane water content to changes in feed gas humidification and fuel cell current. Geiger et al.[29] reported the gas/liquid two-phase flow patterns in the flow fields by neutron imaging for the first time. However, the inadequate spatial resolution (500 μm/pixel) and low image acquisition rates (5 s per image), which included the exposure and read-out time for the CCD camera, in the presented radiographs limited the investigation of the two-phase flow to real time. In Satija et al.'s experimental design,[30] spatial and temporal resolutions for CCD images were improved to 160 μm/pixel and 2 s per image, respectively. The water content in a PEM fuel cell as a function of time over 2,000 s was presented by using masking techniques to differentiate the location of the water in the cell and performing tomography to create a digital three-dimensional representation of the fuel cell stack. Figure 1 shows a typical setup for neutron imaging of a fuel cell,[30] in which neutrons are converted to light using a scintillator screen and the light is focused on the CCD chip. Manke et al.[31] also used neutron tomography to obtain three-dimensional images of a 3- and 5-cell stack. Due to the several hours required to obtain the images, the local water distributions within the stacks were "frozen" by disabling the flow and temperature control. They observed that thinner membranes induced more homogeneous water distributions, but the technique is limited by the long acquisition times required.

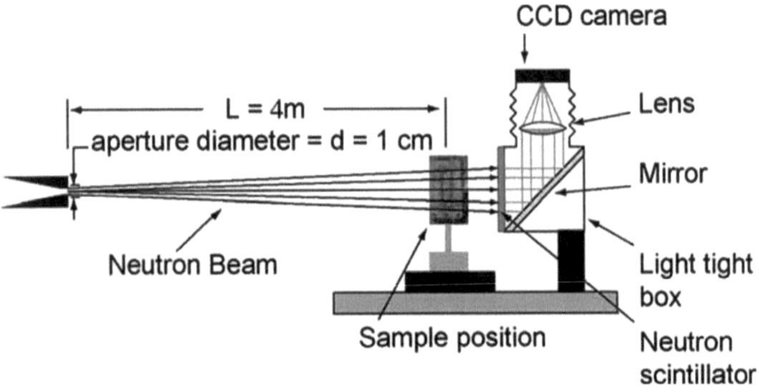

Figure 1. Typical experimental setup for neutron imaging of a fuel cell (reprinted from[30] Copyright (2004), with permission from Elsevier).

Recently, a neutron imaging technique developed at the Penn State Breazeale Nuclear Reactor has significantly improved the spatial resolution to 129 μm/pixel and the temporal resolution to 30 images/s.[32,33] Their results showed that liquid water had a tendency to accumulate at specific locations, depending on operating conditions. Kramer et al.[34] employed the neutron imaging technique to reveal the relationship between water content in a PEM fuel cell and operating current density with a spatial resolution of 115 μm/pixel and exposure times ranging from 0.5 to 15.0 s and, for the first time, quantified liquid water in the GDL was reported with proper algorithms. More recently, rather than a CCD camera, a flat panel amorphous silicon detector was utilized by Hickner et al.[35] to measure water content in a PEM fuel cell under varying current densities and operating temperatures. The images obtained by this detector could reach 127 μm/pixel in spatial resolution and 3 images/s in temporal resolution.

A major issue with two-dimensional neutron imaging is the difficulty in distinguishing between water content in the anode, cathode, and membrane and within the various components of the anode and cathode. Chen et al.[36] attempted to discriminate between anode and cathode effects at a resolution of 250 μm by using serpentine flow fields with minimal overlap (the anode channel overlapped the cathode land and vice versa, rather than the typical channel/channel and land/land overlaps). While the results proved promising, it is questionable if true discrimination was accomplished and the atypical flow field orientation likely influenced the observed water distribution. Boillat et al.[37] utilized deuterium species ($^2H_2$ and ^2H_2O) for labeling to discriminate between humidification, product water, and anode/cathode effects. Their neutron imaging technique proved to be a valuable method when the numerous exchange steps between the nuclei were correctly accounted for.

Another approach to differentiating between the anode and cathode water contents is to perform through-plane visualization. Until recently, neutron radiography was not able to achieve the resolutions necessary to sufficiently resolve a membrane thickness of 25 μm and GDL thicknesses of 200 μm. However, Hussey et al.[38] developed a new detector technology based on micro-channel plates that allowed in-plane visualization with a resolution of 30 μm (10 μm is feasible with further detector development). An exposure time of 20 min was used and the possibility of increasing temporal resolution was not addressed. Lehmann et al.[39] developed a detector system optimized for high resolutions (50 μm and 10–100 s/image) by improving the scintillator screen, mirror, and lens. By tilting this detector, Boillat et al.[40] were able to increase the resolution and obtain through-plane images of a 0.5 cm × 1 cm active area fuel cell with a 2.5-μm spatial resolution. The temporal resolution was not disclosed.

Other notable neutron imaging investigations include the effects of the following conditions on water distribution: dynamic loading by Park et al.[41]; load cycling with constant gas flow and RH cycling by Owejan et al.[42]; variations in inlet gas flow rate, cell pressure, and inlet gas humidity by Turhan et al.[43]; and locally resolved current density distribution by Hartnig et al.[44] Additional investigations utilizing neutron imaging for cell design are described in Section 2.3.

Neutron imaging, with sufficient spatial and temporal resolution, has proven to be a powerful technique for obtaining qualitative and quantitative information into liquid water content and distribution. However, requirement of a neutron source with a high fluence rate limits the wide application of neutron imaging, for which differentiation between anode and cathode water content continues to be a challenge. In addition, cell rotation is imperative for present tomographic imaging methods to gain three-dimensional information on an operating PEM fuel cell.[35]

2.2.4. Magnetic Resonance Imaging

Magnetic resonance imaging (MRI), based on the nuclear magnetic resonance (NMR) phenomenon, is an imaging technique using gradient radio frequency (RF) pulses in a strong magnetic field. The basic principle of MRI is that certain atomic nuclei within an object, if placed in a magnetic field, can be stimulated by the correct RF pulses. After this stimulation the nuclei relax while energy is induced into a receiver antenna to further obtain a viewable image. The nondestructive nature of MRI enables one to obtain unique in-situ information from a multitude of systems.[45] As for a PEM fuel cell, the hydrogen atom is always observed in MRI experiments due to excellent correlation between its signal intensity and water content in the membrane.[46] According to Tsushima and his coworkers' MRI experiments, the effects of operating condition,[47–49] membrane thickness,[50] and liquid water supply[51] on water distribution within PEMs in fuel cells were demonstrated. An example of the experimental setup used by the previous studies can be observed in Fig. 2. Bedet et al.[52] demonstrated non-homogeneous hydration due to membrane swelling in a plastic flow field with a perimetric gold wire for current collection at resolutions of 100 μm and 60 s/image. By fabricating flow fields from Delrin cylinder halves, Feindel et al.[53–56] also used MRI to obtain liquid water distribution in the flow field.

A few studies have used unique approaches to obtain increasing amounts of information. ^2H has been used as a labeling species in MRI, which is of particular interest because it does not produce a signal. Feindel et al.[57] switched between ^2H_2O and H_2O humidification and between H_2 and $^2H_2$ fuel supply to observe the interactions and transport throughout the PEM (based on the decrease or increase of signal strength) at resolutions of 234 μm and 128 s/image. Dunbar and Masel[58,59] obtained three-dimensional images in miniature PEM fuel cells with MRI. Each voxel measured 138 μm × 138 μm × 200 μm and the acquisition time for each sequence was 259 s. Of interest is their observation that liquid water accumulated in stationary waves with occasional slipping, rather than

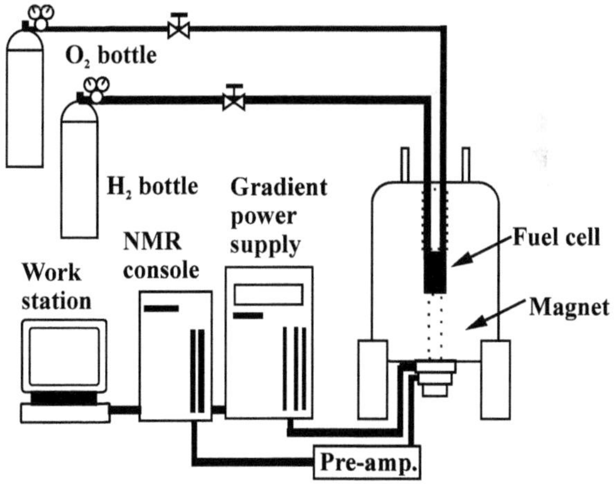

Figure 2. Experimental setup for MRI tests of a fuel cell (reprinted from[47] Copyright (2004). Reproduced by permission of The Electrochemical Society).

the typical slug-type flow that was expected. However, their unique channel shape (1 mm wide × 3 mm deep) may have been the cause of this phenomenon. Zhang et al.[60] obtained high-resolution MRI images by using gold plated printed circuit board (PCB) flow fields as current collectors and as the resonating plates of the probe. This approach allowed visualization of the through-plane water distribution in the membrane with a resolution of 6 μm and an acquisition time of 360 s/image.

The weakness of MRI technique is mainly in the requirement that the materials have to be nonmagnetic. For this reason, the fuel cell components must be carefully chosen, of which many are not the same size or composition as in industrial cells. Additionally, the water content in the CL and GDL, either made from nonwoven carbon paper or from woven carbon cloth, will be difficult to visualize with MRI.[54]

2.2.5. X-ray Imaging

Due to the inadequate spatial and temporal resolutions of the previously mentioned neutron imaging and MRI techniques, it is imperative to develop a high-resolution in-situ technique for imaging pore-scale flow and multiphase transport in individual components of PEM fuel cells. As a consequence, X-ray imaging has recently gained significant attention in academia and industry for detecting and quantifying water distribution within a fuel cell. The basic principle of X-ray imaging is that the intensity of an X-ray beam is attenuated as it traverses through a material. The transmitted radiation, received by an array of detectors, produces a two-dimensional or three-dimensional map based on the variation in X-ray adsorption throughout the sample.

The X-ray microtomography technique was pioneered by Sinha et al.[61,62] to obtain high-resolution, three-dimensional images of liquid water distribution in a GDL during gas purge. A comparatively high spatial resolution of 10 μm and temporal resolution of 0.07 s for a single image was achieved in their study. The authors successfully measured the liquid water saturation distribution in the cathode GDL during gas purge, as well as fuel cell operation. Synchrotron X-ray radiography was employed by Hartnig et al.[63–66] to investigate water evolution and liquid water transport from the CL through the GDL to the gas channels in an operating fuel cell at a microscopic level of 3 μm and time resolution of 5 s. Mukaide et al.[67] also used synchrotron X-ray radiography to observe water distribution and behavior in the components of an operating PEM fuel cell, with a spatial resolution of ~12 μm. By means of the X-ray radiography technique, Lee et al.[68] quantitatively visualized the water distribution in the region between the flow plate and the GDL in a PEM fuel cell. A high spatial resolution of 9 μm was achieved in their study. Albertini et al.[69] also recently reported the hydration profile of a Nafion® membrane in an operational fuel cell using X-ray radiography. The spatial resolution was 10 μm but the time resolution was relatively low (75 s).

X-ray imaging is a promising technique for the quantification of liquid water in different components of PEM fuel cells due to its resolution of ~10 μm. Though high spatial and temporal resolutions can be realized, X-ray imaging still need much more development and refinement for fuel cell research. For example, further improvement in the spatial resolution down to 1–5 μm is preferred for the quantitative study of liquid water distribution in thin membranes, the CL or the MPL.

2.2.6. Optically Transparent Fuel Cells

In order to delineate the origin and development of flooding with high spatial and temporal resolution in PEM fuel cell, recent research has resorted to the transparent cell design based on optical diagnostics.

Weng et al.[70] designed a transparent PEM fuel cell used to visualize and study the distribution of water and water flooding inside the cathode gas channels (serpentine channels), and also to explain the phenomena of membrane dehydration. This cell was similar in design to the one presented by Yang et al.[71] except it was used for H_2 fueled PEM fuel cells instead of visualizing CO_2 gas bubble behavior in direct methanol fuel cells (DMFC). The cell consisted of two transparent acrylic cover plates and two flow field plates (anode and cathode) made out of brass[70] (see Fig. 3). Liu et al.[72,73] developed a number of transparent fuel cells, each with different flow fields, to study the water flooding, two-phase flow of the reactant and products, and pressure drop in the cathode flow channels. Similarly, Su et al.[74] analyzed flooding in four different cathode flow

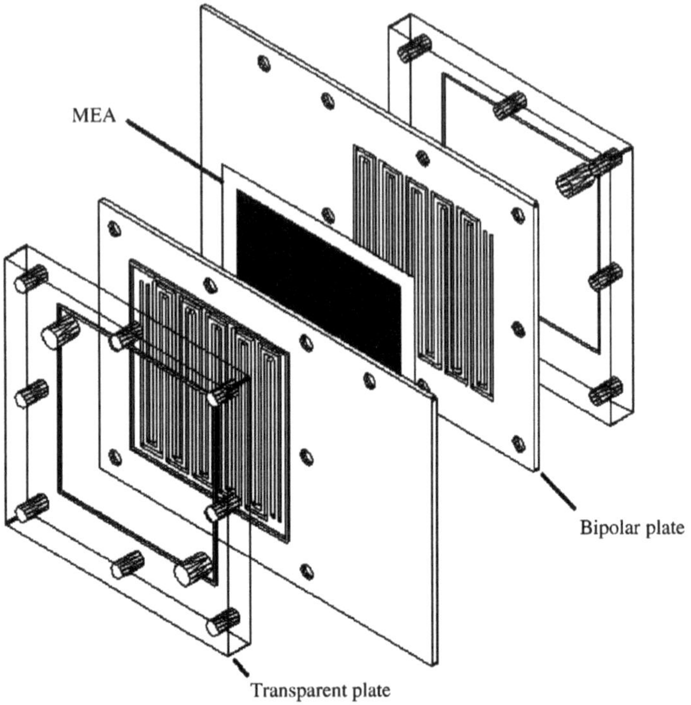

Figure 3. Schematic of a transparent PEM fuel cell (reprinted from[70] Copyright (2006), with permission from Elsevier).

fields and with different operating conditions in order to improve a mathematical model. From these results, it was concluded that a serpentine-interdigitated flow field was the best design for water removal. As mentioned in Section 2.2.1, Ma et al.[11] designed a transparent cell to visualize water flooding in the cathode flow field; this study also used the pressure drop along the cathode side to provide better understanding of water flooding. However, the flow field in this cell was very simple (a single straight channel) and it is not a practical tool for more complex flow fields and cells. Ge and Wang[75] used a transparent cell to visualize the liquid water and ice formation during startup of a fuel cell at subzero temperatures. They also used a transparent cell to visualize water in the anode at low current densities.[76] They observed that water condensed on the channel walls and not inside the hydrophobic GDL and concluded that the water came from the membrane and through the GDL as vapor or condensed due to diminishing anode gas due to hydrogen consumption. Other similar transparent cells used to analyze water distribution and flooding in fuel cells can be found in the literature.[77–80]

In order to study cathode flooding in small fuel cells for portable applications operated at ambient conditions, Tüber et al.[81] designed a transparent cell that was only operated at low current densities and at room temperature. The experimental data was then used to confirm a mathematical model of a similar cell. Fig. 4 describes the schematic top and side view of this transparent fuel cell. The setup was placed between a base and a transparent cover plate. While the anodic base plate was fabricated of stainless steel, the cover plate was made up of plexiglass. A rib of stainless steel was inserted into a slot in the cover plate to obtain the necessary electrical connection. It was observed that clogging of flow channels by liquid water was a major cause for low cell performance. When the fuel cell operated at room temperature during startup and outdoor operation, a hydrophilic carbon paper turned out to be more effective compared with a hydrophobic one.[81]

Theodorakakos et al.[82] developed a computer model that described the detachment of liquid droplets from solid surfaces. In order to validate the model, a transparent fuel cell was used and then an ex-situ experiment was conducted. This test analyzed

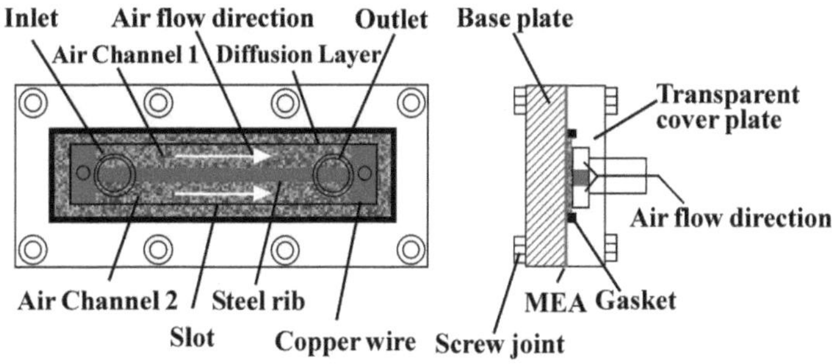

Figure 4. Schematic of the top and side view of a transparent PEM fuel cell (reprinted from[81] Copyright (2003), with permission from Elsevier).

a single-droplet pushed by air along a channel. Chen et al.[83] also developed an ex-situ experiment to validate a model of the transport mechanism of liquid water through the GDL after it is formed at the CL. A similar study was also performed by Kumbur et al.[84] Yang et al.[85] presented another ex-situ experiment in order to have a better understanding of two-phase flows, especially in the anode flow fields of DMFCs.

Transparent cells were first introduced for flow visualization by Argyropoulos et al.[86,87] who designed a cell with an acrylic anode to observe two-phase flow phenomena in DMFCs. Current was drawn from the anode side of the cell using a peripheral stainless steel strip embedded into the acrylic block, which contacted the MEA. With the aid of a high-speed video camera, appropriate computer software, and transparent acrylic cells, gas evolution was recorded in a fuel cell working environment. The flow visualization technique was revisited by Lu and Wang[88] and Nordlund et al.[89] where Lu and Wang[88] applied the technique to the cathode side for the first time to observe the cathode flooding phenomena in DMFCs. Extension of the work to PEMFCs was conducted by their colleagues, Yang et al.[77] who designed a transparent fuel cell with two gold-plated stainless steel plates that acted as current collectors. Before being gold plated, the plates were machined through to form seven straight flow channels. Two clear polycarbonate plates were placed outside the current collector plates to constrain the gas flow. A series of physical processes under different operating conditions were described, including water droplet emergence from the GDL, growth and detachment from the GDL, and distribution and removal from the flow channel.[77]

The effects of species other than water have also been investigated through the use of transparent cells. Inukai et al.[90] conducted novel work with an oxygen sensitized dye complex dispersed in an oxygen permeable polymer matrix. By irradiating the dye with a laser and recording the image with a CCD camera, they were able to visualize oxygen depletion from inlet to outlet. With the use of other dye materials, the cell could also be used to visualize temperature, CO_2 concentration and water vapor concentration, etc. Murahashi et al.[91] used a transparent cell in combination with electrochemical impedance spectroscopy (EIS) to study the effects of CO poisoning on an operating fuel cell. In addition to a near fivefold increase in impedance after 300 h at 100 ppm CO, the transition point to a two-phase flow moved towards the cathode exit for low humidity and high CO concentration conditions.

There are a few transparent cell techniques for investigating species distribution that, for the most part, have been conducted on mock fuel cells. However, techniques such as laser induced fluorescence (LIF) and particle image velocimetry (PIV) have provided valuable information and may eventually prove suitable for in situ PEM fuel cell investigations. In basic terms, PIV utilizes seeding particles that are illuminated or fluoresced by a laser source to capture discrete images that are representative of the bulk flow. Algorithms are then used to determine the displacement of the fluid flow and consequently the velocity field. In LIF, portions of the flow are fluoresced with a laser light source to visualize flow propagation within the system. More thorough details regarding PIV and LIF can be found in the book by Raffel et al.[92] and the review by Hanson et al.,[93] respectively. A schematic of a PIV experimental setup is provided in Fig. 5.

The earliest application of PIV to fuel cells was by Martin et al.[94] in which a 180° sharp turn was used to simulate the first two passes of a serpentine channel. Water flow in a Plexiglas mock cell (with 12.7 mm square cross section) was visualized at the proper Reynolds numbers to ensure hydraulic similarity with operating fuel cells. The study visualized flow instabilities in the stream-wise and cross-flow distributions caused by the bend. Sharp et al.[95] and Yoon et al.[96] utilized PIV to visualize flow distribution in mm-sized mock channels using smoke and water droplets as the seeding particles. They proposed water droplets because of the compatibility with operating fuel cells (smoke will quickly contaminate the cell); however, they found that water particles <1 μm in diameter are required to accurately follow the flow at all flow rates. Feser et al.[97] investigated parallel, serpentine, and interdigitated flow fields with a mock cell utilizing water flow. However, by utilizing a glass fiber paper, they were able to conduct experiments with a system that was hydraulically similar to a fuel cell with GDL. Grega et al.[98] visualized air flow seeded with oil-based aerosol particles to investigate flow distribution in a mock fuel cell manifold with 21 straight-channeled cells. They observed turbulent flow in the manifold and unsteady flow in the cells, although the seeding particles would not be compatible with an operating fuel cell. Sugii and Okamoto[99] used fluorescent oil particles to visualize flow in a non-operational fuel

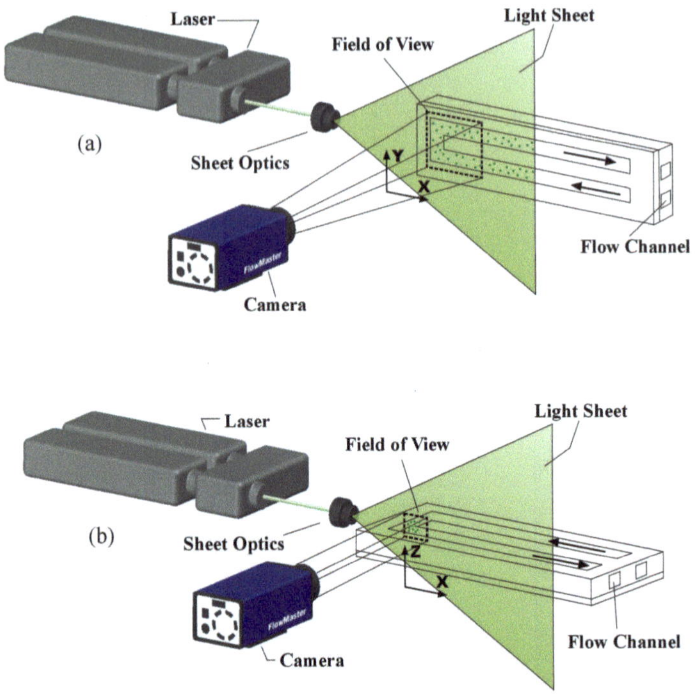

Figure 5. Schematic of a PIV setup for image acquisition in streamwise (a) and cross-flow (b) planes (reprinted from[94] Copyright (2005), with permission from ASME).

cell. They introduced the seeding particles into an operational fuel cell for 60 min before the performance decreased due to fouling. Salloum et al.[100] conducted the only study in an operational fuel cell, using fluorescent polystyrene microspheres in a membrane-less formic acid/potassium permanganate micro-fuel cell. They were able to visualize flow and study the effective hydraulic height of the cell to determine gasket compression and uniformity.

The use of LIF in fuel cells is less extensive than PIV, but has produced valuable results. Bazylak et al.[101] used ex-situ testing to quantify the effects that cell compression have on GDL hydrophobicity by damaging the GDL and PTFE layer. Lozano et al.[102,103] and Barreras et al.[5] conducted a series of LIF experiments on mock fuel cells with various flow field designs (diagonal, serpentine, and interdigitated) both with and without a GDL. The experiments utilized both water and air as the fluid and indicated that the flow through the GDL was controlled more by the pressure gradient across the GDL than the uniformity of the velocity distribution in the flow field. Their work implies that it may be more important for a flow field to create a smooth pressure field throughout the active area than to necessarily have uniform flow distribution, although the two are often correlated.

2.2.7. *Embedded Sensors*

Another method capable of mapping the distribution of species within PEM fuel cells is to embed sensors within the fuel cell to directly measure the presence of the species of interest. One promising technology is the use of fiber optic sensors, with a signal that is unaffected by electric fields and a structure that should minimally affect the performance of the cell components. As part of the DOE Hydrogen Program, McIntyre et al.[104] developed two distinct fiber optic temperature probes (free space and monolithic) and a gas species sensor based on capillary mass spectrometer probe measurements (for H_2O, H_2, and O_2 concentrations) for deployment in fuel cells. While Tao et al.[105] did not test their fiber optic sensors in a fuel cell, they did verify the performance of various sensors with plans to integrate them into fuel cells. The developed fiber optic sensors included a moisture sensor sensitive to RH = 4–95%, a temperature sensor sensitive to T = 22–90°C, and a hydrogen peroxide sensor sensitive down to 10 ppb.

Various other sensor technologies have been successfully embedded in operating PEM fuel cells. Takaichi et al.[106] placed platinum wires in between eight 25-μm thick sheets of Nafion® membrane to measure the mixed potential, throughout the resulting 200-μm thick membrane, caused by reactant gas permeation. Takaichi et al.[107] also used Pt-black coated Au wires to determine the changes in water content through the membrane in response to abrupt load changes. Nishikawa et al.[108] developed a cell with six humidity sensors in the cathode plate. The plates also included three current sensors and they were able to correlate humidity and current distribution changes to the current density, humidity, and oxygen utilization. The results showed that high utilization levels shifted the highest performance towards the cathode inlet, while low humidity shifted the highest performance towards the outlet. Büchi and Reum[109] placed 10-μm thick gold wires in between the CL and GDL. The wires, spaced 0.2 mm apart, acted as potential sensors to calculate local current density and membrane resistance (influenced by water content) at sub-channel resolutions.

2.2.8. *Other Methods*

There are a variety of additional methods not falling within the previous sections that have been utilized to determine the species distribution in fuel cells. A commonly used approach has been to utilize a cell consisting of electrically segmented components to calculate species distribution from local measurements. Brett et al.[110,111] utilized a segmented single channel to obtain the distribution of CO poisoning and membrane water content. The CO poisoning distribution was determined by localized stripping voltammetry, while the membrane water content distribution was determined by local EIS. Tingelöf et al.[112] used a cathode with 34 segments to study CO and CO_2 poisoning and the effect of an air bleed. They found that the CO poisoning effects were higher at the inlet, while the air bleed (which did not affect the performance of a non-poisoned electrode) induced a recovery of cell performance that occurred evenly over the electrode surface. Reiser et al.[113] conducted one of the earliest experimental studies on the effects that startup, shutdown, and hydrogen starvation have on accelerating performance decay. The anode can be only partially filled with hydrogen during startup and shutdown (due to air entering the cathode) or during fuel undersupply. The effect is that the electrolyte potential drops to –0.59 V, which increases the cathode potential to ~1.44 V, leading to electrode corrosion. They found that even oxygen crossover from the cathode can be sufficient to cause this reverse-current condition. Baumgartner et al.[114] studied the effects of dead-end and undersupply conditions at the anode with a cell that incorporated four reference electrodes. Increasing concentrations of nitrogen and water in the dead-end anode resulted in high overpotential and an abrupt increase in hydrogen supply after a period of undersupply can result in very high cathode potentials. In similar work, Siroma et al.[115] used a cell with 97 segments to examine current distribution during a simulated startup. As the O_2 in the anode was replaced with H_2, large potentials developed between the inlet and outlet causing carbon corrosion.

Water distribution has also been investigated by various other techniques. Fabian et al.[116] measured water vapor distribution, along with oxygen concentration and temperature, with a point-based technique on an air breathing fuel cell. An x-y-z translation stage was used to obtain measurements throughout the fuel cell, but it is a time-consuming technique that cannot be used for transient data. Basu et al.[117] simultaneously measured temperature and the partial pressure of water vapor using tunable diode laser absorption spectroscopy. Fiber optic cables were used to direct the beam through the channel to photodetectors on the opposite end of the cell.

2.3. Design Implications

The ultimate goal of any diagnostic technique is to improve the design of PEM fuel cells to achieve the desired performance, durability, reliability, and cost. Thus far most species distribution research has focused on technique development and verification. A summary of the previously presented techniques is shown in Table 1. However, as the techniques continue to mature, they are providing increasing insight into improved fuel cell designs.

Several species distribution mapping techniques have been utilized for fuel cell design studies. Gebel et al.[118] utilized neutron imaging to investigate the benefits of porous gas distributors (with both high and low porosity) for improved water management and the possibility of eliminating the GDL. They were able to obtain water concentration profiles that can be used to validate mass transfer models. Turhan et al.[119] studied water transport in seven different flow field geometries (with varying landing-to-channel ratios) with neutron imaging. They observed that similar performance was possible between the flow fields with vastly different stored water profiles. It was also found that large landing-to-channel ratios were good for low current, dry inlet gas operation while small landing-to-channel ratios were good for wet conditions and high current density. Yoshizawa et al.[120] used neutron imaging to examine water distribution variations for cloth versus paper GDL and straight versus serpentine flow fields. Cloth showed higher performance, which was most likely due to wider pore distribution, while the

Table 1
Species Distribution Mapping Techniques.

Techniques		Measured species	Pros	Cons
Pressure drop		H_2O (liquid)	Simple technique, real-time measurements	Indirect measurement, pressure drop not solely caused by liquid water
Gas composition analysis	Gas chromatography	H_2O (vapor), H_2, O_2, N	Provides molar percent level accuracy, can measure gas crossover	Measures data only at discrete points
	Mass spectrometry	H_2, O_2, N	Temporal resolution of 104 ms^{-1} s is possible	High humidity can cause interference, measures data only at discrete points
Neutron imaging	Two-dimensional	H_2O (liquid)	Relatively high resolution technique, real-time data, applicable to typical fuel cell designs	Need a powerful neutron source, very difficult to separate anode/cathode effects
	Three-dimensional	H_2O (liquid)	Relatively high resolution technique ($\sim$100–150 μm), applicable to typical fuel cell designs	Need a powerful neutron source, extremely long data acquisition time (up to hours) limits data to steady-state
Magnetic resonance imaging		H_2O (liquid)	High spatial resolution possible ($\sim$25 μm)	Low temporal resolution ($\sim$2–6 min), magnetic materials interfere with signal

Table 1
Continuation.

Techniques		Measured species	Pros	Cons
X-ray radiography		H_2O (liquid)	High spatial (~10 μm) and temporal resolutions are possible	Less sensitivity to water, technique still needs refinement
Optically transparent fuel cells		H_2O (liquid), O_2, O_2 and H_2 flow distribution	Full flow field flow visualization, real-time data acquisition	Modified cells have lower performance than traditional fuel cells
Embedded sensors	Fiber optic sensors	H_2O, H_2, O_2,	Signal is unaffected by electric fields, should minimally affect cell performance	Currently an immature technology
	Metal wires	H_2O (in the membrane), H_2 and O_2 crossover	Capable of sub-mm resolutions, minimally affects cell performance	Obtains one-dimensional data
	Humidity sensors	H_2O (vapor)	Fairly established technology	Low spatial resolution, requires significant cell modification

straight flow field showed more uniform water distribution and higher performance. However, through the use of additional manifolds and seals between bends in the serpentine flow field, the performance was increased to that of the straight flow field. Owejan et al.[121] also used neutron imaging to assess the effects of GDL type and flow field channel properties on liquid water accumulation. The study produced the following observations: low in-plane gas permeability increases water retained under the lands; a PTFE coating on the channel surface reduces water accumulation; and triangular channels retain less water, while rectangular channels produce larger water droplets.

Studies have also combined species distribution with other measurement techniques to improve fuel cell design. Wen and Huang[122] used an array of 11 thermocouples and an optically accessible cell to determine the ability of a pyrolitic graphite sheet (PGS) to improve performance and to reduce and homogenize temperature distribution. The PGS was found to effectively increase heat transfer and lower the cell temperature. This led to improved performance at high flow rates (although lower performance at low flow rates) and improved performance for low humidity conditions. Hsieh and Huang[123] used a transparent cell with a segmented plate to investigate the differences between mesh, interdigitated, parallel, and serpentine flow fields in a micro-PEM. Current density and water variation was observed over extended operation times and the interdigitated flow field showed the highest and most uniform current density.

3. TEMPERATURE DISTRIBUTION MAPPING

The generation of heat always accompanies the operation of a fuel cell. The heat is due to inefficiencies in the basic fuel-cell electrochemical reaction, crossover (residual diffusion through the fuel-cell solid-electrolyte membrane) of fuel, and electrical heating of interconnection resistances. Spatial temperature variation can occur if any of these heat-generating processes occur preferentially in different parts of the fuel cell stack. For example, non-uniform distribution of fuel across the surfaces of electrodes, different resistances between the interconnections in a stack, and variations among cells within a stack resulting from variables in fabrication, can all cause local temperature variation.

3.1. Methodology and Results

3.1.1. IR Transparent Fuel Cells

In recent years, there have been a number of studies on temperature distribution along the active area of a fuel cell, using infrared (IR) cameras. Wang et al.[124] developed a fuel cell with a window made out of barium fluoride (transparent to infrared light) at the anode side. This fuel cell can be seen in Fig. 6. After producing the IR images, they measured the temperature at certain points along the flow channel; thus, plots of temperature as a function of the position in the flow field were developed. Similarly, Hakenjos et al.[125] designed a cell for combined measurement of current and temperature distribution (see Section 4.1.2.2). For the temperature distribution an IR transparent window made out of zinc selenide was located at the cathode side of the fuel cell. Other similar studies were presented by Shimoi et al.[126] which used a narrow straight channel flow field and sapphire as the IR transparent window, and Kondo et al.[127] which studied different flow field designs and used a calcium fluoride crystal as the IR transparent window. NASA's Jet Propulsion Laboratory[128] has also employed IR thermography to record the temperature distribution of fuel cells and stacks to

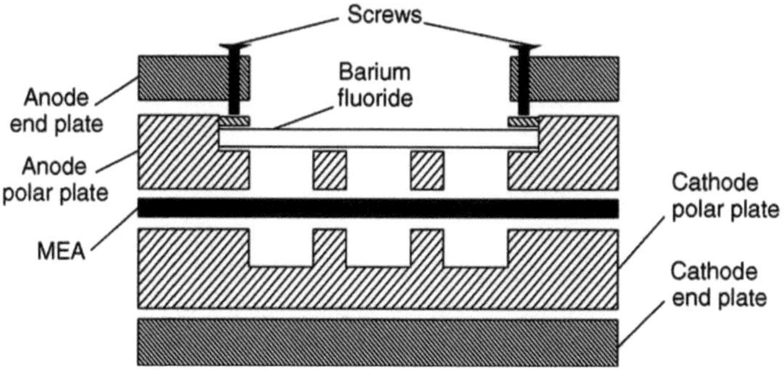

Figure 6. Schematic of a PEM fuel cell for temperature measurements through IR techniques (reprinted from,[124] Copyright (2006), with permission from Elsevier).

further optimize their designs by using an IR camera equipped with quantum-well infrared photodetectors (QWIPs), which can detect temperature differences as small as 0.005 K. Sailler et al.[129] conducted combined measurements of current density and temperature distribution. The current density was measured with magnetic sensors (see Section 4.1.3), while the temperature distribution was measured with an array of nine thermocouples and an IR camera. The current density and temperature distributions highlighted heterogeneous distribution influenced by cell geometry and collector location.

3.1.2. Embedded Sensors

The most commonly utilized embedded sensor for temperature distribution mapping is the thermocouple. Wilkinson et al.[130] developed a simple, in-situ, and noninvasive method of measuring the temperature distribution of a fuel cell with micro-thermocouples. In this study, thermocouples were located in the landing area of the flow field plates (in contact with the GDL of the MEA) of a fuel cell. The temperature data taken at different locations along the flow channel was then used to find each temperature slope, which in turn were related through mathematical equations to the local current density of each location. Thus, the current density distribution in the fuel cell was determined by simple temperature measurements. The results of this approach are discussed in more detail in Section 4.1.3. Mench et al.[131] also presented a method of measuring temperature distribution in a PEM fuel cell using micro-thermocouples. The thermocouples were placed inside the MEA between two sheets of Nafion® and a different approach was utilized for the analysis. In an extension of the work, Burford and Mench[132] showed that temperature variation within the 50 cm^2 could be greater than 10°C at current densities above 1 A/cm^2.

Two combinatorial studies also utilized thermocouple arrays to measure temperature distribution. Yan et al.[133] used an electrically segmented plate with three thermocouples to investigate the effects of a dynamic load on current and temperature distribution. The results are described further in Section 3.2. Maranzana et al.[134] developed a transparent cell with 20 thermocouples and 20 electrically isolated segments of gold wire to monitor performance and the effects that liquid water had on performance instabilities. Temperature and current density distribution were strongly correlated and both depended on water management.

Fiber optic sensors are an alternative to thermocouples as embedded temperature distribution mapping sensors. As described in Section 2.2.7, McIntyre et al.[104] developed two distinct fiber optic temperature probe technologies for fuel cell applications (free space probes and optical fiber probes). Both sensor technologies showed similar trends in fuel cell temperature and were also used to study transient conditions.

3.2. Design Implications

Temperature distribution mapping has not yet seen significant implementation toward improved cell designs. However, as previously mentioned in Section 2.3, Wen and Huang[122] used an array of 11 thermocouples in combination with a visualization cell to evaluate the effectiveness of a PGS sheet to improve performance and to reduce and homogenize temperature distribution in a PEM fuel cell. Also, in the study by Yan et al.[133] they investigated dynamic load conditions by studying the effects of air stoichiometry, rate

Table 2
Temperature Distribution Mapping Techniques.

Techniques		Pros	Cons
IR transparent fuel cells		Full flow field flow visualization, suitable for real-time data acquisition	Modified cells have lower performance than traditional fuel cells
Embedded sensors	Thermocouples	Suitable for real-time data acquisition, high measurement accuracy	Measures data only at discrete points, electrical insulation from other components is necessary
	Fiber optic sensors	Suitable for real-time data acquisition, high measurement accuracy	Currently an immature technology

of load application, magnitude of load increase, and humidity on temperature and current density distribution. It was observed that increasing the air stoichiometry or decreasing the rate of load application improved the uniformity of temperature and current density distribution. A summary of the previously presented techniques is shown in Table 2.

4. CURRENT DISTRIBUTION MAPPING

In PEM fuel cells, uniformity of the current density across the entire active area is critical for optimizing the fuel cell performance. A non-uniform current density in the fuel cell can drastically affect different parameters of the fuel cell, such as reduced reactant and catalyst utilization along the active area, decrease in total efficiency and lifetime, and durability failure modes. Thus, determination of the current density distribution information is vital for designing PEM fuel cells that achieve higher performance and longer life.[130] A number of methods for measuring current distribution in PEM fuel cells have been demonstrated; the following sections discuss some of these methods in further detail.

4.1. Methodology and Results

4.1.1. Partial MEA

The partial MEA approach involves the use of several MEAs, each with different sections of the catalyzed active area covered, thus, reducing the total active area of each MEA,[135] as shown in Fig. 7. Appointed sectional performance could be achieved by subtracting one steady state polarization curve from another one. One of the main advantages of this method is that it is relatively simple to implement with respect to other techniques. In addition, this technology allows the cell performance to be analyzed in a steady state, with low spatial resolution. However, as pointed out by Natarajan et al.,[136] though the resolution can be improved by increasing the number of segments or portions to be tested, significant errors would arise due to inherent variations in electrical, transport, and kinetic properties between different MEAs.

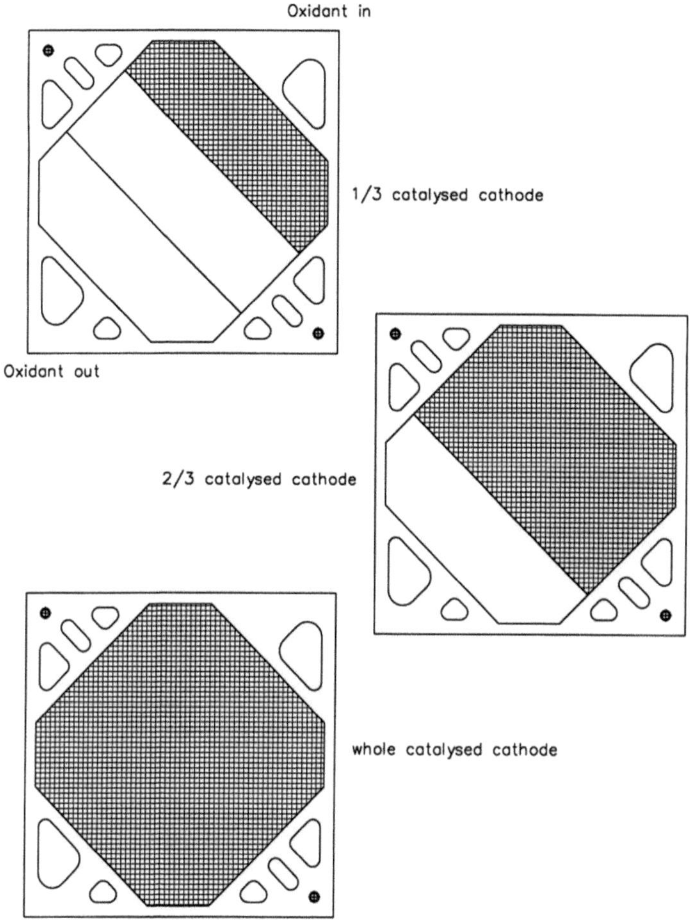

Figure 7. Sketch of MEA design showing active cathode area for the partial MEA method (reprinted from[135] Copyright (1998), with permission from Elsevier).

4.1.2. Segmented Cells

A number of research groups have presented segmented cell approaches and combined them with electrochemical methods, e.g., EIS and MRED (MEA resistance and electrode diffusion). These diagnostic approaches provide direct information on not just the current distribution of the cell, but also on other phenomena that are occurring inside the cell at various operating conditions. On the other hand, an ultimate understanding of the physicochemical phenomena within the PEM fuel cell system is extraordinarily helpful for optimizing electrode and flow plate design.

In addition to the use of conventional MEAs, good spatial resolution and the ability to determine transient effects resulting from sudden changes in operating conditions are desired. To date, a parallel effort has been made to achieve sufficient spatial and temporal resolution by designing segmented flow field plates. The basic concept of the segmented cell (or segmented flow field plate) approach is to divide the anode or/and cathode plate into conductive segments that are electrically isolated from each other.

4.1.2.1. Subcell Approach Stumper et al.[135] presented the subcell approach to measure localized currents and localized electrochemical activity in a fuel cell. In this method a number of subcells were situated in different locations along the cell's active area and each subcell was electrically isolated from each other and from the main cell. Separate load banks controlled each subcell. Figure 8 shows the subcells in both the cathode and anode flow field plates (the MEA also had such subcells). The current–voltage characteristics for the

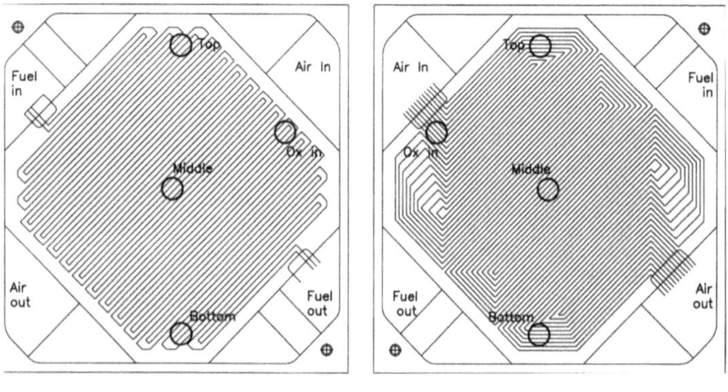

Figure 8. Schematic diagram of subcells method, showing fuel and oxidant flow fields separately (reprinted from[135] Copyright (1998), with permission from Elsevier).

subcells, when compared to those of the main cell, are indicative of local fuel cell performance.[135] Although this method provides a good understanding of the current distribution along a flow field, the manufacturing of the modified flow field plates and MEA makes it very complex and difficult.

Nopenen et al.[137] and Hottinen et al.[138] adopted a similar approach to study steady state and transient current density distribution in free-breathing PEM fuel cells. Liu et al.[139] used the subcell approach to determine current density distribution in a single cell, which was eventually expanded to investigating the cell behavior under starvation conditions.[140] Strickland et al.[141] examined the effects of an electroosmotic pump on a straight flow field in a porous plate with the subcell approach. They conducted both steady state (with and without pump) and transient (turning on the pump) measurements and concluded that the pump improved water management with a pump power consumption of less than 1% of the cell output. Wang and co-workers[16,142,143] and Rajalakshmi et al.[144] also used the subcell approach to determine the current distributions under various operating conditions in PEM fuel cells with a single serpentine flow field. Lastly, as previously mentioned in Section 2.2.7, Nishikawa et al.[108] combined the subcell approach with humidity sensors to examine the effect of load, humidity, and reactant utilization on performance.

4.1.2.2. Segmented Current Collector/Flow Field In order to have spatially resolved performance data, Stumper et al.[135] used a passive resistor network made from resin-isolated graphite blocks located between the flow field plate and current collector. The potential drops across these blocks were monitored to establish the current flowing through them via Ohm's law. By scanning the entirety of the graphite blocks, the current distribution mapping of the electrode was obtained. One of the main advantages is that time-dependent phenomena can be monitored in real time in order to observe sudden changes after certain parameters have been modified. One of the main issues of this configuration arises from the fact that the graphite blocks are not part of the flow field plate, thus, with these passive resistors there is the possibility of having low spatial resolution due to the lateral in-plane current through the flow field plate. The setup of these resistors also makes this approach quite complex and tedious. Hakenjos et al.[145] designed a similar passive resistor network by inserting 45 graphite segments into a plastic plate. In their experiment, the flow field plate was segmented at the same time to avoid lateral flow of current. Ghosh et al.[146,147] designed an approach similar to that of Stumper et al., in which a segmented expanded graphite plate was used as a passive resistor network (each segment represented a resistor), except that this plate was inserted between two adjacent plates for sealing purposes. This sandwich of three thin plates was then placed at the back of the cathode plate (see Fig. 9 to observe how this cell was assembled).

Noponen et al.[148] designed a fuel cell with a segmented anode flow field and anode current collector to be used for high power densities and flow rates. Similarly, Yoon et al.[149] investigated the influence of flooding and stoichiometry variation of the feed gas through the use of a cell composed of 81 segments in both the flow field plate and current collector plate. Each segment was insulated from each other and the current distribution was measured using Hall effect sensors that were connected to each segment. Natarajan

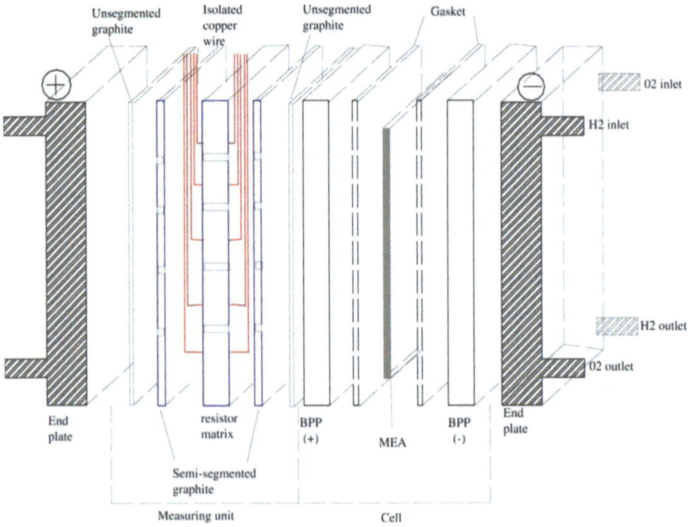

Figure 9. Schematic diagram of the segmented cell approach inside the fuel cell (reprinted from[146] Copyright (2006), with permission from Elsevier).

and Nguyen[136,150,151] developed a segmented fuel cell (both anode and cathode plates) with a segmented MEA used to study a number of parameters with different operating conditions. As mentioned in Section 2.2.2, Araki et al.[17] and Onda et al.[18] used the segmented cell approach to measure the current at different locations in the active area by shunt resistances. This data was used to confirm the reliability of a mathematical model. Lastly, both Sun et al.[152] and Weng et al.[153] used a segmented cell to study the effect of humidity on the local current characteristics.

Another study was the one by Wieser et al.[154,155] in which a magnetic loop array was first employed to determine the currents generated by segmented fuel cells. A flow field plate, which was divided into 40 electrically isolated segments, was used to avoid lateral currents between adjacent segments. The current sensors, which consisted of Hall sensors fixed in the air gaps of annular soft magnetic ferrites, were placed around a gudgeon on each flow field segment. The advantage of this method is that a combination of high spatial and time resolution can be achieved, and in principle, the integration into fuel cell stacks is possible. However, the installation of the Hall sensors into the flow field plate makes maintenance difficult and the investigation of different flow field structures expensive. Instead of installing the magnetic loop array inside the fuel cell, Geiger et al.[156] developed another approach based on externally measuring the currents from different segments. The improvement makes this approach easier to adjust to different cell designs.

Galvanostatic discharge of a fuel cell (MRED method) provided information related to liquid water in a fuel cell in a minimally invasive manner.[157] Stumper et al.[158] showed that through a combination of this MRED method with a current mapping (segmented fuel cell similar to the one discussed in Stumper et al.[135]), it was possible to obtain the local membrane water content distribution across the cell area. The test cell was operated with a current collection plate segmented on the cathode along the reactant flow direction. In addition to the pure ohmic resistance, this experimental setup allowed the determination of the free gas volume of the unit cell (between the inlet and outlet valves). Furthermore, the total amount of liquid water presented in the anode or cathode compartment was obtained.

Segmented cells are the most commonly utilized technique in combinatorial cell investigations. Segmented cells that have been used for combined studies of species and current distribution include

Onda et al.,[18] Lu et al.,[19] Liu et al.,[20] Brett et al.,[111] Tingelöf et al.,[112] Siroma et al.,[114] and Hakenjos et al.[125] Additionally, Maranzana et al.[134] combined a segmented cell with thermocouples and a transparent cell to correlate current density distribution, current distribution, and water management.

4.1.2.3. Printed Circuit Board Cleghorn et al.[159] performed pioneering work using PCB technology to create a segmented anode current collector and anode flow field to measure current distribution in PEM fuel cells. In this study, the gas diffusion backing and CL in the anode side were also segmented and each segment had an active area of 4.4 cm^2. Fig. 10 shows this segmented fuel cell designed using PCB technology. Through the use of current distribution they were able to investigate different flow field designs and optimize the utilization of the active electrode area with the best humidification conditions and reactant stoichiometries. A combination of two load units and a specially designed patch board, which acted as a multiplexer to control the voltage at various segments, were employed. This only allowed analysis of steady-state behavior. Brett et al.[160]

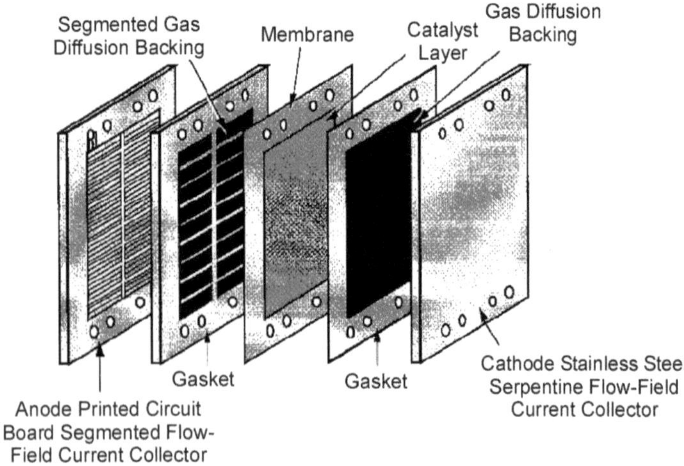

Figure 10. Schematic of the segmented fuel cell design using printed circuit board (PCB) technology (with kind permission from Springer Science+Business Media:[159] Fig. 1).

also used this approach to measure the current distribution along a single flow channel with a relatively high spatial resolution (0.3 cm^2 active area per contact). The difference between their experiments and those conducted by Cleghorn et al.[159] was that they used individual loads to control each segment and monitored the current flowing through that segment. All segments were driven at the same potential to avoid lateral current. In this way, there was no requirement to segment the MEA into electrically isolated sections. Current profiles at different cell polarizations and reactant flow rates were determined. This segmented cell was the one used by Brett et al.[161] to study the distribution of CO within the fuel cell. Hicks et al.[162] utilized PCB technology to create a single serpentine 50-cm^2 flow field with 121 segments. However, constraints with the current collection equipment required rows of segments to be shorted, which only allowed data collection from 11 larger segments. In a combinatorial study, Hartnig et al.[44] utilized neutron imaging to correlate current density with water content.

It is important to note that some of the mentioned segmented cell approaches have also been used to develop other diagnostic tools. For example, Brett et al.[163] used a segmented cell[160] to demonstrate localized EIS response over a frequency range of 0.1 Hz–10 kHz as a function of position in PEM fuel cell. An impedance analyzer and multiplexor were used during the experiment to avoid lateral current flow between adjacent segmented cells. Their results proved that integral EIS measurement alone was not sufficient to properly characterize the operation of a fuel cell. Hakenjos et al.[164] set up test equipment containing a multichannel frequency response analyzer system together with a multichannel potentiostat that allowed measurements of the local electrical impedance spectra of segmented fuel cells simultaneously. Their results showed good agreement between the high frequency and simulated protonic resistivity realized with a PEM model based on CFD software.[164] Bender et al.[165] used a similar method to measure the cell current and high-frequency resistance (HFR) distribution.

4.1.3. *Other Methods*

Current distribution in PEM fuel cell can be determined not only by setting up segmented cells, electrically isolated from each other, but also indirectly, by measurement of the induced magnetic field. A superconducting quantum interference device (SQUID), fluxgate magnetometer, and magnetoresistive sensor were employed respectively to measure the locally induced magnetic field by Claycomb et al.[166] and Candusso et al.[167] Consequentially, the magnetic field was calculated throughout the cell using the Maxwell equations and the current density distribution was evaluated indirectly. In Candusso et al.'s experiment, several magnetoresistive sensors combined with thermocouples were used as probes to correct the results at different temperatures. Furthermore, Hauer et al.[168] used this technology to detect the current distribution in a fuel cell stack. As mentioned in Section 3.1.1, Sailler et al.[129] utilized a magnetic sensor in combination with thermocouples and an IR camera to correlate current and temperature distribution. The advantage of electromagnetic approach is that its implementation causes only limited perturbation, or even no perturbation, to the operation of the fuel cell.

Another method to determine current distribution in a PEM fuel cell was presented by Sun et al.[169] in which they designed a current distribution measurement gasket that can be placed anywhere in the fuel cell (usually at the back of the cathode side) and can measure the local current density at various point along the active area of the cell. The advantage of this approach is that it can be used without having to modify any component of the cell. The same technique was also used by Zhang et al.[170] to compare the performance of interdigitated and serpentine flow fields.

As mentioned previously in Section 2.6, Wilkinson et al.[130] demonstrated that local temperature measurements at specific experimental conditions correlate well with local current densities determined through other published current density approaches. Their results suggest that current mapping can be indirectly conducted through local temperature measurements using an array of microthermocouples in the active area.

Freunberger et al.[171,172] developed a current distribution technique with sub-mm resolution. The technique utilized thin gold wires (diameter of 10 or 25 μm), placed between the CL and GDL, and calculated the current density based on the known resistance between the probe and the current collector. The technique was also utilized in the study by Buchi and Reum[109] described in Section 2.2.7.

4.2. Design Implications

As current distribution mapping techniques have matured, similar to species distribution techniques, they have been increasingly utilized for improving fuel cell design. A summary of the previously presented techniques is shown in Table 3. As previously mentioned in Section 3.2, Yan et al.[133] used a cell combining a segmented plate and an array of thermocouples to determine how variations in the application of a dynamic loading profile affected current and temperature uniformity. Regarding flow field design, Hwang et al.[173] utilized a segmented plate with Hall-effect sensors to study serpentine, parallel, interdigitated, and biomimic configurations. They investigated the effect of cathode stoichiometry and cathode humidity on current distribution and observed more uniform distribution at higher voltages and that the serpentine flow field demonstrated the highest performance. Similarly, as mentioned in Section 2.3, Hsieh and Huang[123] used neutron imaging and a segmented micro-PEM fuel cell to compare mesh, interdigitated, parallel, and serpentine flow fields.

Current distribution studies have also been used to develop optimal loading gradients for compounds within a PEM fuel cell. Santis et al.[174] utilized a segmented plate design with Hall-effect sensors

Table 3
Current Distribution Mapping Techniques.

Techniques		Pros	Cons
Partial MEA		Implementation is relatively simple	Data limited to steady state, low spatial resolution
Segmented cells	Subcell approach	Suitable for real-time data acquisition	Implementation is complicated, low spatial resolution
	Segmented current collector/flow field	Suitable for real-time data acquisition, well-suited for combinatorial cells	Typically spatial resolution is low, implementation can be complicated
	Printed circuit board	Relatively high spatial resolution ($0.3\ cm^2$), relatively simple implementation	High spatial resolutions require expensive and complicated current collection equipment

to determine the capacity for catalyst loading gradients to homogenize current distribution. They found that increasing catalyst loading from inlet to outlet led to uniform distribution. It was also observed that a steep loading profile is best at lower air stoichiometries, while a smoother profile was more beneficial at higher air stoichiometries. While Schulze et al.[175] conducted current distribution studies with both segmented cells and the subcell approach, it was the subcell approach that was used to investigate the effect of a hydrophilically graded GDL on current distribution in the cathode of a DMFC. A GDL with low hydrophibicity treatment at the inlet and high hydrophobicity treatment for the remainder produced more uniform current density at a wider range of humidity conditions.

5. CONCLUDING REMARKS

The developing level of relevant equipment and materials plays an important role in the application of diagnostic techniques to PEM fuel cells. As more detailed information about the numerous species and their interactions throughout a fuel cell is obtained, it will be increasingly possible to predict the response of a fuel cell in real world applications. For example, neutron radiographic imaging has already proven to be a unique experimental method to visualize and quantify liquid water inside an operating PEM fuel cell. However, it is currently limited in both spatial and temporal resolution, making it difficult to capture transient two-phase flow phenomena. Further, it is difficult to differentiate between the liquid transport inside the anode and cathode and their individual components and is limited to locations with applicable neutron sources. Other visualization approaches that help understanding phenomena within a fuel cell have been discussed (IR tests for thermal imaging and transparent cells for liquid water visualization). However, these methods also fail to provide a better understanding of the transport mechanisms of liquid water in microstructures (i.e., inside and through the GDL).

Current density distribution along the active area in a fuel cell is a very critical parameter that needs to be understood completely in order to improve the design of each component of a cell. So far, one limitation prevailing in much of the published work regarding segmented cells is the difficulty in producing a similar level of current density as non-segmented cells. The complexity of the design, fabrication, and assembly of the segmented cells may result in higher and non-uniform contact resistance between components of the fuel cell, which leads to lower performance. Another limitation is that the test conditions of many studies are far from a typical automotive drive cycle. Additionally, since the parameters studied are so highly dependent on the design of the fuel cell, different researchers can obtain contradictory results. A detailed understanding of the various influences is critical and must be accounted for when developing conclusions. Only when these limitations are conquered, the fuel cell redesign approaches can have the potential to become a standard diagnostic tool for automotive fuel cell research and development.

As shown in this review, test equipment integrated with several diagnostic techniques is preferred for a deeper insight into the mechanisms that cause performance losses and spatial non-uniform distribution. As a consequence, more information, which is simultaneously obtained with these diagnostic tools, will strongly support development of empirical models or validate theoretical models predicting performance as a function of operating conditions and fuel cell characteristic properties.

REFERENCES

[1] L. Ma, S. Warthesen, A. Shores, J. New Mater. Electrochem. Syst. **3**, 221 (2000)

[2] B. Du, Q. Guo, R. Pollard, D. Rodriguez, C. Smith, J. Elter, JOM **58**, 45(2006)

[3] J. Tan, Y. J. Chao, J. W. Van Zee, W. K. Lee, Mater. Sci. Eng. A **445–446.** 669(2007)

[4] X. Cheng, Z. Shi, N. Glass, L. Zhang, J.J. Zhang, D. Song, Z.S. Liu, H. Wang, J. Shen, J. Power Sources **165**, 739 (2007)

[5] F. Barreras, A. Lozano, L. Valiño, R. Mustata, C. Marin, J. Power Sources **175**, 841 (2008)

[6] W. He, G. Lin, T.V. Nguyen, AIChE J. **49**, 3221 (2003)

[7] H. Görgün, M. Arcak, F. Barbir, J. Power Sources **157**, 389 (2006)

[8] F. Barbir, H. Gorgun, X. Wang, J. Power Sources **141**, 96 (2005)

[9] A.D. Bosco M.H. Fronk, "Fuel cell flooding detection and correction," U.S. Patent 6, **103**, 409 (2000)

[10] P. Rodatz, F. Büchi, C. Onder, L. Guzzella, J. Power Sources **128**, 208 (2004)

[11] H.P. Ma, H.M. Zhang, J. Hu, Y.H. Cai, B.L. Yi, J. Power Sources **162**, 469(2006)

[12] P. Pei, M. Ouyang, W. Feng, L. Lu, H. Huang, J. Zhang, Int. J. Hydrogen Energy **31**, 371 (2006)

[13] M.M. Mench, Q.L. Dong, C.Y. Wang, J. Power Sources **124**, 90 (2003)

[14] H.S. Kim, T.H. Ha, K. Min, *Proceedings of FUELCELL2006*. Irvine,CF, 2006, 431

[15] Q. Dong, J. Kull, M.M. Mench, J. Power Sources **139**, 106 (2005)
[16] X.G. Yang, N. Burke, C.Y. Wang, K. Tajiri, K. Shinohara, J. Electrochem. Soc. **152**, A759 (2005)
[17] T. Araki, H. Koori, T. Taniuchi, K. Onda, J. Power Sources **152**, 60 (2005)
[18] K. Onda, T. Araki, T. Taniuchi, D. Sunakawa, K. Wakahara, M. Nagahama, J. Electrochem. Soc. **154**, B247 (2007)
[19] G.Q. Lu, F.Q. Liu, C.Y. Wang, J. Power Sources **164**, 134 (2007)
[20] F.Q. Liu, G.Q. Lu, C.Y. Wang, J. Membr. Sci. **287**, 126 (2007)
[21] S. Cleghorn, J. Kolde, W. Liu, in *Handbook of Fuel Cells: Fundamentals, Technology and Applications*, ed. by W. Vielstich, H.A. Gasteiger, A. Lamm Vol. 3, (Wiley, West Sussex, 2003) p. 566
[22] K. Broka, P. Ekdunge, J. Appl. Electrochem. **27**, 117 (1997)
[23] F. Liu, B. Yi, D. Xing, J. Yu, H. Zhang, J. Membr. Sci. **212**, 213 (2003)
[24] W.P. Partridge, T.J. Toops, J.B. Green, T.R. Armstrong, J. Power Sources **160**, 454 (2006)
[25] X. Ren, S. Gottesfeld, J. Electrochem. Soc. **148**, A87 (2001)
[26] G.J. Janssen, J. Electrochem. Soc. **148**, A1313 (2001)
[27] A.E.–G. El-Abd, A. Czachor, J.J. Milczarek, J. Pogorzelski, IEEE Trans. Nucl. Sci. **52**, 299 (2005)
[28] R.J. Bellows, M.Y. Lin, M. Arif, A.K. Thompson, D. Jacobson, J. Electrochem. Soc. **146**, 1099 (1999)
[29] A.B. Geiger, A. Tsukada, E. Lehmann, P. Vontobel, A. Wokaun, G.G. Scherer, Fuel Cells **2**, 92 (2002)
[30] R. Satija, D.L. Jacobson, M. Arif, S.A. Werner, J. Power Sources **129**, 238 (2004)
[31] I. Manke, C. Hartnig, M. Grünerbel, J. Kaczerowski, W. Lehnert, N. Kardjilov, A. Hilger, J. Banhart, W. Treimer, M. Strobl, Appl. Phys. Lett. **90**, 184101(2007)
[32] N. Pekula, K. Heller, P.A. Chuang, A. Turhan, M.M. Mench, J.S. Brenizer, K. Ünlü, Nucl. Instrum. Methods Phys. Res., Sect. A **542**, 134 (2005)
[33] J.J. Kowal, A. Turhan, K. Heller, J. Brenizer, M.M. Mench, J. Electrochem. Soc. **153**, A1971 (2006)
[34] D. Kramer, J. Zhang, R. Shimoi, E. Lehmann, A. Wokaun, K. Shinohara, G. Scherer, Electrochim. Acta **50**, 2603 (2005)
[35] M.A. Hickner, N.P. Siegel, K.S. Chen, D.N. Mcbrayer, D.S. Hussey, D.L. Jacobson, M. Arif, J. Electrochem. Soc. **153**, A902 (2006)
[36] Y.S. Chen, H. Peng, D.S. Hussey, D.L. Jacobson, D.T. Tran, T. Abdel-Baset, M. Biernacki, J. Power Sources **170**, 376 (2007)
[37] P. Boillat, G.G. Scherer, A. Wokaun, G. Frei, E.H. Lehmann, Electrochem. Commun. **10**, 1311 (2008)
[38] D.S. Hussey, D.L. Jacobson, M. Arif, J.P. Owejan, J.J. Gagliardo, T.A. Trabold, J. Power Sources **172**, 225 (2007)
[39] E.H. Lehmann, G. Frei, G. Kühne, P. Boillat, Nucl. Instrum. Methods Phys. Res., Sect. A **576**, 389 (2007)
[40] P. Boillat, D. Kramer, B.C. Seyfang, G. Frei, E. Lehmann, G.G. Scherer, A. Wokaun, Y. Ichikawa, Y. Tasaki, K. Shinohara, Electrochem. Commun. **10**, 546 (2008)
[41] J. Park, X. Li, D. Tran, T. Abdel-Baset, D.S. Hussey, D.L. Jacobson, M. Arif, Int. J. Hydrogen Energy **33**, 3373 (2008)
[42] J.P. Owejan, T.A. Trabold, D.L. Jacobson, R.D. Baker, D.S. Hussey M. Arif, Int. J. Heat Mass Transfer **49**, 4721 (2006)

[43] A. Turhan, K. Heller, J.S. Brenizer, M.M. Mench, J. Power Sources **160**, 1195 (2006)
[44] C. Hartnig, I. Manke, N. Kardjilov, A. Hilger, M. Grünerbel, J. Kaczerowski, J. Banhart, W. Lehnert, J. Power Sources **176**, 452 (2008)
[45] P.T. Callaghan, *Principles of Nuclear Magnetic Resonance Microscopy* (Oxford University Press, New York, NY, 1991)
[46] N.J. Bunce, S.J. Sondheimer, C.A. Fyfe, Macromolecules **19**, 333 (1986)
[47] S. Tsushima, K. Teranishi, S. Hirai, Electrochem. Solid State Lett. **7**, A269 (2004)
[48] S. Tsushima, K. Teranishi, S. Hirai, Energy **30**, 235 (2005)
[49] K. Teranishi, S. Tsushima, S. Hirai, J. Electrochem. Soc. **153**, A664 (2006)
[50] K. Teranishi, S. Tsushima, S. Hirai, Electrochem. Solid-State Lett. **8**, A281 (2005)
[51] S. Tsushima, K. Teranishi, K. Nishida, S. Hirai, Magnet. Reson. Imaging **23**, 255 (2005)
[52] J. Bedet, G. Maranzana, S. Leclerc, O. Lottin, C. Moyne, D. Stemmelen, P. Mutzenhardt, D. Canet, Int. J. Hydrogen Energy **33**, 3146 (2008)
[53] K.W. Feindel, J.P.-A. LaRocque, D. Starke, S.H. Bergens, R.E. Wasylishen, J. Am. Chem. Soc. **126**, 11436 (2004)
[54] K.W. Feindel, S.H. Bergens, R.E. Wasylishen, Chem. Phys. Chem. **7**, 67 (2006)
[55] K.W. Feindel, S.H. Bergens, R.E. Wasylishen, J. Am. Chem. Soc. **128**, 14192 (2006)
[56] K.W. Feindel, S.H. Bergens, R.E. Wasylishen, Phys. Chem. Chem. Phys. **9**, 1850 (2007)
[57] K.W. Feindel, S.H. Bergens, R.E. Wasylishen, J. Power Sources **173**, 86 (2007)
[58] Z. Dunbar, R.I. Masel, J. Power Sources **171**, 678 (2007)
[59] Z. Dunbar, R.I. Masel, J. Power Sources **182**, 76 (2008)
[60] Z. Zhang, J. Martin, J. Wu, H. Wang, K. Promislow, B.J. Balcom, J. Magn. Reson. **193**, 259 (2008)
[61] P.K. Sinha, P. Halleck, C.Y. Wang, Electrochem. Solid-State Lett. **9**, A344 (2006)
[62] P.K. Sinha, P.P. Mukherjee, C.Y. Wang, J. Mater. Chem. **17**, 3089 (2007)
[63] I. Manke, C. Hartnig, M. Grünerbel, W. Lehnert, N. Kardjilov, A. Haibel, A. Hilger, J. Banhart, H. Riesemeier, Appl. Phys. Lett. **90**, 174105 (2007)
[64] C. Hartnig, R. Kuhn, P. Krüger, I. Manke, N. Kardjilov, J. Geobbels, B.R. Müller, H. Riesemeier, Mater. Testing **50**, 609 (2008)
[65] C. Hartnig, I. Manke, R. Kuhn, S. Kleinau, J. Goebbels, J. Banhart, J. Power Sources, **188**, 468 (2009)
[66] C. Hartnig, I. Manke, R. Kuhn, N. Kardjilov, J. Banhart, W. Lehnert, Appl. Phys. Lett. **92**, 134106 (2008)
[67] T. Mukaide, S. Mogi, J. Yamamoto, A. Morita, S. Koji, K. Takada, K. Uesugi, K. Kajiwara, T. Noma, J. Synchrotron. Rad. **15**, 329 (2008)
[68] S.J. Lee, N.Y. Lim, S. Kim, G.G. Park, C.S. Kim, J. Power Sources, **185**, 867 (2008)
[69] V.R. Albertini, B. Paci, F. Nobili, R. Marassi, M.D. Michiel, Adv. Mater. **21**, 578 (2009)
[70] F.B. Weng, A. Su, C.Y. Hsu, C.Y. Lee, J. Power Sources **157**, 674 (2006)
[71] H. Yang, T.S. Zhao, Q. Ye, J. Power Sources **139**, 79 (2005)
[72] X. Liu, H. Guo, C.F. Ma, J. Power Sources **156**, 267 (2006)
[73] X. Liu, H. Guo, F. Ye, C.F. Ma, Electrochim. Acta **52**, 3607 (2007)

[74] A. Su, F.B. Weng, C.Y. Hsu, Y.M. Chen, Int. J. Hydrogen Energy **31**, 1031 (2006)
[75] S. Ge, C. Y. Wang, Electrochem. Solid-State Lett. **9**, A499 (2006)
[76] S. Ge, C.Y. Wang, J. Electrochem. Soc. **154**, B998 (2007)
[77] X.G. Yang, F.Y. Zhang, A.L. Lubawy, C.Y. Wang, Electrochem. Solid-State Lett. **7**, A408 (2004)
[78] H.S. Kim, T.H. Ha, S.J. Park, K. Min, M. Kim, *Proceedings of FUELCELL2005*. Ypsilanti, 2005, 57
[79] K. Sugiura, M. Nakata, T. Yodo, Y. Nishiguchi, M. Yamauchi, Y. Itoh, J. Power Sources **145**, 526 (2005)
[80] F.Y. Zhang, X.G. Yang, C.Y. Wang, J. Electrochem. Soc. **153**, A225 (2006)
[81] K. Tüber, D. Pócza, C. Hebling, J. Power Sources **124**, 403 (2003)
[82] A. Theodorakakos, T. Ous, M. Gavaises, J.M. Nouri, N. Nikolopoulos, H. Yanagihara, J. Colloid Interface Sci. **300**, 673 (2006)
[83] K.S. Chen, M.A. Hickner, D.R. Noble, Int. J. Energy Res. **29**,1113 (2005)
[84] E.C. Kumbur, K.V. Sharp, M.M. Mench, J. Power Sources **161**, 333 (2006)
[85] H. Yang, T.S. Zhao, P. Cheng, Int. J. Heat Mass Transfer **47**, 5725 (2004)
[86] P. Argyroulos, K. Scott, W.M. Toama, Electrochim. Acta **44**, 3575 (1999)
[87] P. Argyroulos, K. Scott, W.M. Toama, J. Appl. Electrochem. **29**, 663 (1999)
[88] G.Q. Lu, C.Y. Wang, J. Power Sources **134**, 33 (2004)
[89] J. Nordlund, C. Picard, E. Birgersson, M. Vynnycky, G. Lindbergh, J. Appl. Electrochem. **34**, 763 (2004)
[90] J. Inukai, K. Miyatake, K. Takada, M. Watanabe, T. Hyakutake, H. Nishide, Y. Nagumo, M. Watanabe, M. Aoki, H. Takano, Angew. Chem. Int. Ed. **47**, 2792 (2008)
[91] T. Murahashi, H. Kobayashi, E. Nishiyama, J. Power Sources **175**, 98 (2008)
[92] M. Raffel, C.E. Willer, J. Kompenhans, *Particle Image Velocimetry: A Practical Guide* (Springer, New York, 1998)
[93] R.K. Hanson, J.M. Seitzman, P.H. Paul, Appl. Phys. B **50**, 441 (1990)
[94] J. Martin, P. Oshkai, N. Djilali, J. Fuel Cell Sci. Technol. **2**, 70 (2005)
[95] K.V. Sharp, S.Y. Yoon, J.W. Ross, M.M. Mench, ECS Trans. **1**, 571 (2006)
[96] S.Y. Yoon, J.W. Ross, M.M. Mench, K.V. Sharp, J. Power Sources **160**, 1017 (2006)
[97] J.P. Feser, A.K. Prasad, S.G. Advani, J. Fuel Cell Sci. Technol. **4**, 328 (2007)
[98] L. Grega, M. McGarry, M. Begum, B. Abruzzo, J. Fuel Cell Sci. Technol. **4**, 272 (2007)
[99] Y. Sugii, K. Okamoto, *Proceedings of ICNMM2006*. Limerick, 2006, 533
[100] K.S. Salloum, J.R. Hayes, C.A. Friesen, J.D. Posner, J. Power Sources **180**, 243 (2008)
[101] A. Bazylak, D. Sinton, Z.S. Liu, N. Djilali, J. Power Sources **163**, 784 (2007)
[102] A. Lozano, F. Barreras, L. Valiño, C. Marin, Exp. Fluids **42**, 301 (2007)
[103] A. Lozano, L. Valiño, F. Barreras, R. Mustata, J. Power Sources **179**, 711 (2008)
[104] T.J. McIntyre, S.W. Allison, L.C. Maxey, W.P. Partridge, M.R. Cates, R. Lenarduzzi, C.L. Jr. Britton, D. Garvey, T.K. Plant, DOE Hydrogen Program FY 2005 Progress Report, 2005, 989
[105] S. Tao, J.C. Fanguy, X. Hu, Q. Yan, *Proceedings FUELCELL2005*. Ypsilanti, 2005, 231
[106] S. Takaichi, H. Uchida, M. Watanabe, Electrochem. Commun. **9**, 1975 (2007)
[107] S. Takaichi, H. Uchida, M. Watanabe, J. Electrochem. Soc. **154**, B1373 (2007)

[108] H. Nishikawa, R. Kurihara, S. Sukemori, T. Sugawara, H. Kobayasi, S. Abe, T. Aoki, Y. Ogami, A. Matsunaga, J. Power Sources **155**, 213 (2006)
[109] F.N. Büchi, M. Reum, Meas. Sci. Technol. **19**, 085702 (2008)
[110] D.J.L. Brett, P. Aguiar, N.P. Brandon, A.R. Kucernak, Int. J. Hydrogen Energy **32**, 863 (2007)
[111] D.J.L. Brett, S. Atkins, N.P. Brandon, N. Vasileiadis, V. Vesovic, A.R. Kucernak, J. Power Sources **172**, 2 (2007)
[112] T. Tingelöf, L. Hedström, N. Holmström, P. Alvfors, G. Lindbergh, Int. J. Hydrogen Energy **33**, 2064 (2008)
[113] C.A. Reiser, L. Bregoli, T.W. Patterson, J.S. Yi, J.D. Yang, M.L. Perry, T.D. Jarvi, Electrochem. Solid-State Lett. **8**, A273 (2005)
[114] W.R. Baumgartner, P. Parz, S.D. Fraser, E. Wallnöfer, V. Hacker, J. Power Sources **182**, 413 (2008)
[115] Z. Siroma, N. Fijiwara, T. Ioroi, S.I. Yamazaki, H. Senoh, K. Yasuda, K. Tanimoto, J. Power Sources **172**, 155 (2007)
[116] T. Fabian, R. O'Hayre, F.B. Prinz, J.G. Santiago, J. Electrochem. Soc. **154**, B910 (2007)
[117] S. Basu, M.W. Renfro, B.M. Cetegen, J. Power Sources **162**, 286 (2006)
[118] G. Gebel, O. Diat, S. Escribano, R. Mosdale, J. Power Sources **173**, 132 (2008)
[119] A. Turhan, K. Heller, J.S. Brenizer, M.M. Mench, J. Power Sources **180**, 773 (2008)
[120] K. Yoshizawa, K. Ikezoe, Y. Tasaki, D. Kramer, E.H. Lehmann, G.G. Scherer, J. Electrochem. Soc. **155**, B223 (2008)
[121] J.P. Owejan, T.A. Trabold, D.L. Jacobson, M. Arif, S.G. Kandlikar, Int. J. Hydrogen Energy **32**, 4489 (2007)
[122] C.Y. Wen, G.W. Huang, J. Power Sources **178**, 132 (2008)
[123] S.S. Hsieh, Y.J. Huang, J. Power Sources **183**, 193 (2008)
[124] M. Wang, H. Guo, C. Ma. J. Power Sources **157**, 181 (2006)
[125] A. Hakenjos, H. Muenter, U. Wittstadt, C. Hebling, J. Power Sources **131**, 213 (2004)
[126] R. Shimoi, M. Masuda, K. Fushinobu, Y. Kozawa, K. Okazaki, J. Energy Res. Tech. **126**, 258 (2004)
[127] Y. Kondo, H. Daiguji, E. Hihara, *Proceeding of the International Conference on Power Engineering*. Kobe, 2–463 (2003)
[128] NASA's Jet Propulsion Laboratory Pasadena, California. Thermal imaging for diagnosing fuel cells. http://www.nasatech.com/Briefs/Nov01/NPO21177.html. Accessed Feb 2007)
[129] S. Sailler, S. Rosini, M.A. Chaib, J.Y. Voyant, Y. Bultel, F. Druart, P. Ozil, J. Appl. Electrochem. **37**, 161 (2007)
[130] M. Wilkinson, M. Blanco, E. Gu, J.J. Martin, D.P. Wilkinson, J.J. Zhang, H. Wang, Electrochem. Solid-State Lett. **9**, A507 (2006)
[131] M.M. Mench, D.J. Burford, T.W. Davis, *Proceedings of IMECE2003*. Washington, DC, 2003, 415
[132] D.J. Burford, M.M. Mench, *Proceedings of IMECE2004*. Anaheim, 2004, 317
[133] X. Yan, M. Hou, L. Sun, H. Cheng, Y. Hong, D. Liang, Q. Shen, P. Ming, B. Yi, J. Power Sources **163**, 966 (2007)
[134] G. Maranzana, O. Lottin, T. Colinart, S. Chupin, S. Didierjean, J. Power Sources **180**, 748 (2008)
[135] J. Stumper, S.A. Campbell, D.P. Wilkinson, M.C. Johnson, M. Davis, Electrochim. Acta **43**, 3773 (1998)

[136] D. Natarajan, T.V. Nguyen, J. Power Sources **135**, 95 (2006)
[137] M. Noponen, T. Mennola, M. Mikkola, T. Hottinen, P. Lund, J. Power Sources **106**, 304 (2002)
[138] T. Hottinen, M. Nopenen, T. Mennola, O. Himanen, M. Mikkola, P. Lund, J. Appl. Electrochem. **33**, 265 (2003)
[139] Z.X. Liu, Z.Q. Mao, B. Wu, L.S. Wang, V.M. Schmidt, J. Power Sources **141**, 205 (2005)
[140] Z. Liu, L. Yang, Z. Mao, W. Zhuge, Y. Zhang, L. Wang, J. Power Sources **157**, 166 (2006)
[141] D.G. Strickland, S. Litster, J.G. Santiago, J. Power Sources **174**, 272 (2007)
[142] M.M. Mench, C.Y. Wang, J. Electrochem. Soc. **150**, A79 (2003)
[143] M.M. Mench, C.Y. Wang, J. Electrochem. Soc. **150**, A1052 (2003)
[144] N. Rajalakshmi, M. Raja, K.S. Dhathathreyan, J. Power Sources **112**, 331 (2002)
[145] A. Hakenjos, C. Hebling, J. Power Sources **145**, 307 (2005)
[146] P.C. Ghosh, T. Wüster, H. Dohle, N. Kimiaie, J. Mergel, D. Stolten, J. Power Sources **154**, 184 (2006)
[147] P.C. Ghosh, T. Wüster, H. Dohle, N. Kimiaie, J. Mergel, D. Stolten, J. Fuel Cell Sci. Technol. **3**, 351 (2006)
[148] M. Noponen, J. Ihonen, A. Lundblad, G. Lindbergh, J. Appl. Electrochem. **34**, 255–262 (2004)
[149] Y.G. Yoon, W.Y. Lee, T.H. Yang, G.G. Park, C.S. Kim, J. Power Sources **118**, 193–199 (2003)
[150] D. Natarajan, T.V. Nguyen, AIChE J. **51**, 2587–2598 (2005)
[151] D. Natarajan, T.V. Nguyen, AIChE J. **21**, 2599–2608 (2005)
[152] H. Sun, G. Zhang, L.J. Guo, S. Dehua, H. Liu, J. Power Sources **168**, 400 (2007)
[153] F.B. Weng, B.S. Jou, C.W. Li, A. Su, S.H. Chan, J. Power Sources **181**, 251 (2008)
[154] C. Wieser, A. Helmbold, E. Gülzow, J. Appl. Electrochem. **30**, 803 (2000)
[155] C. Wieser, A. Helmbold, "Method for determining materials conversion in electrochemical reactions and electrochemical unit," WO Patent 9,926,305, (1999)
[156] A.B. Geiger, R. Eckl, A. Wokaun, G.G. Scherer, J. Electrochem. Soc. **151**, A394 (2004)
[157] J. Stumper, H. Haas, A. Granados. J. Electrochem. Soc. **152**, A837 (2005)
[158] J. Stumper, M. Lohr, S. Hamada, J. Power Sources **143**, 150 (2005)
[159] S.J.C. Cleghorn, C.R. Derouin, M.S. Wilson, S. Gottesfeld. J. Appl. Electrochem. **28**, 663 (1998)
[160] D.J.L. Brett, S. Atkins, N.P. Brandon, V. Velisa, N. Vasileiadis, A.R. Kucernak, Electrochem. Commun. **3**, 628 (2001)
[161] D.J.L. Brett, S. Atkins, N.P. Brandon, V. Velisa, N. Vasileiadis, A.R. Kucernak, J. Power Sources **133**, 205 (2004)
[162] M. Hicks, K. Kropp, A. Schmoeckel, R. Atanasoski, ECS Trans. **1**, 605 (2006)
[163] D.J.L. Brett, S. Atkins, N.P. Brandon, V. Vesovic, N. Vasileiadis, A. Kucernak, Electrochem. Solid-State Lett. **6**, A63 (2003)
[164] A. Hakenjos, K. Tüber, J.O. Schumacher, C. Hebling, Fuel Cells **4**, 185 (2004)
[165] G. Bender, M.S. Wilson, T.A. Zawodzinski, J. Power Sources **123**, 163 (2003)
[166] J.R. Claycomb, A. Brazdeikis, M. Le, R.A. Yarbrough, G. Gogoshin, J.H. Miller, IEEE Trans. Appl. Supercond. **13**, 211 (2003)
[167] D. Candusso, J.P. Poirot-Crouvezier, B. Bador, E. Rulliere, R. Soulier, J.Y. Voyant, Eur. Phys. Appl. Phys. **25**, 67 (2004)

[168] K.H. Hauer, "Method for detecting the current density distribution in a fuel cell stack," WO Patent 0,155,735 (2001)
[169] H. Sun, G. Zhang, L.J. Guo, H. Liu, J. Power Sources **158**, 326 (2006)
[170] G.S. Zhang, B. Ma, D.H. Shang, L.J. Guo, H. Sun, H.T. Liu, ECS Trans. **11**, 1545 (2006)
[171] S.A. Freunberger, M. Reum, J. Evertz, A. Wokaun, F.N. Büchi, Electrochem. Commun. **8**, 1435 (2006)
[172] S.A. Freunberger, M. Reum, J. Evertz, A. Wokaun, F.N. Büchi, J. Electrochem. Soc. **153**, A2158 (2006)
[173] J.J. Hwang, W.R. Chang, R.G. Peng, P.Y. Chen, A. Su, Int. J. Hydrogen Energy **33**, 5718 (2008)
[174] M. Santis, S.A. Freunberger, A. Reiner, F.N. Büchi, Electrochim. Acta **51**, 5383 (2006)
[175] M. Schulze, E. Gülzow, S. Schönbauer, T. Knöri, R. Reissner, J. Power Sources **173**, 19 (2007)

5

High-Resolution Neutron Radiography Analysis of Proton Exchange Membrane Fuel Cells

D.S. Hussey and D.L. Jacobson

National Institute of Standards and Technology, 100 Bureau Dr., MS 8461, Gaithersburg, MD 20899-8461, USA

1. INTRODUCTION

Proper water management in proton exchange membrane fuel cells (PEMFCs) is critical to PEMFC performance and durability. PEMFC performance is impaired if the membrane has insufficient water for proton conduction or if the open pore space of the gas diffusion layer (GDL) and catalyst layer (CL) or the gas flow channels becomes saturated with liquid water, there is a reduction in reactant flow to the active catalyst sites. PEMFC durability is reduced if water is left in the CL during freeze/thaw cycling which can result in CL or GDL separation from the membrane,[1] and excess water in contact with the membrane can result in accelerated membrane thinning.[2]

One challenge to developing water management strategies is visualizing and accurately measuring the relatively small mass of water (10 mg cm^{-2}) typically present in the PEMFC, since standard PEMFC hardware is metallic and conducting, which limits inspection by standard imaging techniques, such as optical, X-ray, or magnetic resonance imaging (MRI). Optical techniques have been used to study slug and droplet flow in the gas channels, as well as water escape from the GDL surface.[3,4] X-ray imaging at

U. Pasaogullari, C.-Y. Wang (eds.), *Modeling and Diagnostics of Polymer Electrolyte Fuel Cells*, Modern Aspects of Electrochemistry 49,
DOI 10.1007/978-0-387-98068-3_5, © Springer Science+Business Media, LLC 2010

synchrotron sources have thinned out endplates to examine small regions of interest, over short periods of time, as the beam intensity causes radiation damage to the polymer membrane.[5] MRI requires reducing the amount of electrically conducting material to overcome the PEMFC self-shielding of the MRI signal. With these specially designed test sections, MRI is able to rapidly measure the through-plane water content with spatial resolution in the through-plane direction of order 10 μm, though on the order of several 100 μm in the in-plane directions.[6]

The only measurement technique that is able to accurately measure and visualize the water content in PEMFCs using standard materials of construction for fuel cells is neutron radiography, owing to the nature of the interaction between the neutron and material being dominated by the strong nuclear force, rather than the electromagnetic interaction. In particular, neutrons strongly interact with hydrogen, while only weakly interacting with aluminum and carbon. This sensitivity to hydrogen enables accurate measurement of the water content typical of standard PEMFCs with water volume resolution of about 10^{-7} cm^3, limited by neutron counting statistics from an image acquisition time of 60 s in a 15-μm pixel (see Section 3.2.1). The very first application of neutron radiography[7] to PEMFC research was of the through-plane water content in a fuel cell with an artificially thick membrane electrode assembly (three thermally bonded Nafion 117[1] membranes). The image detector had a spatial resolution of about 50 μm and achieved this spatial resolution by centroiding an intensified emission from a scintillator. The centroiding algorithm had a long dead time, with a maximum event rate of about 3 kHz; hence each image was acquired over a period of about a day. Many interesting features of the membrane water content were observed, including an expected gradient from anode to cathode and depressions in the membrane water content at the interfaces between the thermally bonded membranes. These oscillations were attributed to an incomplete rearrangement of the hydrophobic surfaces of the thermally bonded membranes. As this was the first

---

[1] Certain trade names and company products are mentioned in the text or identified in an illustration in order to adequately specify the experimental procedure and equipment used. In no case does such identification imply recommendation or endorsement by the National Institute of Standards and Technology nor does it imply that the products are necessarily the best available for the purpose

measurement, the water quantification utilized the gravimetric measurements of the water uptake in Nafion to convert neutron intensity to membrane water content. Further investigations of the through-plane water content were not pursued until later due to the event rate limitation and lack of suitable high-resolution imaging detectors. Since conventional neutron scintillator-based detectors had a spatial resolution of about 250 μm, the focus of further PEMFC neutron radiography research was the in-plane water content and the differences between water in the channels versus water in the land regions. A comprehensive review of this initial neutron radiography work has been given by Trabold et al.[8] Neutron detector technology has progressed significantly in the past decade. As of publication of this article, the state-of-the-art neutron imaging detector spatial resolution is of the order 20 μm, allowing direct measurement of the through-plane (anode versus cathode) water content in standard GDL. This progress has been achieved through both dramatic event processing speed enabling high count rate and high-resolution microchannel plate detectors and in developments to fabricating thinner scintillators. There have been a few recent articles that have focused on this new development, but without a detailed discussion of the measurement performance of contributions to measurement uncertainties.[9–15] This article will discuss the technique of neutron radiography with a focus on high spatial resolution 50 μm to 20 μm measurements of the through-plane water content in PEMFCs.

2. NEUTRON RADIOGRAPHY FACILITY LAYOUT AND DETECTORS

2.1. Neutron Sources and Radiography Beamlines

Neutron sources fall into one of two categories, reactor or spallation. In a fission reactor, 2 to 3 neutrons are produced from the fission of uranium, and only one neutron per fission must be captured to sustain the chain reaction. The neutrons from the fission reaction have energy of order 2 MeV and must be slowed to meV energies in order to be used for neutron scattering beamlines. The slowing occurs via inelastic scattering in a moderator material, such as water or heavy water. The resulting neutrons have an energy spectrum that is described by a Maxwell–Boltzmann distribution, the peak energy and width of which are determined by the temperature of the moderator. Neutron spectra are often referred to by the mean temperature of the moderator material, two relevant designations being thermal and cold (http://en.wikipedia.org/wiki/Thermal_neutron). Thermal neutrons have a characteristic temperature of about 40°C to 50°C, yielding a mean neutron energy of about 25 meV, while cold neutrons are moderated in a cryogenic liquid and have an energy of about 3 meV. In a typical spallation source, a high energy (GeV) proton beam collides with a neutron-rich nucleus target, such as mercury, tungsten, or uranium, knocking out or spallating neutrons from the nucleus. The neutrons from the spallation process are high energy (MeV to GeV) and must also be slowed down by scattering with a moderator material. Two advantages of a reactor source are that it is a continuous source of neutrons of all energies with a high average fluence rate. The advantage of a spallation source is that the neutrons are produced in short (<1 μs) bursts with a large peak fluence rate, but have a smaller integrated fluence rate. Scattering techniques that can make use of the pulsed structure are best suited to a spallation source. However, typical neutron radiography experiments use the entire neutron spectrum, and therefore a continuous, high integrated fluence rate of neutrons is optimal for studying PEMFC. One exception to this generalization is the SINQ spallation source at the Paul Scherer Institute in Switzerland, which is a continuous spallation neutron source, with both cold and thermal neutron radiography facilities.[16]

Neutron radiography beamlines are relatively simple compared to most scattering instruments, see Fig. 1 which depicts the NIST neutron imaging facility layout. An aperture is placed some distance from the neutron source. Moderated neutrons diffuse away from the source, through the aperture to form a beam. The aperture forms a "pin-hole" image of the neutron source downstream at the sample and detector position. Typical dimensions are aperture with diameter, D, 1 cm, aperture to detector distance, L, 3 m to 15 m. In order to minimize beam intensity loss from scattering in air (thermal neutrons have about a 97% transmission in air), the distance between the aperture and the sample position is evacuated. At some facilities, filter material such as single crystal Bi is placed between the aperture and neutron source to reduce the intensity of the gamma-rays and unmoderated fast neutrons both for radiation safety

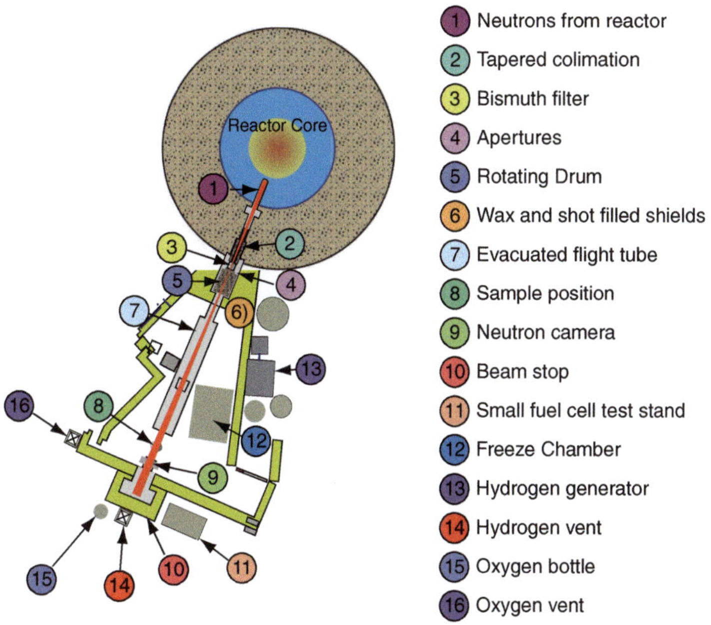

Figure 1. Schematic layout of the NIST BT-2 neutron imaging facility, including the main neutron optic components as well as the location of the fuel cell test and control infrastructure.

and to reduce image background noise. Bi has a similar electron density as lead and is effective as a gamma-ray filter, but for every 5 cm of filter length, scatters about 50% of the thermal neutron intensity. The Bi filter also modifies the neutron spectrum due to Bragg diffraction, as shown in Fig. 2. The two primary figures of merit for a neutron radiography facility are the *L*/*D* ratio and the neutron fluence rate (neutrons cm^{-2} s^{-1}). The neutron fluence rate determines the image acquisition time needed to reach a given signal-to-noise rate and is discussed in more detail in Section 3.1.1. The *L*/*D* ratio, along with the sample-to-detector separation, z, determines the geometric blur, λ_g, which is calculated by

$$\lambda_g = zD/(L - z) \approx zD/L, \tag{1}$$

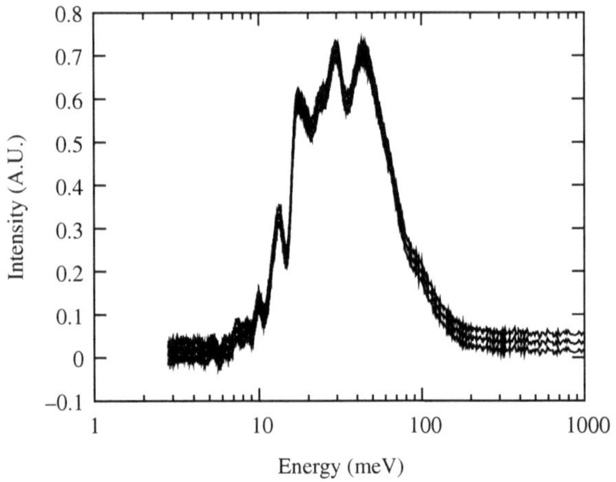

Figure 2. The neutron spectrum at the NIST BT2 neutron imaging facility. The spectrum is not corrected for the neutron detection efficiency, which increases as the square root of the energy.

where λ_g defines the minimum spatial resolution (full width half maximum) achievable for a beamline, assuming that the detector is also capable of a similar or smaller resolution. A larger L/D ratio means that an object can be placed further from the detector, and the image will still be sharp. Since the neutron fluence rate scales with the square of both $1/L$ and D, increasing L/D comes at the expense of decreasing the neutron fluence rate and, therefore, an increase in the image acquisition time. In high-resolution PEMFC imaging, the interest is the through-plane water content. Thus using a slit rather than circular aperture reduces the blur along the dimension of interest, while maintaining a reasonable neutron fluence rate for image acquisition. Most imaging facilities have a set of apertures so as to optimize beam conditions for the imaging application. As an example, the current choice of apertures at the NIST neutron imaging facility, along with the corresponding fluence rates are shown in Table 1. The other means to reduce λ_g is to minimize the dimension of the object along the beam path direction, which means reducing the width of the active area. An additional way to minimize λ_g is to design the test section such that gas and electrical connections are made on one side, allowing the other side of the test section to be mounted flush with the face of

Table 1
The Fluence Rate and Approximate *L/D* for the Five Apertures Available at the NIST Neutron Imaging Facility. The Slits in Positions 1 and 2 Reduce the Geometric Blur Along One Direction by About a Factor of 5 over the Previous Configuration for Use in High-Resolution Imaging of the Through-Plane Water Content.

Aperture number	Aperture dimension	≈ *L/D* (*x*,*y*)	Fluence rate (cm^{-2} s^{-1})
5	15 mm	450	1.38×10^7
4	10 mm	600	4.97×10^6
3	3 mm	2000	5.23×10^5
2	10 × 1 mm	600, 6000	6.54×10^5
1	1 × 10 mm	6000, 600	7.17×10^5

the detector. Typical PEMFCs for high-resolution neutron imaging are about 1 cm in width and can be placed about 2 cm from the neutron detector and are imaged with an $L/D = 6000$. This means that λ_g is about 5 μm or about 5 times smaller than the intrinsic detector spatial resolution.

2.2. Neutron Imaging Detectors

Since neutrons are neutral particles, they cannot be directly detected. Instead neutron detectors are composite devices. The first detector component is a thin neutron sensitive region, where a neutron is captured. This nuclear reaction results in the emission of energetic charged particles. The second detector component detects the emitted charged particles, either from scintillation light or amplifying the charge onto a position sensitive electrode. Two figures of merit for a neutron imaging detector are the spatial resolution and the detection efficiency. Typical imaging detectors have a thermal neutron detection efficiency of at least 15%. The spatial resolution of the detector is described by a point spread function (PSF), which is often approximated by a Gaussian. A common definition of spatial resolution is the frequency at which the Fourier transform of the PSF reaches 10% of its maximum value. The reciprocal of this frequency is the spatial resolution, which for a Gaussian PSF, with standard deviation σ_d

$$\delta_d = \frac{\sigma_d \pi}{\sqrt{2 \ln(10)}}. \quad (2)$$

The spatial resolution can be measured in a number of ways, including the contrast between two absorbing bars as the bar width and separation is varied, or by measuring the width of a sharp edge.[17] The effects of the PSF, including λ_g, on measuring the water profile are discussed in more detail in Section 3.1.5.

The most common neutron imaging detector uses a scintillator viewed by a charge-coupled device (CCD) camera. Because the scintillation light is emitted in all directions, the spatial resolution of a scintillator is approximately the thickness of the scintillator. Since the neutron detection efficiency is an asymptotic function of the thickness, reducing the scintillator thickness in order to increase spatial resolution comes at the cost of counting statistics (water volume resolution). CCDs are sensitive to radiation damage. Thus, CCDs must be located outside of the main beam, typically with a mirror reflection, resulting in reduced light collection. The high-resolution imaging system developed at PSI uses a variety of scintillators to optimize the spatial resolution and image acquisition time for a given application. A mirror reflects the light out of the main beam path. A special lens with an f-stop of 1, magnification of 1 to 1 with no vignetting, focuses the scintillation light onto a CCD sensor.[18] With this system, a variety of scintillators were tested, and with a 10-μm thick gadolinium oxysulfide scintillator a resolution approaching 20 μm was achieved.

Another approach to high-resolution neutron imaging is based on microchannel plates (MCP) shown schematically in Fig. 3. In these devices, the neutron is captured in the wall of the MCP. The charged particle products from the nuclear reaction after neutron capture exit into the pores of the MCP, knocking out electrons at the wall–pore interface. A negative high voltage (about –5 kV) gives rise to an electron avalanche, which amplifies the initial charge by a factor of order 10^6. This charge cloud is imaged with a position sensitive anode. The spatial resolution depends on the pore-to-pore separation and the range of the charged particle. Since the charged particles are emitted in all directions, the limit to the spatial resolution is the charged particle range in the glass. The range of the charged particle is determined by its initial kinetic energy and the

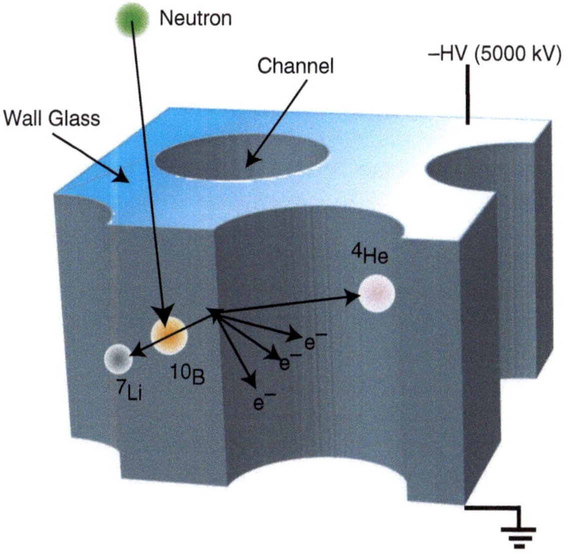

Figure 3. Schematic drawing of a microchannel plate detector for neutron radiography. Channel diameters range from 6 to 12 μm, and typical wall thickness are about 2 μm. The high voltage depends on the particular MCP used and can range from 2 to 6 kV.

electron density of the medium in which it moves. Current technology has a pore diameter of 10 μm, a pore-to-pore separation of 12 μm, and a charge particle range of about 5 μm, which yields a spatial resolution of about 25 μm.[19] Future detectors are anticipated with a MCP pore diameter of 6 μm, which should result in a detector resolution less than 15 μm.[20] Since the MCP system is an event-based detector (as opposed to CCDs which are integrating detectors), the electronic dead-time limits the overall event rate, which for a given neutron fluence rate limits the field of view (FOV). In the case of cross-delay-line anode readout electronics, there is a 10% dead-time for an overall event rate of about 200 kHz. For an incident fluence rate of about 10^6 cm^{-2} s^{-1}, the FOV is limited to about 1 cm^2. Future readouts using a cross-strip anode are expected to have a factor of ten higher event rate and higher spatial resolution than the cross-delay-line anode.[21]

In high-resolution imaging, sample mounting and alignment are critical to ensure that the test section is stable and the through-plane direction is well-aligned with the neutron beam axis. (This axis is the only axis that must be aligned, as other misalignments introduce a relative uncertainty in the measured water thickness that scale as $1/\cos(\theta)$ or about 10^{-4}.) For instance, a 1-cm wide active area and a pixel pitch of 15 μm requires an angular alignment to about 0.1°, which is easily achieved with standard motorized rotation and tilt stages. The alignment is performed by minimizing the measured width of the through-direction of the fuel cell by rotating the test section. Typical test sections for high-resolution imaging are composed of gold-coated aluminum flow fields, which have a high neutron transmission. There is a sharp, nearly step-like change in the neutron transmission going from the flow fields to the gas diffusion layer and membrane electrode assembly. The width of the through-plane direction is determined from the position of the half-height of this step change. The measured width as a function of angle is a parabola, and the minimum of the parabola is the optimal angle. With high spatial resolution, small changes in the test section position can result in poor image registration. Sample position changes can result from thermal expansion, changing a fitting on the cell, or stress in the gas or electrical connections. Incorporating robust mounting into the test section design can reduce many of these issues.

3. WATER METROLOGY WITH NEUTRON RADIOGRAPHY

3.1. Neutron Attenuation Coefficient of Water, μ_w

For thermal and cold neutron radiography, the neutron can be treated as a point particle, and the interaction with material can be characterized by a collision cross-sectional area or total neutron scattering cross section, σ_{tot}, encompassing coherent scattering (i.e., Bragg diffraction), incoherent scattering (resulting from isotopic variation or nuclear spin-flip scattering) and absorption. In a composite material, σ_{tot} is the weighted average over the species present. However, for some composite materials, especially hydrogenous liquids, the cross section for neutron scattering from hydrogen can depend on the vibrational and rotational excitation modes. These effects can be calculated and properly modeled.[22] There is also an energy dependence to the incoherent scattering and absorption processes, with σ_{tot} increasing with decreasing neutron energy. Water exhibits both properties, energy dependent σ_{tot} as shown in Fig. 4 and a deviation from the weighted average cross sections.[23] The impact of this energy dependence will be discussed in Section 3.2.2.

Neutron radiography measures the spatially resolved neutron transmission through a sample. The neutron transmission is derived from the Lambert–Beer Law of attenuation, which in differential form is

$$\mathrm{d}I = -IN\sigma_{tot}\,\mathrm{d}t, \qquad (3)$$

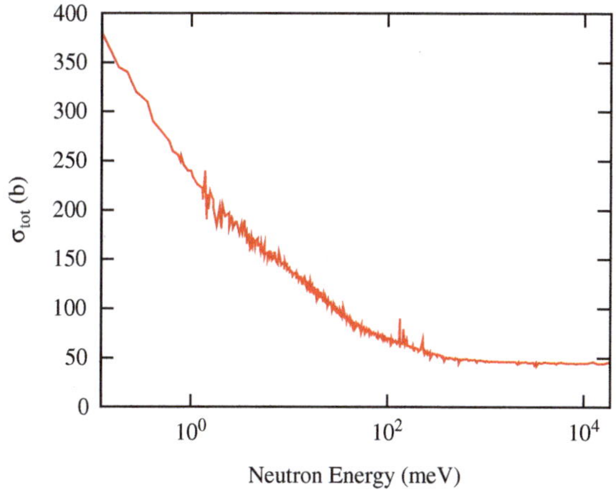

Figure 4. The measured total neutron scattering cross section for water at 300 K.[23] The cross section diverges as the neutron energy approaches 0. Due to this, the transmission through a thick section of water results in the transmitted beam having a modified spectrum, shifted to higher energy. This modification is referred to as beam hardening. Beam hardening is a potential source of systematic uncertainty, as the measured water thickness from applying the Lambert–Beer Law would yield a systematic thinner section of water. In addition, since the measurement resolution is inversely proportional to the σ_{tot}, a neutron beam with a cold neutron energy distribution will have a higher sensitivity than a thermal neutron beam, as shown in Fig. 5.

where I is the incident intensity, N is the number density, and dt is a differential unit of length. In connection with other transmission methods, the product $N\sigma_{\text{tot}}$ is referred to as the attenuation coefficient, μ, with dimension inverse length. For an operating fuel cell, the transmitted intensity is

$$I_{\text{w}} = I_0 \exp(-\mu_{\text{dry}} t_{\text{dry}} - \mu_{\text{w}} t_{\text{w}}), \tag{4}$$

where I_0 is the incident neutron intensity, $\mu_{\text{dry}} t_{\text{dry}}$ is the attenuation due to the dry components, and $\mu_{\text{w}}\ t_{\text{w}}$ is the attenuation due to the water in the test section. With a calibration measurement of μ_{w}, the water thickness, t_{w}, can be obtained by taking the natural logarithm of the ratio of the operating intensity to the dry intensity. As discussed in Section 2.1, typical active area widths along the neutron beam path are about 1 cm. Assuming a water uptake of 30% for Nafion indicates that a minimum thickness for the calibration is about 3 mm. Shown in Fig. 5 is the calibration measurement performed at a cold and thermal neutron imaging setup with an MCP detector. As one can see from the plot and Table 2, the cold neutron

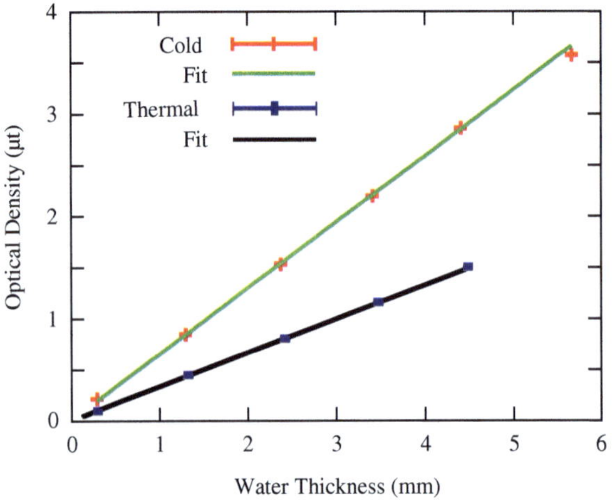

Figure 5. The measured attenuation coefficient for the NIST MCP detector at the BT2 thermal neutron beam and the NG1 cold neutron beam.

Table 2
Attenuation Coefficient Measured in the Calibration Experiment. The Uncertainty is the One-Sigma Fit Error. The Reduced-Chi-Square for Each Fit was of Order 1.

Spectrum type	Mean energy (meV)	μ_W (mm^{-1})	$\delta\mu_W$ (mm^{-1})
Thermal (BT-2)	25	0.338	0.001
Cold (NG-1)	3	0.645	0.002

beam has about a factor of two larger attenuation coefficients than the thermal beam, as expected from the increase in σ_{tot} shown in Fig 4.

The water content in an operating fuel cell is obtained from the ratio of the background (B) corrected images of the operating fuel cell ($I_w = \tilde{I}_w - B$) to the dry fuel cell ($I_d = \tilde{I}_d - B$), where the tilde indicates the raw, measured image:

$$t_W = \frac{-1}{\mu_W} \ln\left(\frac{I_w}{I_d}\right). \tag{5}$$

In order to account for variations in incident neutron intensity, such as reactor power fluctuations or dead-time corrections, one must scale two images based on the observed intensity in an unchanging region of the image, for instance where there is no sample (open beam) or in the end-plates away from the flow channels. The background in CCD systems is due primarily to the electronic read noise and the dark current. MCP detectors have negligible electronic read noise (which is the advantage of an event-based detector) but have a relative larger sensitivity to gamma-rays from the reactor core that are not completely stopped by the Bi filter. This gamma-ray background can be measured by blocking the thermal neutron beam with a thin plate of borated aluminum (5 mm) and Cd (1 mm), which negligibly reduces the gamma-ray flux from the reactor.

3.2. Sources of Uncertainties in Neutron Radiography

The method of neutron radiography is simple to implement. Since water has a large neutron attenuation coefficient, the technique of radiography has a high sensitivity for small water mass changes of order 100 ng achievable, as determined from Poisson counting statistics. There are several contributions to the overall water measurement uncertainty that limit this sensitivity. The primary sources of uncertainty for high-resolution imaging of the through-plane water content in PEMFCs are neutron counting statistics, departure from the Lambert–Beer law of attenuation due to beam hardening, uncertainty in the background subtraction of the reactor core and sample prompt gamma-ray radiation, reduction in the measured water content in rapidly changing regions due to image blur, and change in the total neutron scattering from water that is absorbed in the membrane.

3.2.1. Counting Statistics

The water thickness measurement uncertainty due to neutron counting statistics can be calculated from Poisson counting statistics. For a random process, the standard deviation, Δ_I, in the observed counts I is $\Delta_I = \sqrt{I}$. The number of neutrons in the incident, or open beam, I_0, is the product of the neutron fluence rate Φ (cm^{-2} s^{-1}), integration time T (s), integration area A (cm^2), and neutron detection efficiency, η:

$$I_o = \Phi \times \eta \times T \times A. \tag{6}$$

Typical values for through-plane water measurements are $\Phi = 1\times10^6$ cm^{-2} s^{-1}, $T = 100$ s, $A = 2.25\times10^{-6}$ cm^2 (pixel area), $\eta = 0.2$, giving a per pixel intensity of $I_0 = 45$, so that $\Delta_I \approx 6.7$, or a 15% relative uncertainty. Since the dry and background images are static conditions, the counting statistics of the wet image typically determine the liquid water uncertainty. Taking the partial derivative of Eq. (5) with respect to I_w gives the water thickness uncertainty as follows:

$$\Delta_t = \frac{1}{\mu_\mathrm{w}}\sqrt{\frac{1}{I_\mathrm{w}}} = \frac{1}{\mu_\mathrm{w}}\sqrt{\frac{\exp(\mu_\mathrm{w} t_\mathrm{w})}{\phi A T \eta}}. \tag{7}$$

As shown in Table 2 and Fig. 5, $\mu_w = 0.338$ mm^{-1} for the NIST 25 μm MCP detector system, which means that for a 100 s exposure, the per pixel one-sigma water thickness uncertainty due to Poisson counting statistics is $\Delta_t \approx 0.45$ mm, or a water volume of about 10^{-7} cm^3. In a PEMFC, the minimum sensitivity may be larger than this, as the membrane and the GDLs attenuate the beam. During an experiment, the water thickness sensitivity can be improved by integrating for a longer period of time and integrating the water content along the in-plane direction (for instance, using a "differential cell"), as shown in Fig. 6. The sensitivity can also be improved by using a cold neutron beam. Cold neutrons provide two benefits. The first is that the neutron detection efficiency (η) is about a factor of two larger for typical detectors. The second is that the total neutron scattering cross section (Fig. 4) increases with decreasing neutron energy, such that the attenuation coefficient (μ) also increases by about a factor of two, as shown in Fig 5. Thus, for equal thermal and cold neutron fluence rates, the standard uncertainty in the measured water thickness for cold neutron radiography is about 35% smaller than that for thermal neutron radiography.

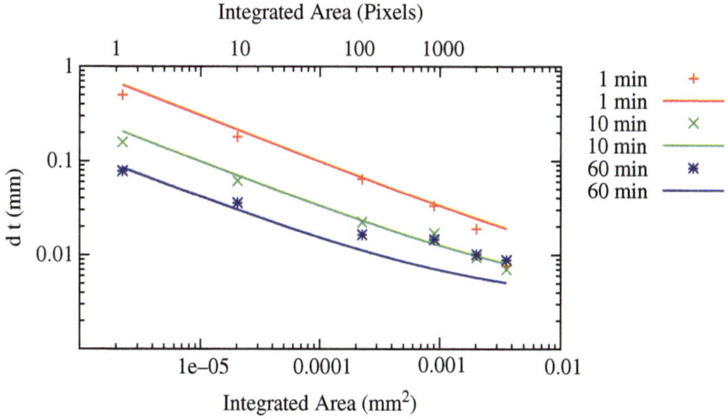

Figure 6. One-sigma measurement uncertainty measured for the 25 μm MCP detector, a fluence rate of 2×10^6 cm^{-2} s^{-1} as a function of integration area for three different integration times. The solid curves are estimates based on Eq. (7). The measured and predicted uncertainties are in reasonable agreement.

3.2.2. *Beam Hardening*

As mentioned in Section 3.1, the total neutron scattering of water depends on the incident neutron energy. Thus, for sufficiently thick sections of water and high counting statistics data, beam hardening can be observed. Beam hardening results in lower energy neutrons being preferentially scattered or absorbed, and the spectrum of the emerging neutron beam has a higher mean energy. As a result, the object can appear to have a higher transmission and applying Eq. (5) to a thick region section of water would result in measuring a thinner section of water. In the calibration measurements, this would be manifested by a non-linear relationship between the neutron optical density and water thickness. As shown in Fig. 5, the optical density versus water thickness up to 4.5 mm resulted in a linear fit with a reduced χ^2 of about 1, with no evidence for beam hardening beyond the uncertainty due to counting statistics of the measurement. Thus, analyzing typical PEMFC images with Eq. (5) will introduce a negligible uncertainty compared to that of the counting statistics.

3.2.3. *Background Subtraction*

A changing background is a source of measurement uncertainty. The background can change from varying CCD temperature, changes in reactor gamma-ray fluence rate that is related to power fluctuations, prompt gamma-rays from the sample, or scattered neutrons. If in Eq. (5) there is a different background for the wet and dry images that is not properly accounted for, the resulting fractional error in the measured water thickness is

$$\frac{\Delta t_{\mathrm{w}}}{t_{\mathrm{w}}} = -\frac{1}{\mu_{\mathrm{w}} t_{\mathrm{w}}} \ln\left(1 + \frac{(1-P)B}{\exp(-\mu_{\mathrm{w}} t_{\mathrm{w}})/I_{\mathrm{d}}}\right), \tag{8}$$

where $P = B_{\mathrm{w}}/B$ is the ratio of the wet and dry backgrounds. Shown in Fig. 7 is the fractional error for $I_{\mathrm{d}} = 50$ and $B = 0.5$. The resulting systematic uncertainty is of the order of or less than the uncertainty due to counting statistics, meaning that varying backgrounds, including that due to scattered neutrons will introduce negligible uncertainty into the measurement of the membrane and GDL water content.

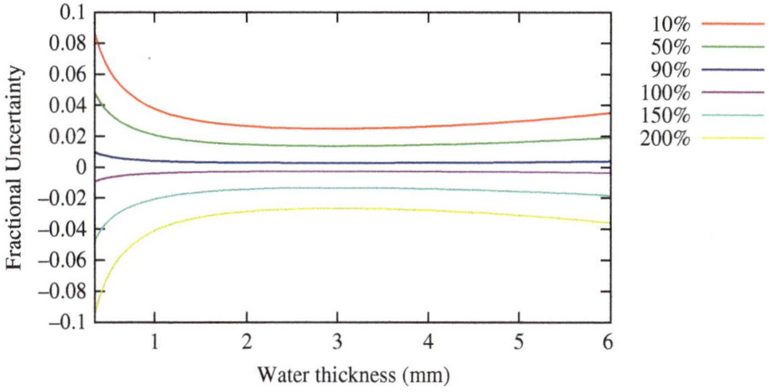

Figure 7. The relative systematic error due to improper background subtraction for an open beam signal-to-noise ratio of 100. The uncertainty is plotted for the wet image background being a relative percentage of the dry background, $B_W = PB$.

3.2.4. Changes in the Total Neutron Scattering from Water Absorbed in the Membrane

As discussed in Section 3.0, the total neutron scattering cross section of hydrogenous materials often depend on the rotational and vibrational modes of the molecule. Thus, when water is absorbed in the PEMFC membrane, it is possible that there could be a change in the neutron scattering cross section. To determine if there is a change in the scattering cross section, the water sorption of a Nafion 117 membrane as a function of water activity was measured. A humidified nitrogen gas stream, with a flow of 100 sccm was introduced on either side of a bare membrane at 40 °C and 80 °C. For each relative humidity (RH) condition, the gas was flowed for 30 min, and then the 15 min of images were acquired for a total of 45 min at each RH condition. (This will result in a lower water sorption value at 100% RH due to insufficient time to reach equilibrium.) The membrane water content was determined by the value at the center of the membrane. The systematic uncertainty due to the image resolution (as discussed in Section 3.2.5) was estimated by convolving the PSF due to a 50-μm blur with a model function of a uniform water thickness across the membrane. The simulation was performed for each saturation condition and demonstrated that the relative uncertainty due to spatial resolution was less than 2%. The water content of the membrane, λ (reported in moles of water per moles of sulfonic acid), is shown in Fig. 8, and there is no evidence for a change in the total neutron scattering cross section that is larger than the uncertainty due to counting statistics, or of that due to the RH control.

3.2.5. Image Spatial Resolution

A large potential source of systematic uncertainty is due to the finite image resolution. The measured image can be described as the convolution of the true sharp image with the imaging system point spread function, PSF. The PSF is reasonably approximated by a Gaussian,

$$\mathrm{PSF}(x - x') = \frac{1}{\sqrt{2\pi\sigma_i^2}} \exp\left(-\frac{(x - x')^2}{2\sigma_i^2}\right), \quad (9)$$

where σ_i is related to the overall spatial resolution δ_i by Eq. (2). There are two independent contributions to δ_i, the geometric blur, $\lambda_g = 2.35\,\sigma_g$ and the neutron detector spatial resolution, δ_d, so that

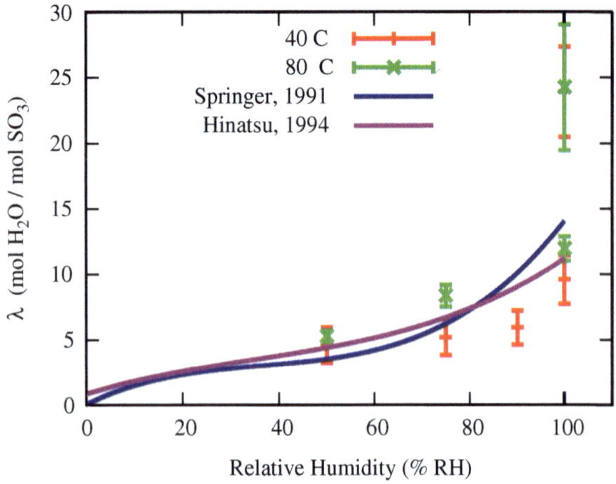

Figure 8. Water sorption as a function of inlet gas humidity. The uncertainty plotted is derived from counting statistics alone. The two solid lines are the correlation for λ as a function of activity derived from Springer et al.[26] and Hinatsu et al.[27]

$$\delta_i = \frac{\pi}{\sqrt{2\ln(10)}}\sqrt{\sigma_d^2 + \sigma_g^2} \tag{10}$$

As shown in Fig. 9, the spatial resolution effects high-resolution PEMFC imaging in two ways. The first effect is that the measured water profile will not contain discontinuous jumps or step-like image features, for instance at the boundary of the catalyst and microporous layer (MPL). Rather, an image of a step change will be described as an error function, with the width determined by δ_i, as is demonstrated in Figure 9(a), where the transmission through a

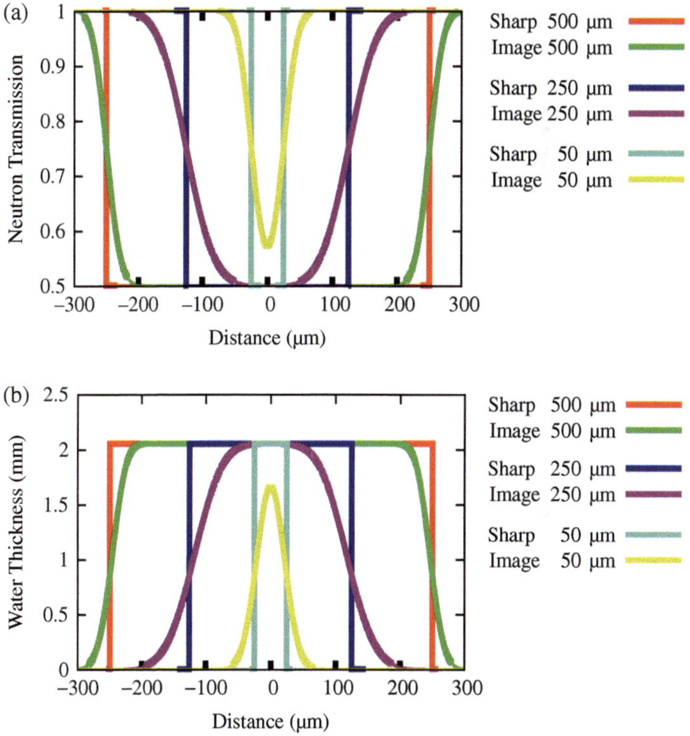

Figure 9. A two-dimensional, noiseless image was simulated on a slab of water that was 2-mm thick along the beam path ($t_W = 2$ mm) with a variable width transverse to the beam direction. The average transmission along the width of slab was from the simulated, blurry image is shown in (a) and is referred to as a through-plane transmission profile. Taking the natural logarithm of the transmission profile yields the through-plane water profile as shown in (b).

slab of water is convolved with a PSF described by a $\delta_{\mathrm{i}} = 25\ \mu\mathrm{m}$. At the edges of the transition, the neutron transmission significantly deviates from that of the sharp profile. The second effect is a lower observed water content in the water profile in the through-plane direction. The size of the deviation depends both on the thickness of the water and the width. As shown in Figure 9(b), as the width of the region approaches $2\delta_{\mathrm{i}}$, the measured neutron transmission at the center of the slab ceases to reach the true transmission value. In through-plane imaging of PEMFCs, the water content is concentrated in the membrane electrode assembly, which is a thin region compared to the image spatial resolution. This situation is similar to the simulated water profile with a 50-μm width in Figure 9(a). As shown in Figure 9(b), significantly less water is observed in the water profile.

This means that if the total water content of the test section is measured from the image

$$V_{\mathrm{t,I}} = -\frac{A_{\mathrm{p}}}{\mu_{\mathrm{w}}} \sum_{j=1}^{N} \ln\left(\frac{I_{\mathrm{w},j}}{I_{\mathrm{d},j}}\right), \tag{11}$$

the measured water content will be systematically smaller. However, the correct total water content can be obtained from the image because the total water content attenuates the neutron beam as expected, and by calculating the total water content from the average transmission,

$$V_{\mathrm{t,a}} = -\frac{NA_{\mathrm{p}}}{\mu_{\mathrm{w}}} \left(\ln \sum_{j}^{N} I_{\mathrm{w},j} - \ln \sum_{j}^{N} I_{\mathrm{d},j}\right) \tag{12}$$

will yield an accurate measurement of the water content. In Eqs. (11) and (12), the sum is taken over the entire region of interest of N pixels, and A_{p} is the area of a single pixel. In other words, the overall transmission of the neutron beam only depends on the amount of water in the test section. This systematic effect is reduced in in-plane PEMFC imaging, as the water thickness that the neutron beam passes through is typically a factor of 10 smaller, and the natural logarithm is nearly linear in transmission.

While the PSF can be modeled, it is not straightforward to deconvolve the images with standard Fourier transform methods to obtain sharpened water content. Due to small pixel areas and low fluence rates ($\sim 10^6$ cm^{-2} s^{-1}) that are due to the necessary collimation, high-resolution neutron images tend to demonstrate high shot noise, with a relative uncertainty on the order of a few percent. This shot noise results in a small signal-to-noise ratio for high-frequency components. Deconvolved images using Wiener filtering show little improvement in the quantitative water content over the unsharpened data, and suffer visually. Other algorithms, such as the one based on Maximum Likelihood methods, may result in sharper images, with a large computation cost. Instead, rather than attempting to sharpen the measured data, a more robust method of comparison is to simulate the neutron image based on a model prediction of the water content.

4. RECENT IN SITU HIGH-RESOLUTION NEUTRON RADIOGRAPHY EXPERIMENTS OF PEMFCS

The improved neutron detector spatial resolution has been a recent advance, with user instruments first available at the end of 2006 and 2007 at NIST and PSI, respectively. The work can be classified as proof-of-principle,[9,10] in situ measurement of the steady-state through-plane water content during fuel cell operation,[11–13] and dynamic through-plane mass transport measurements.[14,15]

4.1. Proof-of-Principle Experiments

The primary results of these first experiments from each facility were primarily proof-of-principle. Hussey et al.[9] used a MCP detector with intrinsic spatial resolution of 25 μm and overall image resolution of about 30 μm. With this setup, a test section was operated in a differential cell mode, with a minimum stoichiometric ratio of about 50 on both the anode and cathode. Due to cell motion, it was not possible to quantify the total water content in the cell, but relative changes in the through-plane water content were observed from open circuit voltage, and the water content increased as a function of current density.

Boillat et al.[10] imaged a 0.15-mm wide slit with three different detector setups. By using a 10-μm Gadox scintillator, tilted at a high angle to the beam direction, a spatial resolution approaching 20 μm along one direction was reported. Using this system, changes in the through-plane water content in a test section maintained at 70°C and 1 A cm^{-2} for four different anode and cathode inlet RH conditions were observed. The primary observation was that for undersaturated cathode inlet RH water was collected in the ribs in the cathode GDL.

4.2. In Situ, Steady-State Through-Plane Water Content

In the work of Hickner et al.,[11] an operating PEMFC, with a Nafion 112 membrane, was imaged with 25 μm detector resolution in situ, and a wide range of observations were made. The center of the test section was about 4 cm from the detector face and the beam *L*/*D* was 1200 resulting in an overall spatial resolution of about 50 μm. The through-plane water profile of the cell was measured at different cell temperatures, current densities, and anode and cathode gas feed flow rates, and the counting statistics uncertainty corresponded to a water thickness resolution of about 3 μm for the 800 pixel-wide regions used for the line profiles. Detailed information was obtained on the cross-sectional water content of the gas flow channels and membrane electrode assembly. At low current densities, water was observed to remain on the cathode. Significant water in the anode gas flow channel was observed when the waste heat and water production of the cell were moderate. No significant gradient in water content within the membrane was observed from anode to cathode, which is due to the image spatial resolution. The total liquid water content of the membrane electrode assembly was fairly stable between current densities of 0.25 A cm^{-2} and 1.25 A cm^{-2} even though the water in the gas flow channels changed drastically over this current density range. At 60°C, the water content in the center of the gas diffusion layer was depleted compared to the membrane or gas flow channel interfaces. This phenomenon was not observed at 80°C where evaporative water removal is prevalent. In a second related article, the measured water content from three operating temperatures at nominally the same current density (0.75 mA cm^{-2}) was compared with calculations from a pseudo two-dimensional model.[12] The quantitative agreement was not very good due in part to model simplification and not incorporating the instrumental broadening. However, many qualitative trends could be identified, including water transport via the phase change induced flow (heat pipe effect). When the instrumental broadening was incorporated, the model prediction at 80°C was in reasonable agreement with the measured data.[25] However, at lower temperatures, a large discrepancy persists between the measurement and model prediction. This first comparison provided useful insight into required model refinement and image analysis requirements.

In the work of Spendelow et al.,[13] high-resolution neutron images of water profiles in an operating fuel cell were obtained under varying cell conditions, including temperature, current, inlet humidity, concurrent flow and counter-flow, orientation of the cell to examine effects due to gravity, and the anode flow rate was varied to simulate the effects of a recycle loop. Significantly more water was observed in the membrane electrode assembly during counter-flow operation than under co-flow. This led to a lower high-frequency resistance, as well as better cell performance for counter-flow. Rotating the cell vertically such that the anode and cathodes are flipped has a significant effect on the water density, and it was observed that positioning the anode on top decreases the tendency towards flooding, leading to improved performance.

4.3. Dynamic Through-Plane Mass Transport Measurements

The focus of Boillat et al.[14] was examining the replacement of ^{1}H (hydrogen) with ^{2}H (deuterium) in the membrane, which exploits the ≈10x smaller neutron cross section for ^{2}H. The test section was operated with light water and $^1H_2$ fuel on the anode at two current densities, 0.02 and 0.80 A cm^{-2}. After stabilizing, the fuel (but not the humidifying water) was switched to $^2H_2$. The authors observed a very rapid increase in the uptake of ^{2}H, which was faster than that expected solely from the applied current, and further somewhat independent of the current. This rapid exchange indicates a high exchange current density of the hydrogen oxidation reaction. They did not observe a complete replacement of all ^{1}H in the membrane, only 25% of the membrane ^{1}H was exchanged for ^{2}H, and the authors state that this might be due to the use of light water for humidification.

Kim and Mench[15] used high-resolution neutron imaging to visualize the liquid water flow in the through-plane direction under different thermal gradients, and two sets of GDLs, one with only a substrate loading of polytetraflouroethylene (PTFE) of 5% by weight and a MPL, and the other with no PTFE content. This visualization work was in support of an extensive series of ex situ measurements to determine the water transport rates for thermo-osmosis and phase change induced flow. The GDL with no PTFE showed a slight water flow with no thermal gradient due to the piezometric head. Applying a thermal gradient, so that the liquid side was cooler, prevented the slow leakage water flow. Reversing the thermal gradient resulted in significantly higher fluid flow than in the case of that driven by the piezometric head. Images of the GDL containing PTFE showed that with no thermal gradient the piezometric head was insufficient to create a flow through the diffusion media. For all cases investigated, when the gas-phase side of the test section was colder, the channels were observed to fill rapidly (few minutes) due to phase change induced flow.

5. CONCLUSIONS

High-resolution neutron radiography has enabled the direct measurement of the through-plane water content in PEMFCs. This in turn allows the measurement of a range of water transport properties in the GDL and in thick membranes. Measurement uncertainties are dominated by counting statistics and the blurring effects due to the finite spatial resolution. Typical uncertainties due to counting statistics contribute to the membrane water content uncertainties at the 5% level. Depending on the operating conditions of the PEMFC, primarily at low temperature and high current density, the spatial resolution can result in a large uncertainty. The image spatial resolution can be modeled, and image simulations based on through-plane water content can be easily performed in order to properly account for the instrumental broadening. New detector technology with 10 μm resolution is anticipated to be available in the near future and will further extend the range of investigation. The currently published results have demonstrated the utility of the technique, and it is anticipated that several new reports utilizing the current technology will be published in the near future.

ACKNOWLEDGMENTS

This work was supported by the US Department of Commerce, the NIST Ionizing Radiation Division, the Director's office of NIST, the NIST Center for Neutron Research, and the Department of Energy interagency agreement No. DEAI01-01EE50660. The authors wish to acknowledge R. Mukundan, J. Spendelow, R. Borup, and J. Davey from Los Alamos National Laboratory for assistance with the water sorption measurements, and E. Baltic at the National Institute of Standards and Technology for technical support.

REFERENCES

[1] S.H. He, M.M. Mench, One-dimensional transient model for frost heave in polymer electrolyte fuel cells. J. Electrochem. Soc. **153**, A1724–A1731 (2006)

[2] X. Huang, X. Wang, J. Preston, L. Bonville, H.R. Kunz, M. Perry, D. Condit, effect of water management schemes on the membrane durability in PEMFCs. ECS Trans. **16,** 1697 (2008)

[3] D. Spernjak, S.G. Advani, A.K. Prasad, Simultaneous neutron and optical imaging in PEM fuel cells. J. Electrochem. Soc. **156**, B109–B117 (2009)

[4] Y. Wang, C.Y. Wang, K.S. Chen, Elucidating differences between carbon paper and carbon cloth in polymer electrolyte fuel cells. Electrochimica. Acta. **52**, 3965–3975 (2007)

[5] C. Hartnig, I. Manke, R. Kuhn, S. Kleinau, J. Goebbels, J. Banhart, High-resolution in-plane investigation of the water evolution and transport in PEM fuel cells. J. Power Sources **188**, 468–474 (2009)

[6] S. Tsushima, S. Hirai, K. Kitamura, M. Yamashita, S. Takasel, MRI application for clarifying fuel cell performance with variation of polymer electrolyte membranes: Comparison of water content of a hydrocarbon membrane and a perfluorinated membrane. Appl. Magn. Reson. **32**, 233–241 (2007)

[7] R.J. Bellows, M.Y. Lin, M. Arif, A.K. Thompson, D. Jacobson, Neutron imaging technique for in situ measurement of water transport gradients within nafion in polymer electrolyte fuel cells. J. Electrochem. Soc. **146**, 1099–1103 (1999)

[8] T.A. Trabold, J.P. Owejan, J.J. Gagliardo, D.L. Jacobson, D.S. Hussey M. Arif, *Use of Neutron Imaging for Proton Exchange Membrane Fuel Cell (PEMFC) Performance Analysis and Design*, ed. by W. Vielstich, H. Yokokawa, H.A. Gasteiger. Handbook of Fuel Cells – Fundamentals, Technology and Applications, vol. 6 (Wiley, New York, NY, 2009)

[9] D.S. Hussey, D.L. Jacobson, M. Arif, J.P. Owejan, J.J. Gagliardo, T.A. Trabold, neutron images of the through-plane water distribution of an operating PEM fuel cell. J. Power Sources **172**, 225–228 (2007)

[10] P. Boillat, D. Kramer, B.C. Seyfang, G. Frei, E. Lehmann, G.G. Scherer, A. Wokaun, Y. Ichikawa, Y. Tasaki, K. Shinohara, In situ observation of the water distribution across a PEFC using high resolution neutron radiography. Electrochem. commun. **10**, 546 (2008)

[11] M.A. Hickner, N.P. Siegel, K.S. Chen, D.S. Hussey, D.L. Jacobson, M. Arif, In situ high-resolution neutron radiography of cross-sectional liquid water profiles in proton exchange membrane fuel cells. J. Electrochem. Soc. **155**, B427 (2008)
[12] A.Z. Weber, M.A. Hickner, Modeling and high-resolution-imaging studies of water-content profiles in a polymer-electrolyte-fuel-cell membrane-electrode assembly. Electrochimica. Acta. **53**, 7668–7674 (2008)
[13] J. Spendelow, R. Mukundan, J. Davey, T. Rockward, D.S. Hussey, D. Jacobson, M. Arif, R.L. Borup, High resolution neutron radiography imaging of operating pem fuel cells: Effect of flow configuration and gravity on water distribution. ECS Trans. **16**(2), 1345 (2008)
[14] P. Boillat, G.G. Scherer, A. Wokaun, G. Frei, E.H. Lehmann, Transient observation of H-2 labeled species in an operating PEFC using neutron radiography. Electrochem. Commu. **10**, 1311 (2008)
[15] S. Kim, M.M. Mench, Investigation of temperature-driven water transport in polymer electrolyte fuel cell: Phase-change-induced flow. J. Electrochem. Soc. **156,** B353 (2009)
[16] W.E. Fischer, Physica B **234,** 1202 (1997)
[17] C. Grunzweig, G. Frei, E. Lehmann, G. Kuhne, C. David, Highly absorbing gadolinium test device to characterize the performance of neutron imaging detector systems. Rev. Sci. Instrum. **78**, 053708 (2007)
[18] E.H. Lehmann, G. Frei, G. Kuhne, P. Boillat, The micro-setup for neutron imaging: A major step forward to improve the spatial resolution. Nucl. Instrum. Methods, Sect. A **576**, 389 (2007)
[19] O.H.W. Siegmund, J.V. Vallerga, A. Martin, et al., A high spatial resolution event counting neutron detector using microchannel plates and cross delay line readout. Nucl. Instrum. Methods, Sect. A **579**, 188–191, (2007)
[20] A.S Tremsin., J.V. Vallerga, J.B. McPhate, et al., On the possibility to image thermal and cold neutron with sub-15 μm spatial resolution. Nucl. Instrum. Methods, Sect. A **592**, 374–384 (2008)
[21] O.H.W. Siegmund, A.S. Tremsin, J.V. Vallerga, R. Abiad, J. Hull, Nucl. Instrum. Methods, Sect. A **504**, 177 (2003)
[22] Y. Edura, N. Morishima, Cold and thermal neutron scattering in liquid water: cross-section model and dynamics of water molecules, Nucl. Instrum. Methods, Sect. A **534**, 531–543 (2004)
[23] E. Melkonian, Slow neutron velocity spectrometer studies of O2, N2A, H2, H2O, and seven hydrocarbons. Phys. Rev. **76**, 1750 (1949); J.L. Russell, J.M. Neill, J.R. Brown, total cross section measurements in H20, General Atomic Div. Reports No.7581 (1966); K. Heinloth, Subthermal neutron scattering on H2O, CH2O2 AND C6H6. Zeitschrift fuer Physik **163**, 218 (1961)
[24] P.R. Bevington, *Data Reduction and Error Analysis for the Physical Sciences*, 3rd edn. (McGraw-Hill, Boston, MA, 2003)
[25] D.S. Hussey, D.L. Jacobson, M. Rangachary, R. Borup, J. Spendelow, Systematic uncertainties in neutron imaging of proton exchange membrane fuel cells, 214th ECS Meeting, Honolulu, HI October 15, (2008)
[26] T.E. Springer, T.A. Zawodzinski, S. Gottesfeld, Polymer electrolyte fuel cell model. J. Electrochem. Soc. **138**, 2334 (1991)
[27] J.T. Hinatsu, M. Mizuhata, H. Takenaka, Water uptake of perfluorosulfonic acid membranes from liquid water and water vapor, J. Electrochem. Soc. **141**, 1493 (1994)

6

Magnetic Resonance Imaging and Tunable Diode Laser Absorption Spectroscopy for In-Situ Water Diagnostics in Polymer Electrolyte Membrane Fuel Cells

Shohji Tsushima and Shuichiro Hirai

Research Center for Carbon Recycling and Energy, Tokyo Institute of Technology, 2-12-1, O-okayama, Meguro-ku, Tokyo 152-8552, Japan

1. INTRODUCTION

Water management is of great importance in polymer electrolyte membrane fuel cells (PEMFCs) with high energy efficiency and high durability.[1–3] Polymer electrolyte membrane plays an important double role as a proton conductor and a gas separator in PEMFC systems. For higher proton conductivity, membrane hydration is essential under fuel cell operation. Humidification of anode and/or cathode gases to the fuel cells is generally needed to keep membrane hydration, even though water is produced in the cathode electrode in PEMFCs. Especially, with higher cell current applied to the cells, dehydration of the membrane proceeds by heat release in the cell and an electro-osmotic drag effect.[4] On the other hand, generated water in the cells possibly plugs pore channels in catalyst layers (CLs) and/or gas diffusion layers (GDLs), that prevents reactant gases from reaching to the reaction sites in the catalyst layers. Thus, control of both retention of water in the membrane and exhaust of

U. Pasaogullari, C.-Y. Wang (eds.), *Modeling and Diagnostics of Polymer Electrolyte Fuel Cells*, Modern Aspects of Electrochemistry 49,
DOI 10.1007/978-0-387-98068-3_6,

water from the CLs and GDLs is needed for wide range of fuel cell operation.

In PEMFC systems, water is transported in both transversal and lateral direction in the cells. A polymer electrolyte membrane (PEM) separates the anode and the cathode compartments, however water is inherently transported between these two electrodes by absorption, desorption and diffusion of water in the membrane.[5,6] In operational fuel cells, water is also transported by an electro-osmotic effect and thus transversal water content distribution in the membrane is determined as a result of coupled water transport processes including diffusion, electro-osmosis, pressure-driven convection and interfacial mass transfer. To establish water management method in PEMFCs, it is strongly needed to obtain fundamental understandings on water transport in the cells.

In this article, magnetic resonance imaging (MRI) technique is described as a diagnostic tool for in-situ visualization of water content in the membrane under fuel cell operation.[7–30] Demonstrative applications and measurement procedure using MRI techniques are presented with discussion on water transport involved in PEMFCs.

In-situ measurement technique of water vapor concentration in gas flow channels in PEMFCs using tunable diode laser absorption spectroscopy (TDLAS)[31–36] is also shown with fundamental descriptions on its measuring principle and validity of a practical system. Localized current density and through-plane water-back transport index are obtained with variation of vapor concentration along the gas channel taken into account. Demonstrative results showing that effect of the micro porous layer (MPL) on variation of through-plane water-back transport index is shown in an operating PEMFC.

2. MAGNETIC RESONANCE IMAGING (MRI): AS A DIAGNOSTIC TOOL FOR IN-SITU VISUALIZATION OF WATER CONTENT DISTRIBUTION IN PEMFCs

2.1. Basic Principle of MRI

If an atom has an uneven number of neutrons or protons, then the atom has a certain angular momentum. The angular momentum is expressed as vector quantity having both magnitude and direction. The nucleus of the hydrogen atom contains a single proton and it thus possesses a significant magnetic moment. The proton behaves as a tiny bar magnet and it is aligned in either a parallel (a low energy state) or anti-parallel (a high energy state) with the direction of an externally applied magnetic field. In the presence of the magnetic field, B_o, the proton precesses at the Larmor frequency, ν_L, expressed as the following equation,

$$\nu_L = \frac{\gamma B_0}{2\pi}. \tag{1}$$

Therein, γ is the gyromagnetic ratio. The energy differential between the high and low energy states is proportional to the strength of the magnetic field, B_o. Therefore, a larger B_o will produce a greater net magnetization vector, M, as shown in Fig. 1a. However the net magnetization does not precess, but is aligned parallel with B_o, because the individual spins creating the net magnetization do not precess in phase. If we apply a pulse of radio frequency (RF) energy into the spins at the Larmor frequency, the individual spins begin to precess in phase. This resonance called nuclear magnetic resonance (NMR) occurs when the frequency of the rotating electromagnetic field matches the Larmor frequency, ν_L. The RF pulse is created by an RF coil that is oriented perpendicular to B_o. As a result, the aligned spins are tipped into the plane perpendicular to the main field to create the transverse magnetization as shown in Fig. 1b. This RF pulse is therefore called a 90° RF pulse. After switching off the RF pulse, the time-varying fields that results from the precession of this transverse magnetization induce

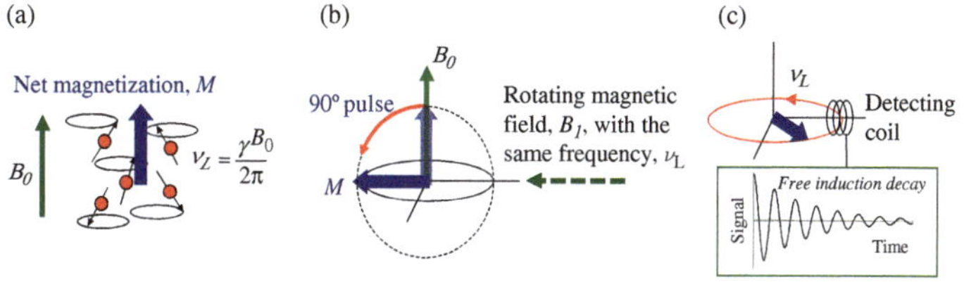

Figure 1. Schematic of NMR signal detection.

voltages in a receiver coil. The induced voltages in the coil are then detected by the MR system. This initial signal is referred to as the Free Induction Decay (FID) as illustrated in Fig. 1c. The signal exponentially decays with a relaxation time as the individual spins contributing to the net magnetization loose their phase coherence, referred to dephasing. The relaxation time is inherently affected by circumstances surrounding the spins.

In order to obtain the spatial distribution of the spins, i.e. NMR imaging (MRI), in a sample of interest, weak magnetic field gradients, **G**, are additionally superimposed to localize the spins. As a result, the NMR signal contains spatial information. The resonance frequency along the sample corresponding to

$$\nu_L(\mathbf{r}) = \frac{\gamma}{2\pi}(B_0 + \mathbf{G} \cdot \mathbf{r}). \quad (2)$$

Temporal variation of magnetic field gradients along the X, Y, and Z axes: a so-called *pulse sequence* for image acquisition is applied for MRI measurements. Several procedures have been developed for an effective determination of the spatial distribution of the spins. A detailed description of MRI pulse sequences is available in explanations by Callaghan,[37] Blűmich,[38] and Kimmich.[39]

Because the induced NMR signal after switching off the 90° RF pulse decays so rapidly as mentioned above, spin echoes are often detected in many practical pulse sequences for MRI visualization. Spin echo is formed by an additional RF pulse applied to the sample.The additional RF pulse is applied after an evolution period, τ. This RF pulse that gives rise to inversion of the magnetization components causes the spins to rephrase and is thus referred to the 180° RF pulse as shown in Fig. 2. This rephrasing process contributes to recover the transverse magnetization which had lost in dephasing process and results in the formation of an "echo". We detected the echo by the receiver coil. The time from when the 90° RF pulse is applied to when the echo forms is referred to as the "echo time", *TE* and is equal to twice the time between the 90° and 180° pulses, i.e. 2τ.

We need to apply temporal variation of magnetic gradient for localization of the spins between the 90° RF pulse and the echo detection, which is typically in a range of millisecond. In MRI experiments, the sample is divided into either a two-dimensional

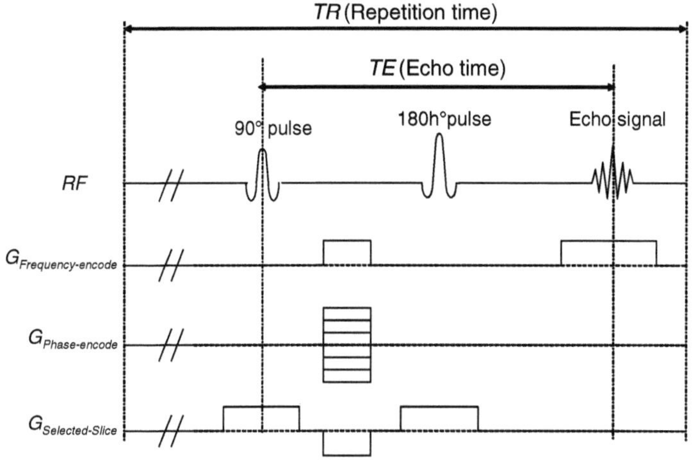

Figure 2. Example of a basic MRI pulse sequence with the spin-echo.

(2D) or three-dimensional (3D) grid (matrix) of voxels. The size of each voxel is given by the imaged region, called field of view (FOV), and thc matrix size. MRI has a unique feature to acquire images in planar slices of preselected thickness and location of the sample. In X-ray tomography, 2D cross sectional images are reconstructed after collecting a series of 2D projection images using different exposure angle.

In typical 2D MRI experiments with the spin echo, the 90° RF pulse, the frequency of which is matched with the Larmor frequency of the spins in a planar slice of the sample under the magnetic field gradient, $\boldsymbol{G}_{Selected\text{-}slice}$, is applied. Hence, the spins in the planar slice of interest in the sample selectively has the transverse magnetization. Then, the phase of the local magnetization is made to vary along one direction by applying a so-called phase-encoding gradient, $\boldsymbol{G}_{Phase\text{-}encode}$. Spatial information is also encoded in the frequencies that comprise the spin echo signal. The frequency of resonance depends on the local value of the magnetic field. Although the main magnetic field is designed to be quite uniform, additional magnetic fields, a so-called frequency-encoding gradient, $\boldsymbol{G}_{Frequency\text{-}encode}$, can be temporarily superimposed on the main static field. This creates spatial variation in the net magnetic field, resulting in a magnetic field gradient. A pulse sequence diagram for MRI visualization using a spin echo is schematically shown in Fig. 2.

Two- or three-dimensional MR images are created by sequential repetitions of the magnetic resonance and signal reception process, i.e., by repeated execution of a combination of RF pulses and time-varying magnetic field gradients in what constitutes a pulse sequence. Each repetition has a length referred to as the sequence repetition time, or *TR*. *TR* intervals are typically on the order of sub-second and are determined by a combination of logistics (the minimum amount of time needed to play out the required set of gradients) and the desire for particular types of contrast in the final images. Spatial resolution is determined by the number of frequency-encoded (or readout) projections and phase-encoded projections for a given FOV. In 2D MRI visualization, the two axes in an image are thus the "readout" (x-gradient) axis and the "phase-encode" axis (y-gradient). The z-gradient is used to select the slice.

During signal acquisition by the receiver coil, the detected spin echo signal is digitized into a certain number of data point, N_F. The number of data point the spin echo is sampled is equal to the number of "projections" along the frequency-encoded (readout) axis, for example, 256. An increase in spatial resolution along the phase encode axis requires an increase in the number of phase-encoded projections, N_P, each with a different strength of the phase encoding gradient. This increases the acquisition time and thus spatial resolution must be set prospectively. This is accomplished by specifying the size of the acquisition matrix ($N_F \times N_P$) and the FOV. Larger matrix (covering the same FOV) results in better spatial resolution, but causes less signal-to-noise ratio (S/N) due to fewer protons in the smaller voxels. In addition, larger matrix requires longer acquisition times since the acquisition time is equal to the product of $TR \times N_P$.

2.2. MRI System Hardware for PEMFC Visualization

Schematics of MRI system hardware and experimental setup for visualization of an operational PEMFC are shown in Fig. 3 reported by Tsushima et al.[27–29]. In MRI system for PEMFC visualization

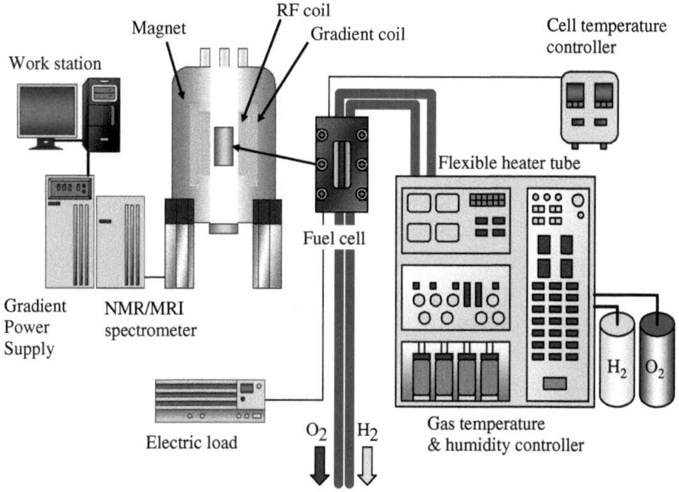

Figure 3. Schematic of MRI system and experimental setup for an operational PEMFC visualization.

found in literature,[7–29] super conducting magnets with its magnet field strength, B_0, from 2.4T to 14.1T are used. The strong magnetic field is advantageous for increasing S/N, although magnetic susceptibility differences between materials gives rise to the frequency shift that might cause image distortions. The gradient coils driven by gradient power supply are installed in the magnet for producing a gradient in B_0 in the x, y, and z directions. The RF coil is located within the gradient coils. The RF coil produces the B_1 magnetic field necessary to rotate the spins by 90°, 180°, or any other value selected by the pulse sequence. The RF coil also detects the signal from the spins within the cell. The work station controls all components on the MRI system, also providing an input/output (I/O) interface in which a pulse sequence is selected and customized.

The fuel cell is positioned within the magnet. Fuel gas (hydrogen) and oxidant gas (oxygen or air) are fed to the cell. For investigation of water transport in PEMFCs by MRI, it is preferable to operate PEMFCs in elevated temperature with humidity control, because both operating temperature and relative humidify (RH) inherently affects water transport in PEMFCs. Tsushima et al.[27–29] introduced a gas temperature and humidity controller. The gas temperature and humidity controller can control gas temperature and humidity literally and also can control gas flow rate and pressure. The fuel and oxidant gases are fed to the cell through flexible heater tube which is used to avoid condensation of water vapor before reaching to the cell.

Because the MRI system generates strong magnetic fields, fuel cells are constructed from nonmagnetic materials. Figure 4 shows an overview and a cross-sectional view of a cell.[29] A polymer electrolyte membrane is sandwiched between two carbon electrodes on which platinum particles are dispersed as a catalyst. It is reported that both perfluorinated sulfonic acid (PFSA) membranes and hydrocarbon (HC) membranes can be visualized by MRI,[26] although a thicker membrane as Nafion® 1110 (E. I. du Pont de Nemours and Company, thickness: 254 μm) was used to examine transversal (through-plane) water content distribution in the membrane.[15,29] Reactive area of the membrane electrode assembly (MEA) has limitation depending on inner diameter of test section in MRI system used for PEMFC visualization. In Fig. 4, the reactive area of the membrane electrode assembly (MEA) is 4.0 cm^2 (GDLs have a width and a length of 1 and 4 cm respectively). The current collectors in the cell, which also offers straight channels as flow path, are made of copper and are coated with gold to prevent the copper from dissolving into the membrane. Silicon gaskets are used in the gaps between the membrane and current collectors to provide a seal around the cell. The cell has a glass heater embedded in both endplates to maintain the temperature of MEA constant. During the

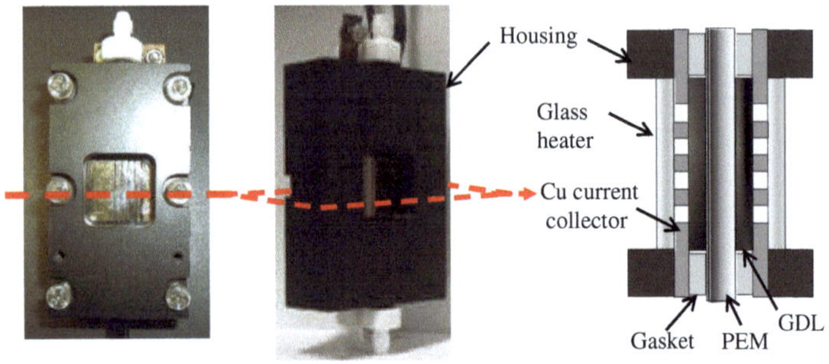

Figure 4. An overview and a cross-sectional view of a PEMFC for MRI visualization.

experiments, current density was controlled by adjusting an external electric load resistance connected to the cell.

2.3. MRI Signal Calibration for Water Content in PEM

Detected MRI signal should be converted for water content in a polymer electrolyte membrane (PEM). Our group performed a simple approach to relate the image intensity to the water content in the membrane by acquiring a series of MR images of an MEA exposed to water vapor activity, α, that are known to result in specific values of λ, as the following equation found in literature.[40]

$$\lambda = 0.043 + 17.81\alpha - 39.85\alpha^2 + 36.0\alpha^3 \qquad (3)$$

Actually, λ is the number of water molecules per SO_3H; it is known to vary up to 22. A calibration curve can then be produced to relate the image intensity to the absolute water content of the PEM.

This method however assumes that a static calibration is sufficient for a dynamic sample. It corresponds that a PEM sample in a fuel cell has well defined and fixed relaxation times. If the PEM however is truly dynamic and undergoes microstructural variation or other evolution then the relaxation times will change and the image will change even though the water content is not changing. For more reliable approach, it is feasible to use mapping of relaxation parameters in the PEM, because relaxation parameters such as T1 and T2 also depends on water contents as performed by Balcom et al.[15]

2.4. In Situ Visualization of Water in PEMFC Using MRI

Figure 5 shows the MRI visualization of the transversal water content distribution in the membrane in an operational PEMFC with variation of the current density.[29] The cell temperature and relative humidify were 70°C and 92%, respectively. The vertical width of the images is about 1.0 cm across the gas channels, which is in the central part of the GDL. The horizontal width of the images is 600 μm. The anode is on the left side of each figure. Figure 6 shows one-dimensional water content profiles in the membrane that were obtained from MRI visualization results at variation of the relative humidity and current density. The horizontal axis and vertical axis respectively indicate the through-plane position of the

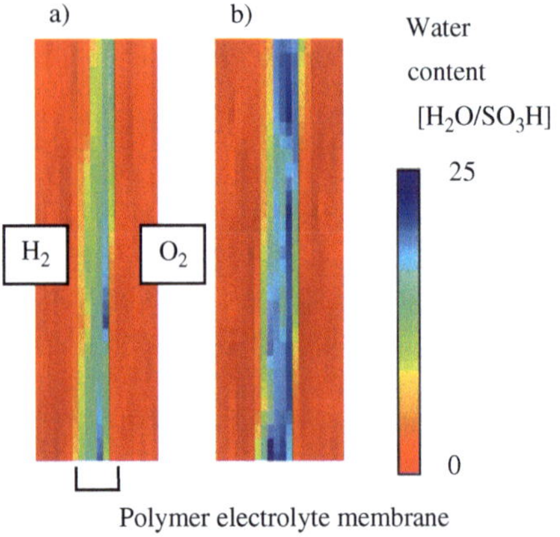

Figure 5. MRI visualization of the transversal water content distribution in a membrane in an operational PEMFC (a) 0.1 A/cm^2 at 92% RH, (b) 0.2 A/cm^2 at 92% RH.

membrane and water content. The position of the membrane is marked by the vertical dashed lines. Small amount of water was detected outside the membrane. This could be explained with contribution by either water inside the catalyst layers located at the both side of the membrane or signal blurring effect at the edge of the membrane.

In the standard condition with 80% RH (Fig. 6b), a water content, λ [H_2O/SO_3H], of around 8 in the membrane and partial dehydration at the anode were observed at a current density of 0.2 A/cm^2. This suggested that electro-osmosis was influential, even at this current density, because the membrane (Nafion® 1110, du Pont, 254 μm) was thicker than usual. In the dry condition (40% RH), the water content in the membrane was around 3 (Fig. 6c) and the water content profile was flat. This can be attributed to molecular diffusion in the membrane, as fast chemical diffusion at this water content has been reported by Zawodzinski et al.[41] It was also noted in both cases of 80% RH and 40% RH that the water generated in the cathode catalyst layer was not sufficiently transported to the membrane. This means that little water generated at the cathode condensed and was exhausted to the cathode gas channel as liquid. In the case of 92% RH shown

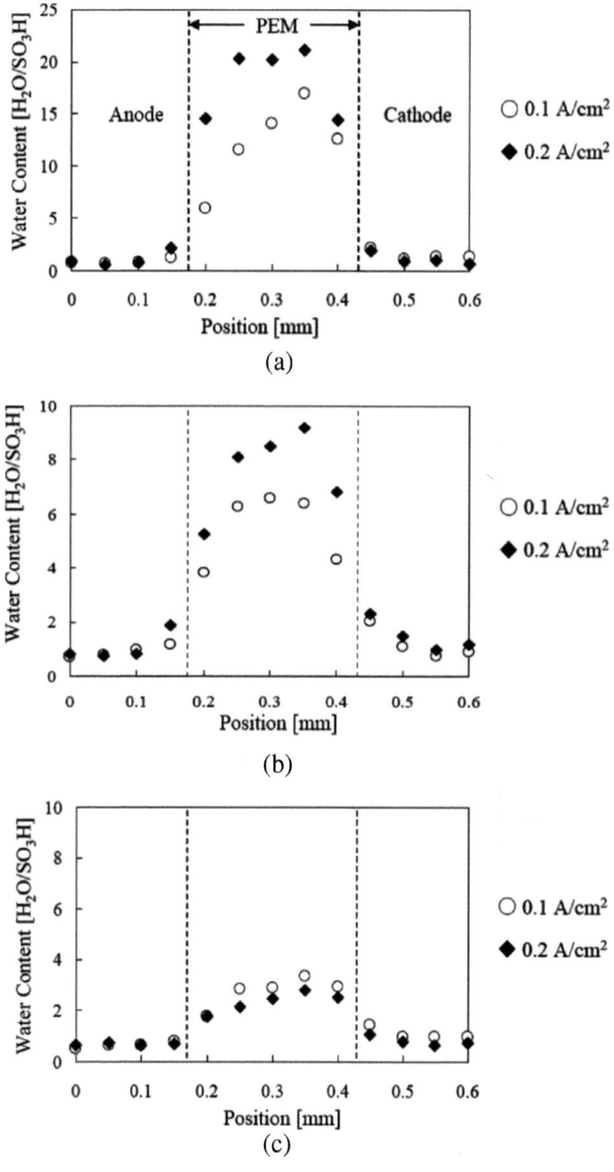

Figure 6. One-dimensional water content profile in the membrane. (a) 92% RH, (b) 80% RH, (c) 40% RH.

in Fig. 6a, the water content on the cathode side of the membrane increased to over 14 under 0.1 A/cm^2 and reached 22 under 0.2 A/cm^2, this being almost equivalent to the value when the membrane was soaked with water. This indicates that the water generated at the cathode was effectively absorbed by the membrane, because of the higher relative humidity that could suppress water discharge as a vapor. It was also noted that the anode side of the membrane at 0.1 A/cm^2 was less hydrated than the cathode by electro-osmosis, while the water content profile at 0.2 A/cm^2 was relatively flat. This suggests that liquid water in the cathode catalyst layer was transported to the membrane, presumably due to a pressure-driven water permeation process that was not apparent under lower humidity. It is noteworthy that a larger water concentration gradient was observed at a higher water content under 0.2 A/cm^2 with 92% RH than that observed with 80% RH. The membrane water content was around 15 at 92% RH which is greater than that obtained at 80% RH. This suggests that the higher water content in the membrane induced greater electro-osmotic drag, because the electro-osmosis coefficient was correlated with the water content.

MRI visualization has been also applied to investigate liquid water in flow fields in operational PEMFCs. The relation between the onset of liquid water accumulation in the cathode flow channel and the power output of the cell has been discussed by Feindel et al.,[8] Minard et al.[11] and Dunbar et al.[12,13] Conventional optical diagnostic can also capture liquid water behaviors in flow fields in operational PEMFCs embedded with transparent glass and can offer real-time visualization, although MRI visualization is time-averaged. This indicates that the most advantage by using MRI for PEMFC is in-situ diagnostic of water in MEAs and therefore recent advances for high-spatial resolution MRI gather much attention to apply a thin film of Nafion® membrane in an operational PEMFC.

Balcom et al. developed a novel diagnostic technique suitable for a thin-film by MRI.[42] The authors obtained one-dimensional water content profiles with nominal resolution of 6 μm using a prototype resonator and double half k-space spin echo single-point imaging (SPI) technique.[15] Even at room temperature, partial dehydration of the membrane in the anode side was captured clearly with high spatial resolution.

MRI visualization with contrast imaging is frequently applied in other applications as described in literature,[37–39] but is not widely regarded in MRI applications for PEMFC. Not only proton density imaging, but also contrast imaging, which is sensitive to both physical and chemical properties of nuclei of interest, is useful for fundamental understanding of water transport and degradation mechanisms in operating PEMFCs. Recently nuclei-labeling MRI (NL-MRI) has been introduced to investigate proton conducting processes in the membrane[25] and to visualize membrane hydration path in the membrane.[27] Spatially resolved nuclear magnetic resonance (NMR) spectroscopy also recently demonstrated its feasibility for probe variation of chemical and/or physical properties in the membrane after an accelerated degradation test of PEMFC.[30]

All demonstrated by using MRI visualization reported so far shows its unique and strong ability as powerful tools for investigation of water in PEMFCs, which is important for evaluating and optimizing fuel cell materials, components, design and operating conditions.

3. TUNABLE DIODE LASER ABSORPTION SPECTROSCOPY (TDLAS): AS A DIAGNOSTIC TOOL FOR IN-SITU DETECTION OF WATER VAPOR CONCENTRATION IN PEMFCs

3.1. Basic Principle of TDLAS

Localized remote sensing for chemical compositions in operating PEMFC can be performed by tunable diode laser absorption spectroscopy (TDLAS). Measurement of water vapor and carbon dioxide has been reported by using TDLAS.[31–36] In TDLAS measurement, the absorption of the laser beam passing through the gas sample is expressed with the partial pressure of the absorbing species, P_a (atm), by Beer's law,

$$\frac{I}{I_0} = \exp\left(-\int_0^L \kappa P_a dl\right) \tag{4}$$

where κ (atm^{-1} cm^{-1}) is the absorption coefficient that depends on the wavelength and temperature. I is the transmitted intensity and I_0 is the incident intensity. The integration is performed over the path length of the gas sample, L. The absorption coefficient shows strong peaks as a function of wavelength due to the rotational and vibrational energy transitions of the absorbing species. Assuming that the absorbing species has uniform concentration and temperature in the path length of the gas sample, the ratio of the transmitted intensity to the incident intensity at a single wavenumber is readily converted to the molar concentration of the absorbing species. However, as absorption bands generally spread over a range of wavelength, quoting the absorption coefficient at a single wavenumber might not give a true indication of the intensity of a transition. Therefore, in practice, preliminarily experiments are performed to obtain a calibration curve to convert the ratio of the transmitted intensity to the incident intensity to the molar concentration of the absorbing species.

For simultaneous measurement of water vapor and gas temperature by TDLAS, effect of water-water collisions on broadening the absorption profiles can be also used. Basu et al.[31–33] investigated the temperature dependence of peak absorption coefficient of water vapor and showed the transition near 1470 nm had the greatest temperature sensitivity with 23% variation, although the water transitions in the vicinity of 1491 nm were only sensitive to partial pressure and not temperature. The authors also showed that the halfwidth of the absorption profile near 1470 nm was not sensitive to temperature, but to the partial pressure. Accordingly, they established simultaneous measurement technique for temperature and water partial pressure by using both the variation of halfwidth of the absorption profile near 1470 nm for partial pressure and the variation of the peak absorption coefficient for temperature. They performed measurements in the fuel cell without external loading to calibrate the measured halfwidths and spectral intensities against the HITRAN simulations and obtained calibration diagrams to determine the partial pressure and the temperature of the supplied gas to the PEMFC.

3.2. TDLAS System Hardware for Water Vapor Measurement

For TDLAS measurement in PEMFCs, the laser diode as the transmitter and the photodiode as the detector are used. In a typical experimental setup with TDLAS as Basu et al. reported, the laser beam from the distributed feedback diode (DFB) laser at a wavelength of 1470 nm was split using a bifurcated optical fiber. One leg of the fiber was fed to a reference photodiode for measuring the laser power without absorption while the other leg was fed into another optical fiber. The output legs of this fiber were placed at the ends of the milled flow channels. They used a specially-designed PEMFC with serpentine flow path in which the selected channels were milled out to the end of the bipolar plate so that the diode laser beam could be transmitted along the channel length. The incident laser light intensity was monitored by the reference photodiode while the attenuated laser signal was measured by other photodetectors at the opposite ends of the PEMFC gas channels.

In other experimental setup with TDLAS as reported by Fujii et al.,[34–36] both the beam transmitter and the detector were packaged in a sensor head, enabling to probe water vapor concentration from the one side of PEMFCs through the optical window as depicted in Fig. 7. For measurement of water vapor in PEMFCs the diode laser at a wavelength of 1392 nm was used. A sensor head projects a laser beam through the glass cover into the fuel cell channel. The back of the fuel cell channel scatters some of the laser light back into the sensor head which is collected onto a photo detector. The photo detector converts the received light into an electrical signal

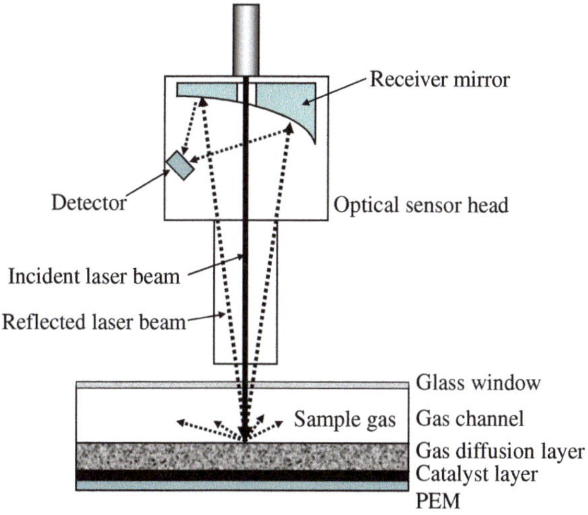

Figure 7. Schematic of optical sensor head of TDLAS.

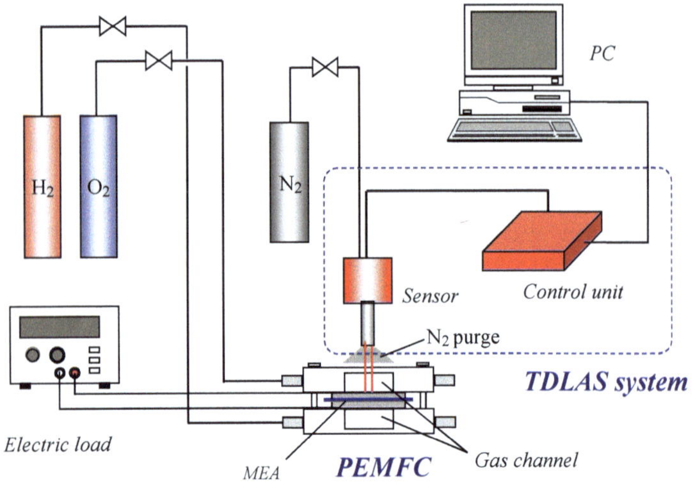

Figure 8. Schematic of an experimental setup with TDLAS for in-situ measurement of water vapor in an operating PEMFC.

and supplies the signal to the control module. Figure 8 shows a schematic of experimental setup for water vapor measurement in an operating PEMFC by TDLAS. The sensors comprise two sections, a control module and an optical sensor head, connected by an optical fiber. Water vapor is purged from the sensor head by N_2 gas flow to eliminate effect of water vapor in the atmosphere.

3.3. TDLAS Signal Calibration for Measurement of Water Vapor Concentration

In practical application of TDLAS to measure water vapor concentration in gas channels in PEMFCs, a signal calibration is needed. In this section, a wavelength modulation spectroscopy (WMS) method that is known to be advantageous on increased sensitivity and reliability than direct absorption spectroscopy[43,44] is described. Figure 9 shows a schematic of signal processing with WMS method. The incident laser wavelength is modulated sinusoidally at a fixed modulation rate as shown in Fig. 9a. The modulation spans a range of wavelengths, called the modulation depth, δ, that is usually comparable to the linewidth of the water vapor. Thus, the wavelength

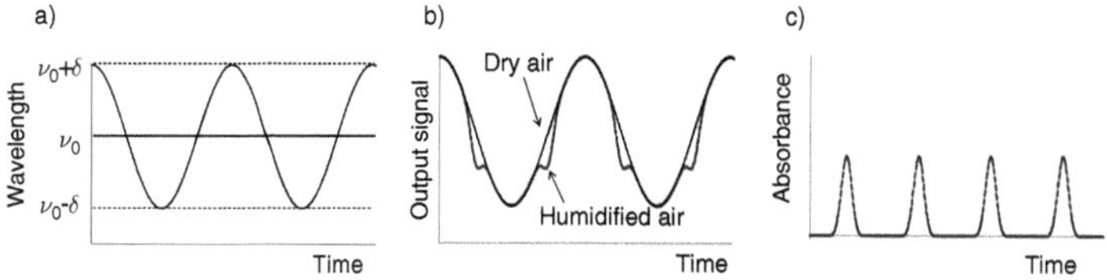

Figure 9. Schematic of signal processing with WMS method in TDLAS.

sweeps periodically back-and-forth across the absorption line. In the absence of water vapor, i.e. dry air condition, the detector output signal is a nearly perfect sinusoid as shown in Fig. 9b. In the presence of water vapor in the optical path, the detector output signal contains harmonic components originated from the absorption of the incident laser beam by water vapor. Figure 9c shows the absorbance versus time deduced from the curves in Fig. 9b.

In these WMS signals, the wavelength crosses the water absorption feature twice for each modulation cycle. Therefore, when water vapor is in the optical path, the amplitude modulation of the received laser power contains a periodic component having precisely twice the wavelength modulation frequency and fixed in phase relative to the wavelength modulation.

In WMS, phase-sensitive (i.e. lock-in) amplification demodulates the periodic signal with a very narrow electrical bandwidth to precisely measure the RMS amplitudes of the fundamental sinusoid and its second harmonic component, averaged over a period of time equal to the inverse of the electrical bandwidth. These are called the $1f$ and $2f$ signals and each amplitude is expressed by the following equations, respectively,

$$P_f = \frac{A_0 \eta \rho}{R^2} S_{DC} m \tag{5}$$

$$P_{2f} = \frac{A_0 \eta \rho}{R^2} S_{DC} k\alpha \times 2C_R \tag{6}$$

where A_0 is the effective area of collection lens, η is the receiver-optics efficiency, ρ is the differential reflectance of target, R is the range from sensor to target, S_{DC} and m are the dc component and the amplitude modulation (AM) ratio of initial laser power, α is the absorption coefficient at the absorption centre, k is a correction factor for a Lorentzian line shape and C_R is the integrated concentration of water vapor. Diving the $2f$ signal by the $1f$ signal, uncertainties of A_0, η, ρ, R and S_{DC} are cancelled out and we obtain a simple relationship between integrated concentration of water vapor, C_R, and the ratio of P_{2f} to P_f, P_{2f}/P_f, used as the detector output.

$$\frac{P_{2f}}{P_f} = \frac{2\,k\alpha}{m} C_R \tag{7}$$

To obtain water vapor concentration, output signal intensity by TDLAS is calibrated under a well-controlled environment in variation of relative humidity and surrounding temperature. Figure 10 shows a relationship between water vapor concentration and output signal intensity detected by TDLAS. It is shown that the

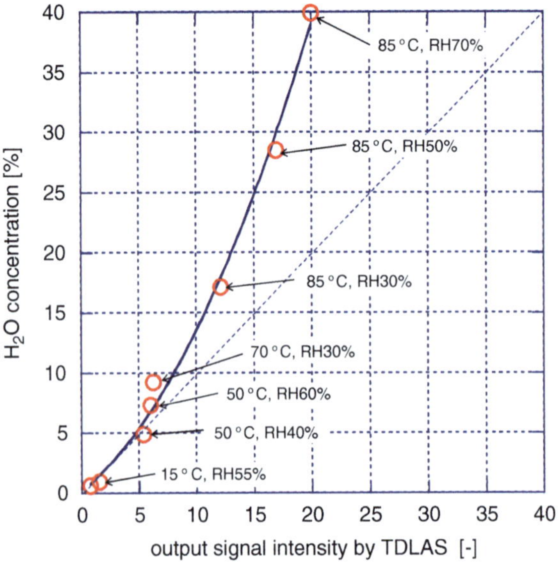

Figure 10. Calibration curve for measurement of water vapor concentration by TDLAS.

output signal intensity is not simply proportional to water vapor concentration, but is well correlated with water vapor concentration in wide range of relative humidity and temperature. Much variation of output signal intensity with relative humidity is observed under elevated temperature condition (85°C), because saturated water vapor pressure considerably increases with temperature. Accordingly, in PEMFC application, measurement of relative humidity of feed gas by TDLAS is advantageous under high temperature condition. On the other hand, application of TDLAS under sub-zero temperature condition is not feasible because of lower concentration of water vapor.

3.4. In Situ Measurement of Water Vapor in PEMFC Using TDLAS

One of the noticeable advantages of TDLAS measurement is to be able to measure local concentration of chemical species and its distribution in operating PEMFCs. Figure 11 shows the variation of water vapor concentration in the cathode and the anode channel of the PEMFC with the MPL at the cathode (®SIGRACET GDL 24BC; SGL Carbon Japan Ltd.) under the counter-flow mode at a setting of 0.5 A/cm^2. In this TDLAS experiment, the GDL without the MPL (®SIGRACET GDL 24BA; SGL Carbon Japan Ltd.) was used at the anode side. Oxygen and hydrogen gases were supplied by a gas supply unit to the single straight channel PEMFC, both the depth and the width of the channel were 3 mm, at a flow rate of 100 mL/min with relative humidity (RH) 20%. The cell temperature was maintained at 80°C using a thermostat. The active area of the membrane electrode assembly (MEA) was 3.5 cm^2 (7 mm wide, 50 mm long). The polymer electrolyte membrane thickness was 50 μm (Nafion® 112, DuPont). The catalyst layers were coated with 0.1 mg/cm^2 of Pt catalyst.

As shown in Fig. 11, water vapor concentration in the oxygen channel increases and the concentration in the hydrogen channel decreases from the oxygen inlet. Near the oxygen inlet region, it is inferred that more water is transported to the cathode channel because the water vapor concentration in the anode channel is higher than that in the cathode channel. On the other hand, near the hydrogen inlet region, more water was transported to the anode

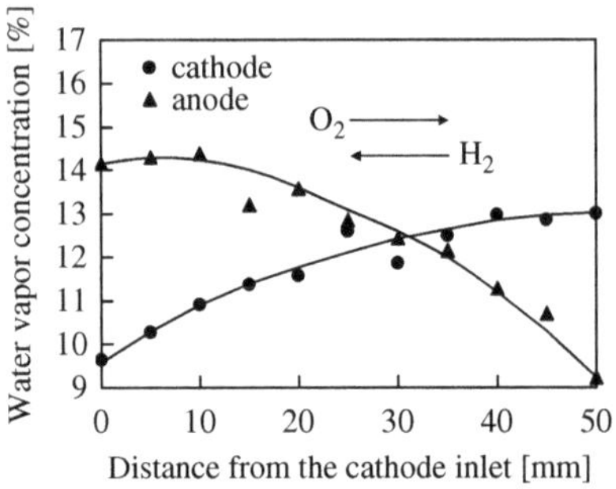

Figure 11. Variation of water vapor concentration in the cathode and the anode channel of the PEMFC with the MPL under the counter-flow mode at a setting of 0.5 A/cm^2 measured using the TDLAS system.

channel because the water vapor concentration in the anode channel is lower than that in the cathode channel.

For quantitative investigation on through-plane water transport, a through-plane water back-transport index can be calculated. The through-plane water back-transport index, χ, is used to evaluate enhancement of water transport from the cathode side to the anode side, although the net electro-osmotic drag coefficient, showing the number of water molecules per proton transports from the anode side to the cathode side, is widely used to investigate water transport through a membrane. Herein, the through-plane water back-transport index is useful because the through-plane water back-transport index shows the ratio of local through-plane water flux across the MEA from the cathode to the anode, $j_{C\to A}(x)$, to the local water production rate, $S_{EC}(x)$, by electrochemical reaction at the cathode catalyst layer, and is expressed by the following equation,

$$\chi(x) = \frac{j_{C\to A}(x)}{S_{EC}(x)} \tag{8}$$

This water back-transport index is expressed as the normalized through-plane water flux from the cathode side to the anode side divided by the water generation rate at the cathode. Consequently, a water back-transport index equal to unity means that all generated water in the cathode catalyst layer backs to the anode channel. Based on measurement of variation of water vapor concentration in both the anode and the cathode channels by TDLAS as shown in Fig. 11, amount of localized water generation by electrochemical reaction can be obtained with hydrogen/oxygen consumption and velocity field variation along the channel taken into account. Accordingly, variation of local current density along the channel can be obtained.

Figure 12 under the counter-flow mode at 0.5 A/cm^2. The index with the MPL is larger than that without the MPL for an index of 0–1 as depicted in Fig. 12. This result also shows that the MPL enhances water back-transport from the cathode side to the anode side. However, when the index is negative, meaning that the internal water circulation from the anode channel to the cathode channel, the index with the MPL is slightly higher than the index without the MPL. Therefore, the MPL at the cathode suppressed water vapor absorption at the anode, which is explainable by membrane hydration attributable to the MPL at the cathode. Consequently, the MPL promotes membrane hydration, leading less internal water circulation from the anode to the cathode side.

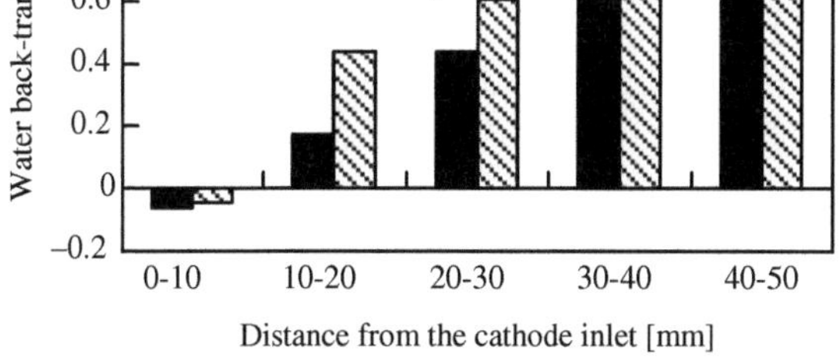

Figure 12. Variation of water back-transport index of the PEMFC without and with the MPL under the counter-flow mode at a setting of 0.5 A/cm^2 based on measurements using the TDLAS system.

As shown above, in-situ TDLAS measurements show its unique potential to investigate localized water transport in operating PEMFCs, leading to fundamental understanding on water management. It is also capable of monitoring localized phenomena with good temporal resolution.

4. SUMMARY

Magnetic resonance imaging (MRI) technique and tunable diode laser absorption spectroscopy (TDLAS) are described as diagnostic tools for in-situ water measurement in operating PEMFCs. Basic principle and system hardware for in-situ MRI visualization of water content in the membrane under fuel cell operation are given. MRI technique has been recognized as a useful tool to investigate water content and its spatial distribution in membranes as well as liquid water accumulation in operating PEMFCs. Further development to achieve higher spatial resolution and other related techniques using isotope-labeling and/or localized spectroscopy have been expected in the future. High spatial resolution MRI possibly leads to probing water content in the catalyst layers. In-situ measurement technique of water vapor concentration in gas flow channels in PEMFCs using TDLAS is described with its measuring principle and validity of a practical system. Localized current density and through-plane water-back transport index are obtained with variation of vapor concentration along the gas channel taken into account, showing its strong potential for fundamental understanding on water transport in PEMFCs. TDLAS technique has already shown its unique potential to simultaneously detect local water vapor concentration and temperature. Detection of contaminant species in both the anode and the cathode channels by TDLAS promisingly contribute for elucidation of degradation mechanism in PEMFC.

REFERENCES

[1] T.A. Zawodzinski, C. Derouin, S. Radzinski, R.J. Sherman, V.T. Smith, T.E. Springer, S. Gottesfeld, J. Electrochem. Soc. **140**, 1041 (1993)
[2] E. Endoh, ECS Trans. **3**(1), 9 (2006)
[3] J.Y. Shim, S. Tsushima, S. Hirai, ECS Trans. **11**(1), 1151 (2007)

[4] Y. Tabuchi, N. Kubo, in *6th International Fuel Cell Science, Engineering and Technology conference (2008)*, FuelCell2008-65201, DENVER, CO, 16–18 June 2008
[5] S.H. Ge, X.G. Li, B.L. Yi, I.M. Hsing, J. Electrochem. Soc. 152 (2005) A1149
[6] P.W. Majsztrik, M.B. Satterfield, A.B. Bocarsly, J.B. Benziger, J. Memb. Sci. 301 (2007) 93
[7] K.W. Feindel, L.P.A. LaRocque, D. Starke, S.H. Bergens, R.E. Wasylishen, J. Am. Chem. Soc. **126**, 11436 (2004)
[8] K.W. Feindel, S.H. Bergens, R.E. Wasylishen, Chem. Phys. Chem. **7**, 67 (2006)
[9] K.W. Feindel, S.H. Bergens, R.E. Wasylishen, J. Am. Chem. Soc. **128**, 14192 (2006)
[10] K.W. Feindel, S.H. Bergens, R.E. Wasylishen, Phys. Chem. Chem. Phys. **9**, 1850 (2007)
[11] K.R. Minard, V.V. Viswanathan, P.D. Majors, L.Q. Wang, P.C. Rieke, J. Power Sources **161**, 856 (2006)
[12] Z.W. Dunbar, R.I. Masel, J. Power Sources **171**, 678 (2007)
[13] Z.W. Dunbar, R.I. Masel, J. Power Sources **182**, 76 (2008)
[14] Z.W. Dunbar, R.I. Masel, ECS Trans. **16**(2), 1001 (2008)
[15] Z. Zhang, J. Martin, J. Wu, H. Wang, K. Promislow, B.J. Balcom, J. Magn. Reson. **193**, 259 (2008)
[16] K. Teranishi, S. Tsushima, S. Hirai, Therm. Sci. Eng. **10**, 59 (2002)
[17] K. Teranishi, S. Tsushima, S. Hirai, Therm. Sci. Eng. **11**, 35 (2003)
[18] K. Teranishi, S. Tsushima, S. Hirai, Therm. Sci. Eng. **12**, 91 (2004)
[19] S. Tsushima, K. Teranishi, S. Hirai, Energy **30**, 235 (2004)
[20] S. Tsushima, K. Teranishi, S. Hirai, Electrochem. Solid-State Lett. **7**, A269 (2004)
[21] T. Mibae, S. Tsushima, S. Hirai, Therm. Sci. Eng. **12**(4), 89 (2004)
[22] S. Tsushima, K. Teranishi, K. Nishida, S. Hirai, Magn. Reson. Imaging **23**, 255 (2005)
[23] K. Teranishi, S. Tsushima, S. Hirai, J. Electrochem. Soc. **153**(4), A664 (2006)
[24] S. Tsushima, T. Nanjo, K. Nishida, S. Hirai, ECS Trans. **1**(6), 199 (2005)
[25] S. Tsushima, K. Teranishi, S. Hirai, ECS Trans. **3**(1), 91 (2006)
[26] S. Tsushima, S. Hirai, K. Kitamura, M. Yamashita, S. Takase, Appl. Magn. Reson. **32**(1), 233 (2007)
[27] T. Kotaka, S. Tsushima, S. Hirai, ECS Trans. **11**(1), 445 (2007)
[28] S. Tsushima, T. Nanjo, S. Hirai, ECS Trans. **11**(1), 435 (2007)
[29] T. Ikeda, T. Koido, S. Tsushima, S. Hirai, ECS Trans. **16**(2), 1035 (2008)
[30] S. Hirai, S. Tsushima, ECS Trans. **16**(2), 1337 (2008)
[31] S. Basu, H. Xu, M.W. Renfro, B.M. Cetegen, J. Fuel Cell Sci. Tech. **3**, 1 (2006)
[32] S. Basu, M.W. Renfro, H. Gorgun, B.M. Cetegen, J. Power Sources **159**, 987 (2006)
[33] S. Basu, M.W. Renfro, B.M. Cetegen, J. Power Sources **162**, 286 (2006)
[34] Y. Fujii, S. Tsushima, K. Fukuzato, S. Hirai, ECS Trans. **16**(2), 1635 (2008)
[35] Y. Fujii, S. Tsushima, S. Hirai, J. Therm. Sci. Tech. **3**(1), 94 (2008)
[36] Y. Fujii, S. Tsushima, S. Hirai, ECS Trans. **11**(1), 451 (2007)
[37] P.T. Callaghan, *Principles of Nuclear Magnetic Resonance Microscopy* (Oxford University Press, New York, 1991)
[38] B. Blűmich, *NMR Imaging of Materials* (Oxford University Press, New York, 2000)
[39] R. Kimmich, *NMR: Tomography, Diffusometry, Relaxometry* (Springer, Berlin, 1997)

[40] T.E. Springer, T.A. Zawodzinski, S. Gottesfeld, J. Electrochem. Soc. **138**(8), 2334 (1991)
[41] T.A. Zawodzinski, M. Neeman, L.O. Sillerud, S. Gottesfeld, J. Phy. Chem, **95**, 6040 (1991)
[42] A.V. Ouriadov, R.P. MacGregor, B.J. Balcom, J. Magn. Reson. **169**, 174 (2004)
[43] J. Reid, D. Labrie, Appl. Phys. B **26**, 203 (1981)
[44] G.B. Riecker, H. Li, X. Liu, J.B. Jeffries, R.K. Hanson, M.G. Allen, S.D. Wehe, P.A. Mullhall, H.S. Kindle, Meas. Sci. Technol. **18**, 1195 (2007)

7

Characterization of the Capillary Properties of Gas Diffusion Media

Jeffrey T. Gostick,[1] Marios A. Ioannidis,[2] Michael W. Fowler,[2] and Mark D. Pritzker[2]

[1]*Department of Chemical Engineering, McGill University, Montreal, QC, Canada*
[2]*Department of Chemical Engineering, University of Waterloo, Waterloo, ON, Canada*

1. INTRODUCTION

The present generation of membrane materials used in polymer electrolyte membrane fuel cells (PEMFCs) requires high humidity to maintain sufficient proton conductivity. Mass transport through the porous electrodes, however, is most effective in dry conditions since the presence of liquid water in the pores reduces effective oxygen diffusivity to the catalytic sites. Management of these competing requirements is further complicated by the production of water inside the cell as a by-product of the cathode reaction. Maximizing fuel cell power density therefore requires effective water management techniques to prevent excessive liquid water from accumulating in the porous electrode components. Liquid water distribution and flow in the cathode gas diffusion media (GDM) of an operating PEMFC is critically affected by capillary forces. Perhaps the most widely employed technique for improving water management is to impregnate the fibrous GDM with a polymer, such as poly-tetra-fluoro-ethylene (PTFE), to coat the carbon fibers and thereby render the GDM more hydrophobic. It is thus important to understand the

U. Pasaogullari, C.-Y. Wang (eds.), *Modeling and Diagnostics of Polymer Electrolyte Fuel Cells*, Modern Aspects of Electrochemistry 49,
DOI 10.1007/978-0-387-98068-3_7,

relationship between wettability and capillary properties of native (i.e., untreated) or PTFE-treated GDMs on the one hand and the relationship between GDM capillary properties and fuel cell performance on the other hand. Until recently, however, few experimental techniques were available to measure the capillary properties of GDMs. This chapter discusses the present understanding of the capillary properties of GDM–water–air systems and provides a critical analysis of reported experimental techniques that have recently contributed to this understanding.

1.1. Motivation

Numerous fuel cell models attempting to incorporate the physics of multiphase flow in the GDM using a continuum framework have been developed. These calculations require accurate knowledge of constitutive relationships for capillary pressure, liquid relative permeability, and effective gas diffusivity, among other things. Such relationships are not well known for GDMs. The first efforts to model two-phase flow in GDMs can be attributed to Wang et al.[1,2] and He et al.[3] The latter employed the capillary diffusivity concept to collect all unknown multiphase transport properties into a single parameter which they treated as a constant. Wang et al.[1,2] adopted a more realistic approach by choosing to account for and estimate each property separately. In the absence of GDM-specific data, they adopted a relationship between air–water capillary pressure and water saturation originally proposed by Leverett[4] and subsequently modified[5] with the intent to capture the effect of wettability on capillary pressure by including the cosine of contact angle θ:

$$P_C\left(S_W\right) = \sigma \cos\left(\theta\right)\left(\frac{\varepsilon}{K}\right)^{0.5} J\left(S_W\right) \tag{1}$$

where S_W is the water saturation, $P_C \equiv P_{NWP} - P_{WP}$ is the capillary pressure (a positive-definite quantity defined as the difference between the pressures of the non-wetting and wetting phases[6]), σ is the surface or interfacial tension of the immiscible fluid pair, ε and K are the porosity and permeability of the porous medium, respectively, and $J\left(S_W\right)$ is the so-called Leverett J-function. When Eq. (1) holds, scaling the capillary pressures measured in different porous media and with different fluid pairs by $\sigma \cos\left(\theta\right)\sqrt{\varepsilon/K}$ causes the data to collapse along a single curve $J(S_W)$ that depends on saturation only. Lacking a GDM specific $J(S_W)$ function, Wang et al.[1,2] used a polynomial fitting of Leverett's experimental data for water imbibition in sand packs[4] given by Udell:[7]

$$J(S_W) = 1.417(S_W) - 2.120(S_W)^2 + 1.263(S_W)^3 \tag{2}$$

Following the original work of Wang et al.,[1,2] a large number of PEMFC continuum models have relied on Eqs. (1) and (2) to describe the dependence of air–water capillary pressure on GDM water saturation. Reliance on these equations has become so customary that some of the more recent works do not even provide a citation for their original use. While it is true that Leverett used Eq. (1) to correlate capillary pressure curves of several different sand packs, it is not true that the same correlation can be used to describe the capillary properties of GDMs, which bear little resemblance to sand packs with regard to structure or wettability. This can be seen in Fig. 1 where Leverett's data (and the fitting by Udell)

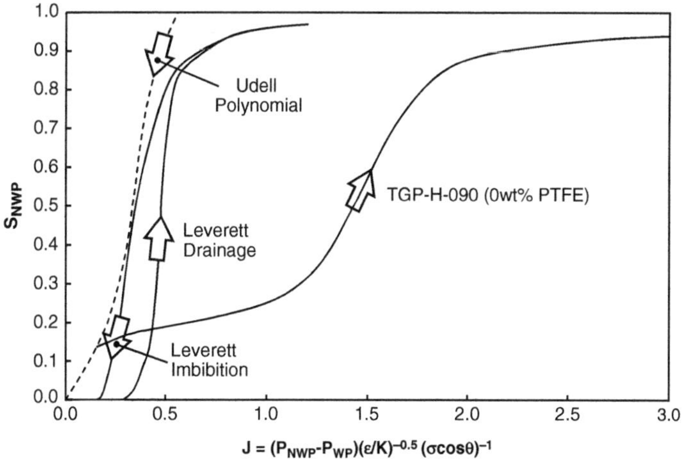

Figure 1. Comparison of Leverett's data[4] and Udell's[7] polynomial fit to measured GDM data.[8] Block arrows indicate direction of saturation change. Physical properties for the GDM are taken from Gostick et al.[9]

are directly compared to a recently reported[8] experimental water–air capillary pressure curve for a GDM, plotted in terms of Eq. (1) using previously reported properties.[9] Clearly, Eq. (2) is not a good description of GDM capillary data; ironically it is not even a good fit to Leverett's original results. It may still be of interest to seek GDM-specific forms of $J\,(S_W)$. This is now possible with the advent of reliable techniques to directly measure the water–air capillary pressure curves of GDMs.

Additional motivation for measuring the capillary pressure-saturation relationships for air and water within GDMs comes from the need to evaluate the effect of hydrophobic treatments on GDM wettability. The coating of GDM fibers with PTFE is based on the premise that the graphitized carbon fibers are hydrophilic in their native state, tending to retain rather than repel water. Thus, the application of a coating of PTFE should render the GDM more water-repellent. A review of contact angle measurements for the water–air fluid pair on graphite and PTFE surfaces seemingly confirms this reasoning, since graphite is reported[10–12] to exhibit contact angles in the range $74° < \theta < 86°$, whereas PTFE is reported[12–14] to exhibit contact angles in the range $108° < \theta < 112°$ (measured through water in all cases). Since it is reasonable to expect that PTFE treatment is non-uniform, it is generally hypothesized that PTFE-treated GDMs have "mixed wettability", in the sense that surfaces exhibiting $\theta \approx 80°$ (i.e., carbon) and $\theta \approx 110°$ (i.e., PTFE) co-exist. The fact that these angles span $\theta = 90°$ has been interpreted to mean that hydrophilic ($\theta < 90°$) and hydrophobic ($\theta > 90°$) pores co-exist in PTFE-treated GDMs.[15–19] For such interpretations, Washburn's equation has been invoked:

$$P_C = -\frac{2\sigma \cos(\theta)}{R} \tag{3}$$

which is valid for straight cylindrical capillaries of radius R.[6] According to Eq. (3), the sign of the capillary pressure is decided by the value of $\cos(\theta)$. Hence, $\theta = 90°$ defines a transition between water being the wetting phase or the non-wetting phase within GDM pores. It has been reasoned on the basis of Eq. (3) that liquid water should spontaneously imbibe into hydrophilic pores in GDMs – a reasoning that has formed the basis of both continuum and pore-network models of water transport in PTFE-treated

GDMs.[15,19] The first reported measurements of air–water capillary pressure in GDMs[16] appeared to confirm this expectation since it was found that $P_G > P_L$ was required to displace water from the GDM pore space. This finding has been interpreted as evidence that the uncoated carbon fibers constitute a hydrophilic pore network.[16,19] Other measurements,[20] however, have indicated that water does not spontaneously imbibe and must be forced ($P_L > P_G$) into dry GDMs even in the absence of PTFE. This observation cannot be explained by Eq. (3) since $\theta < 90°$ for untreated GDMs. Recently reported[21] water–air capillary pressure data for GDMs demonstrate the limitations of using contact angle as a predictor of GDM wettability within the context of Eq. (3) and challenge the concept of mixed-wettability for PTFE-treated GDMs.[15,19]

2. BASIC CONSIDERATIONS

The capillary pressure of interest in water–air–GDM systems is the difference between the pressures of the liquid and gas phases across static air–water interfaces within a GDM. This pressure difference is fundamentally related to the mean curvature H of the air–water interfaces through the well-known Young-Laplace equation:[22]

$$P'' - P' = 2\sigma H \tag{4}$$

Equation (4) shows that, because of the existence of surface tension, an arbitrary surface (interface) of mean curvature H maintains mechanical equilibrium between two fluids at different pressures, P'' and P'.[6] The fluid on the concave side of the surface must have a pressure P'' that is greater than the pressure P' on the convex side and the difference $P'' - P'$ is the capillary pressure P_C. It is known from differential geometry that $2H = \nabla \cdot \vec{n}$, where $\vec{n}$ is the unit normal on the surface.[23] The Young–Laplace equation is therefore a differential equation valid locally at each point on the interface.[24] Strictly speaking, the concept of contact angle is relevant not to the Young–Laplace equation itself, but rather to solutions of this equation representing equilibrium capillary surfaces in contact with a solid phase. In such cases, the contact angle is incorporated in the boundary condition at the three-phase contact line that determines the shape of the interface (and therefore H) in the solution to the

Young–Laplace equation.[6,25] Note that for a drop of liquid placed on a plane surface – a configuration relevant to the measurement of contact angle – the capillary pressure is always positive for any $\theta > 0°$. As pointed out by Dullien,[6] the basic law of capillarity requires that P_C be a positive-definite quantity. Dullien[6] has defined the fluid at higher pressure on the concave side of an interface to be the non-wetting fluid and the fluid at lower pressure on the convex side of an interface as the wetting fluid so that capillary pressure is defined as $P_C \equiv P_{NWP} - P_{WP}$. This corresponds to an operational definition of wettability on the basis of the sign of the mean curvature H, as opposed to whether $\theta < 90°$ or $\theta > 90°$. Whenever the sign of H changes there is an exchange of roles of the two fluids as regards to preferential wettability. It is clear from the preceding discussion that the shape of equilibrium fluid–fluid interfaces in porous media is not determined by θ alone, but by the solution of a differential equation for which θ is incorporated in the boundary condition. Since both pore space geometry and intrinsic contact angle determine the sign of H, preferential wettability is not governed by the intrinsic contact angle θ alone. Note that the conventional definition of air–water capillary pressure as $P_C = P_L - P_G$ allows for both positive and negative capillary pressures. Since $P_C/\sigma \propto H$, it must also be noted that the use of Eq. (1) implies $H \propto \cos(\theta)$. Inclusion of the $\cos(\theta)$ term[5] in Leverett's original correlation[4] is thus motivated by Eq. (3) and is not warranted when Eq. (3) fails to describe the dependence of capillary pressure on contact angle.

Dullien's operational definition of preferential wettability is particularly applicable when the porous medium surface exhibits contact angles in the vicinity of 90°. This is precisely when the sign of mean curvature is most sensitive to the local geometry of the solid structure[6] and Eq. (3) ceases to be valid.[25] For example, in a capillary of varying cross section, relatively small changes in the orientation of the pore walls result in wettability reversal when θ is not very different from 90°.[6] The limitations of classifying porous media as hydrophilic or hydrophobic depending on whether $\theta < 90°$ or $\theta > 90°$, respectively, are well known. According to Anderson,[26] a system is hydrophilic (water-wet) when θ is between 0 and 60–75° and hydrophobic when θ is between 180 and 105–120°. When contact angles lie in the intermediate range, as in the case of native and PTFE-treated GDMs, it is more appropriate to classify the system as being neutrally or intermediately wet. By convention, the process of displacing a wetting phase by a non-wetting one is termed drainage, whereas the reverse process is termed imbibition.[6] A distinguishing feature of systems exhibiting intermediate or neutral wettability is that they do not spontaneously imbibe either fluid.[6,25] Spontaneous imbibition of liquids in porous media has been shown to require contact angles less than 40–50°.[27,28] The first measurements of capillary pressures required to displace water by air[16] and air by water[20] in GDMs are not at odds with this expectation. Subsequent measurements clearly demonstrate that water (air) must be forced into a dry (water-saturated) GDM. The condition of mixed or fractional wettability should be clearly distinguished from that of neutral or intermediate wettability. According to Anderson,[25] fractional wettability describes a porous material in which some pores are strongly wetted by the aqueous phase, whereas other pores are strongly wetted by the non-aqueous phase. The term mixed wettability is used to further describe a material of fractional wettability in which the smaller pores are strongly wetted by the aqueous phase. In both fractional and mixed wettability materials spontaneous imbibition of both fluid phases is expected due to the co-existence of bi-continua of strongly hydrophobic and strongly hydrophilic pore surfaces.[6,25]

Capillary equilibrium of water and air within GDM pores can be established over a broad range of water saturations by changing the phase pressure difference in a step-wise fashion. At each level of saturation, for which capillary equilibrium is established, all air–water interfaces within a GDM have the same mean curvature, provided that hydrostatic gradients are absent. The resulting functional relationship $P_C(S_W)$ between capillary pressure and water saturation is expected to depend on saturation history (i.e., increasing versus decreasing water saturation) and exhibit permanent hysteresis.[6,25] Permanent hysteresis is observed because capillary equilibrium is reached via a sequence of non-equilibrium interfacial configurations that differ depending on the history of saturation change. This is in part due to contact angle hysteresis caused by roughness of the pore surface[12] which modifies the boundary condition at the three-phase contact line, and in part due to the complex geometry and topology of the pore space itself.[6,24,29]

Since water must be forced into a dry GDM regardless of PTFE-content, water is effectively the non-wetting phase. This process has the characteristics of primary drainage, for which invasion percolation[30] is an adequate model.[6,31] Invasion percolation describes the mechanism of fluid invasion into a porous medium whereby a non-wetting invading phase displaces a wetting phase as the capillary pressure ($P_C = P_{NWP} - P_{WP}$) is increased. The two main characteristics of this process are that (a) the invading phase can only enter pores that are penetrable at the applied capillary pressure and (b) the invading phase can only enter pores that are accessible; that is, pores connected directly to the reservoir of invading fluid or indirectly via a path of invaded pores. As the phase pressure difference $P_{NWP} - P_{WP}$ is gradually increased, menisci poised at local constrictions within the pore space (pore throats) become unstable and the invading phase enters adjoining pore bodies in a non-equilibrium piston-type manner (i.e., Haines jumps).[29] Since the defending phase (air) is always able to maintain continuity via corners, crevices, and roughness features of the pore space, air does not become trapped,[8] contrary to the assertions of Sinha and Wang.[19] Chapuis et al.[32] have recently discussed the added complexity of meniscus movements in highly porous, fibrous media of intermediate wettability. In any case, invasion proceeds via a connected path of pore bodies and pore throats. The process is characterized by formation of a sample-spanning cluster of water-invaded pores at the so-called percolation threshold and breakthrough of water at the outlet face of the sample. Clusters of water-invaded pores with no exit to the outlet face are also formed. In samples of infinite size, such dead-end (non-percolating) clusters constitute a negligible fraction of the pore volume. However, in finite size samples, the volume fraction of dead-end clusters can be substantial and affect the percolation threshold. Such effects should be expected to be significant in GDMs which are very thin.

Although water is effectively the non-wetting phase during injection into a dry GDM, spontaneous ejection of liquid from water-saturated GDMs has not been observed. Rather, water must be forced out and therefore is the preferentially wetting phase according to Dullien's operational definition of wettability. According to convention, displacement of a wetting phase by a non-wetting one is classified as a drainage process. Yet, withdrawal of water from GDMs exhibits features consistent with forced imbibition (i.e., lack of finite-size effects and existence of trapped water). Meniscus instabilities controlling imbibition can be much more complex than piston-type invasion of pore throats during drainage.[33–38] Such instabilities include piston-type retraction from pore bodies and snap-off in pore throats, as well as co-operative pore filling by air. Snap-off of liquid threads in pore throats and co-operative pore filling mechanisms involve anticlastic interfaces, i.e., saddle-shaped interfaces with one positive and one negative radius of curvature. At such interfaces, H can assume negative values regardless of θ.[8] Snap-off in pore throats is not favored in porous materials characterized by similar pore body and pore throat sizes and intermediate wettability[39] and should not be expected to play a role in GDMs. Instead, co-operative pore filling is expected to be the dominant mechanism of imbibition in porous materials possessing pore bodies and pore throats of similar size, and narrow pore size distribution, such as GDMs. Importantly, trapping of the defending phase (water) is possible during imbibition, as observed during water withdrawal from GDMs.[8,40]

3. MEASUREMENT OF CAPILLARY PRESSURE CURVES

The most widely accepted method for measuring capillary pressure curves is mercury intrusion porosimetry (MIP). This method, however, is not suitable for GDMs for several reasons. Firstly, since mercury is highly non-wetting to both the graphite substrate and the PTFE coating in GDMs, it is insensitive to changes in the chemical heterogeneity of the solid surfaces, which is precisely what is of greatest interest. Secondly, conversion of mercury intrusion pressure data to an equivalent air–water pressure relies on the validity of Eq. (3) and requires knowledge of the contact angles of both mercury and water on GDM surfaces. Even if a single contact angle can be determined for mercury in the GDM, the water contact angles are not the same for the graphite and PTFE surfaces, making this conversion practically impossible.

These problems can be avoided by measuring capillary pressure curves of GDMs directly with water as the working fluid. A number of techniques have been recently proposed. Of the different types of GDM (carbon paper, carbon cloth, metal foam/mesh, etc.), only carbon paper GDMS have been studied to any appreciable extent using these methods. An ideal capillary pressure measurement technique must be able to vary water saturation in directions of increasing saturation from $S_W = 0$ to $S_W \rightarrow 1$ and decreasing saturation from $S_W = 1$ to $S_W \rightarrow 0$. This requires the application

Table 1
Comparison of Air–Water Capillary Pressure Measurement Techniques.

Method	$-P_C \to 0$	$0 \to -P_C$	$0 \to +P_C$	$+P_C \to 0$	*Variable*
Gostick et al.[16] (2005)		√			S_W
Fairweather et al.[40] (2007)	√	√	√	√	S_W
Rensink et al.[51] (2008)			√		P_C
Sole and Ellis[55] (2008)	√		√		S_W
Gallagher et al.[50] (2008)	√	√			P_C
Koido et al.[58] (2008)			√	√	P_C
Nguyen et al.[57] (2008)			√	√	P_C
Gostick et al.[8] (2008)	√	√	√	√	P_C
Harkness et al.[56] (2009)	√	√	√	√	S_W

of both positive and negative capillary pressures when conventionally defined as $P_C = P_L - P_G$. Table 1 presents a summary of each available technique, listing their abilities and limitations according to the above requirements. A comparison of $P_C(S_W)$ data obtained by different techniques is given in Fig. 2. Figure 2a shows data for untreated Toray TGP-H or similar material and Fig. 2b shows data for PTFE-treated Toray TGP-H or similar material. Clearly, the discrepancies between the data obtained by the various methods are large. The following discussion gives a survey and critical analysis of the water–air capillary pressure measurement techniques reported in the literature. This analysis helps explain the differences shown in Fig. 2 and should aid in the selection of appropriate capillary pressure data and measurement methods.

The method of standard porosimetry (MSP)[41–43] was one of the first approaches used to obtain air–water capillary pressure curves for GDMs.[16] In this test, a GDM sample is initially saturated with water and contacted with a water-saturated porous disc, which is a standard with known $P_C(S_W)$ behavior. The capillary pressure of the sample-standard system is varied by allowing the liquid to evaporate from the standard and sample while in contact. If the two media can be assumed to be in capillary equilibrium, their capillary pressures are equal. Saturation is determined by measuring the weights of the sample and standard periodically. The capillary pressure of the system is found by reference to the known capillary pressure curve of the standard. This method is limited to scanning

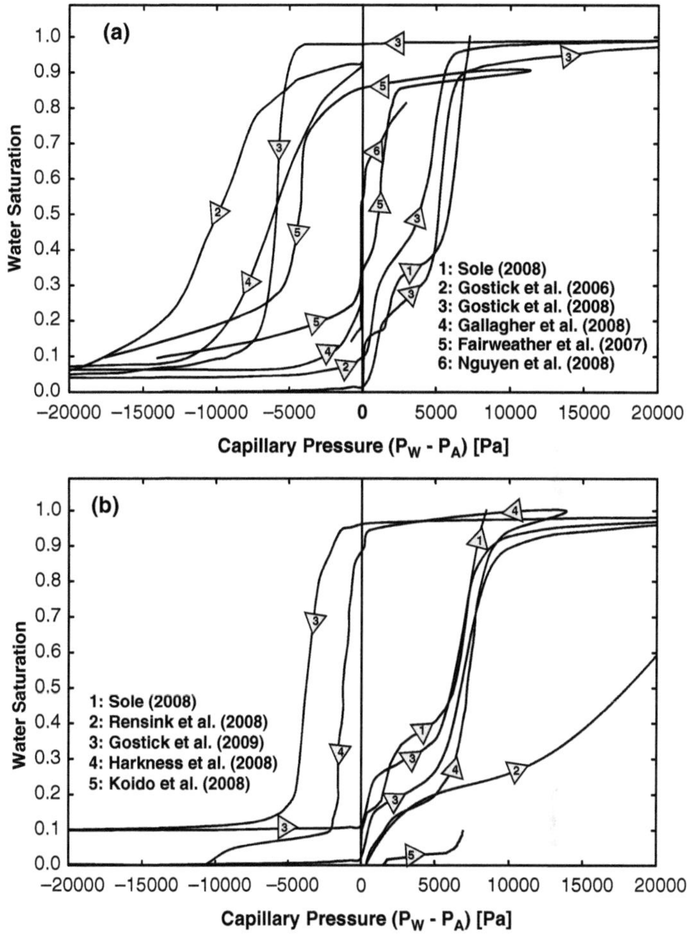

Figure 2. Comparison of air–water capillary pressure curves available in the literature for Toray TGP-H-090 or similar material without PTFE coating (a) and with PTFE coating (b). Triangles indicate direction of saturation change.

only in the direction of decreasing water saturation and $P_C < 0$. The early results of Gostick et al.[16] showed that negative capillary pressures were required to withdraw water from GDMs, an observation attributed to the existence of a network of hydrophilic pores from which displacement of water by air required $P_G > P_L$.

A fundamental assumption behind MSP is that drying and forced displacement by air are equivalent means of reducing water saturation, but this is not necessarily true. For this to be the case, liquid must be able to redistribute throughout the media to maintain capillary equilibrium among all water-filled pores. It has been shown, however, that such redistribution requires that thin films of water exist on the walls, corners, and crevices of air-invaded pores.[44] This is unlikely in GDMs because of the weak affinity of water for either graphite or PTFE surfaces, although it would be the case for evaporation of a strongly wetting fluid such as octane. Instead, disconnection of water is likely to take place as water evaporates during an MSP experiment. Disconnected water can also evaporate, but is not at capillary equilibrium with water in the standard, and this introduces an error of unknown magnitude into the measurement. The MSP data obtained from untreated Toray TGP-H-090 in Fig. 2a clearly show water withdrawal occurring at anomalously large negative capillary pressures. This is expected if large amounts of disconnected water exist, since they would make the saturation higher than the amount of water in capillary equilibrium with the standard. This technique has been utilized by Mench and co-workers to study a wide variety of GDMs with varying hydrophobic polymer contents,[45] under different compressions[46] and over a range of temperatures.[47] They have further attempted to synthesize a single relationship that can describe the capillary properties of any GDM in terms of any combination of the above parameters (temperature, compression, hydrophobic polymer loading),[48] to generate a correlation specific to GDMs in place of Eq. (2). Volfkovich et al.[49] have also employed MSP to study the effect of hydrophobic polymer addition. Capillary pressure curves obtained by MSP, however, are at odds with data obtained by other methods. The MSP method with water as the working fluid is not appropriate for intermediate wettability systems because the extent of water disconnection during evaporation cannot be ascertained.

Gallagher et al.[50] have reported a method that is somewhat similar to MSP. As with MSP, the GDM is initially saturated with water and placed on a porous plate that is also water-saturated. Instead of changing the GDM saturation by drying, as in MSP, the capillary pressure is controlled directly by applying suction to the porous plate. Capillary equilibrium is established between the GDM and the plate as water flows from the GDM into the plate and vice versa.

The GDM saturation is determined by weighing after equilibration at each applied pressure. Since pressure is controlled directly, it is possible to scan along both increasing and decreasing water saturation paths, although this method is limited to $P_C < 0$. Figure 2a shows that the water withdrawal and secondary injection data obtained by Gallagher et al.[50] using this approach are in reasonable agreement with that of Fairweather et al.[40] and Gostick et al.[8] who used different methods (discussed later).

Rensink et al.[51] have developed a technique that involves submerging a sample in water and weighing the water uptake. Since water does not spontaneously imbibe into a GDM, it is forced into the sample in discrete amounts by reducing the gas pressure inside the sample to a value of $P_G < P_{ATM}$, followed by submerging the sample under water and returning the gas pressure above the water to P_{ATM}. This has the effect of forcing water into the sample at a pressure of $P_C = P_{ATM} - P_G$. A capillary pressure curve can be obtained by repeated measurements at successively lower gas pressures. This method only scans in the direction of increasing water saturation and only for $P_C > 0$. A main drawback of this method appears to be that no allowance is made for the compression of gas from P_G to a higher pressure as water invades. This means that the capillary pressure associated with each value of water saturation is overestimated by an amount which increases with water saturation. Figure 2b shows the data presented by Rensink et al.[51] for water injection into treated Toray TGP-H-060. These data agree well with data from other groups for $S_W < 0.2$, but increasingly diverge from the general trend at higher water saturations.

All of the techniques mentioned above determine water saturation by weighing the sample. An alternative approach is to control water saturation directly by injecting and withdrawing water using a syringe pump, while monitoring capillary pressure. This requires a sample holder that allows air to escape as water invades, without allowing water to leave the sample, as shown in Fig. 3. In this arrangement, a porous hydrophobic membrane is used as a capillary barrier to prevent the injected water from escaping out the top of the sample. A porous hydrophilic membrane is located below the sample to prevent air from exiting at the bottom of the sample during water withdrawal. Several variations of this approach are described below.

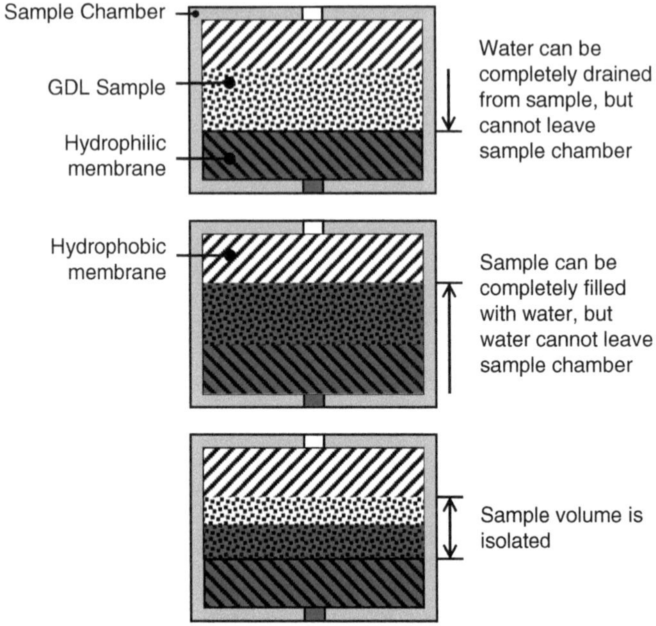

Figure 3. Schematic diagram of a sample holder using capillary barriers for water injection and withdrawal experiments. Liquid enters the sample from the bottom while air escapes from the top.

Fairweather et al.[40] used a syringe pump to inject discrete volumes of liquid into a sample while measuring liquid pressure. In this method, the gas phase is always at atmospheric pressure. Their technique allows the water saturation to be scanned in both increasing and decreasing directions. Since liquid pressure is directly measured, both $P_C > 0$ and $P_C < 0$ are observed. This technique has certain similarities to rate-controlled porosimetry,[52,53] except that water is not injected continuously. Figure 4 (left) shows a schematic diagram of the raw data obtained by this method. During injection, the water pressure rises to a peak pressure P_1, followed by a decay to pressure P_2 while injection is paused. The observed pressure decay is caused by redistribution of water throughout the pore volume accessible at P_1, which is not entirely filled by the available liquid, since otherwise no liquid pressure decay would be observed.[29]

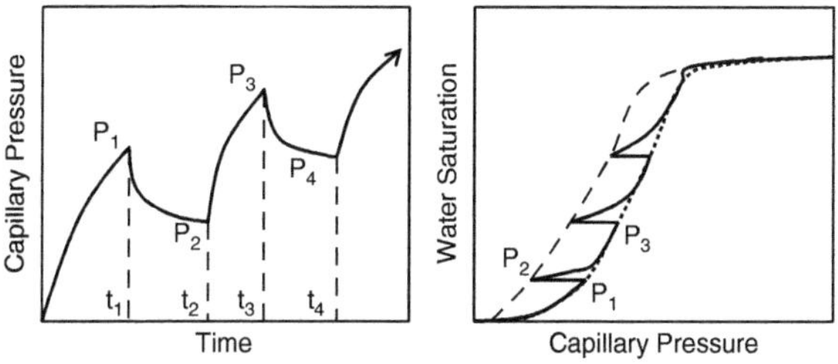

Figure 4. Left: Schematic of pressure profile reported by Fairweather et al.[40] Right: Capillary pressure curve obtained from pressure profile showing pressure decay at constant saturation. Dotted line shows envelope of peak pressures corresponding to pressure-controlled injection[29,53,54] while dashed line shows envelope of decayed pressures corresponding to the data of Fairweather et al.[40]

Figure 4 (right) shows the raw data presented as a capillary pressure curve. The periods of decaying pressure appear as horizontal steps since saturation is unchanged. Fairweather et al.[40] have associated the water saturation reached after each injection event to the decayed capillary pressure, yielding the dashed line in Fig. 4 (right). The dependence of capillary pressure on the volume of liquid injected has been previously studied.[29,53,54] These studies have shown that the envelope of *peak* pressures realized during rate-controlled porosimetry must be used (indicated by the dotted line in Fig. 4 (right)) in order for rate-controlled and pressure-controlled capillary pressure data to be equivalent. Figure 2a shows the water injection and withdrawal data presented by Fairweather et al.[40] for untreated Toray TGP-H-090. The use of the decayed pressure envelope results in a shift of both injection and withdrawal curves toward zero capillary pressure.

Sole and Ellis[55] and Harkness et al.[56] both have used a syringe pump to perform rate-controlled capillary pressure experiments. Unlike Fairweather et al.,[40] water is injected into the GDM at a constant rate while the pressure response is monitored. The water saturation is found from the injection rate and the elapsed time. Both increasing and decreasing water saturation paths can be probed, corresponding to a wide range of both positive and negative capillary pressures. Sole and Ellis[55] did not employ a hydrophilic barrier below the GDM and so could not study water withdrawal. One issue with continuous rate injection is that capillary equilibrium is never truly established in the sample since the fluid interfaces are continuously altered by the addition of fluid. Harkness et al.[56] investigated the effect of water injection rate and found it to have no effect on the results, suggesting that pseudo-equilibrium conditions are achieved with this approach. The data of Harkness et al.[56] display some of the features shown in Fig. 4 (right). However, since injection was continuous, the pressure quickly recovered from such drops and the experiment generally followed the peak pressure envelope.

A practical drawback of rate-controlled experiments is the inability to scan along plateaus in saturation observed during internal scanning loops, as well as the inability to make measurements under conditions of maximum saturation and residual saturation. This is due to the positive displacement of incompressible liquid water by the syringe pump. Maximum saturation must be avoided during injection because liquid pressure will increase abruptly once the sample is filled and the hydrophobic membrane will be breached. Similarly, once full residual water saturation is reached, the pressure will drop sharply, risking breakthrough of the hydrophilic membrane or cavitation and degassing in the water phase.

As an alternative to the rate-controlled methods described above, Gostick et al.[8] have described a method that uses a similar sample holder (Fig. 3) with hydrophobic and hydrophilic capillary barriers, but is pressure-controlled. In this method, the capillary pressure is controlled by adjusting the gas pressure, while water pressure remains atmospheric at all times. Saturation is tracked by monitoring the water uptake from a reservoir on an analytical balance. This method allows scanning of water saturation in both increasing and decreasing directions and over a wide range of positive and negative capillary pressures. The main advantages of this method are the ability to scan saturation plateaus and the assurance that all accessible pore volume is filled at each capillary pressure step. Gostick et al.[8] have also described a unique sample mounting procedure that allows precise control over the starting conditions of the experiment.

Another pressure-controlled method[36] has been adopted by Van Nguyen et al.[57] In this method, capillary pressure is controlled by adjusting the hydrostatic pressure between the sample and a horizontal-graduated tube of known bore diameter. As the static pressure is altered, the liquid saturation in the GDM can be monitored by tracking the movement of a meniscus in the graduated tube. Although their setup included a hydrophobic capillary barrier, no hydrophilic membrane was used. Nonetheless, van Nguyen et al.[57] have managed to perform water withdrawal by carefully reducing the liquid pressure while avoiding gas breakthrough. The capillary loops obtained did not show any hysteresis, which is unexpected and in disagreement with all other results.[8,40,56] Moreover, van Nguyen et al.[57] report anomalously high water saturation values at very low applied capillary pressures. Koido et al.[58] have used a similar approach, but employed neither a hydrophilic nor a hydrophobic barrier, and therefore only reported data for water injection up to the breakthrough point.

Of the proposed methods for measuring the capillary properties of air–water–GDM systems, only the methods of Gostick et al.[8] and Harkness et al.[56] possess all of the desired attributes (see Table 1) and produce accurate results. The results of a rate-controlled approach,[56] however, are less straightforward to interpret than the results of a pressure-controlled approach.[8] Rate-controlled approaches are difficult to implement for conditions near $S_W = 0$ and $S_W = 1$, whereas no such difficulty is encountered with pressure-controlled methods. The rate-controlled approaches have the ability to scan arbitrarily high capillary pressures, limited only by the breakthrough pressure of the hydrophobic membrane. In the negative capillary pressure range, rate controlled approaches are limited by the cavitation of water as liquid pressure is reduced. The pressure-controlled approach described by Gostick et al.[8] has the opposite limitations. The maximum capillary pressure is limited by the vapor pressure of water, while arbitrarily large negative capillary pressures can be scanned.

4. INTERPRETATION OF CAPILLARY PRESSURE CURVES

4.1. Capillary Pressure Hysteresis

It has been widely demonstrated that injection of water into both PTFE-treated and untreated GDMs requires positive capillary pressure while water withdrawal requires negative capillary pressures, where capillary pressure is defined as $P_C = P_L - P_G$. Various explanations for this behavior have been put forward, including contact angle hysteresis,[59] the converging–diverging nature of void space in fibrous media,[56] and cooperative pore filling during water withdrawal.[8] Figure 5 depicts each of the three mechanisms. Each of these effects has the ability to alter the sign of the mean curvature

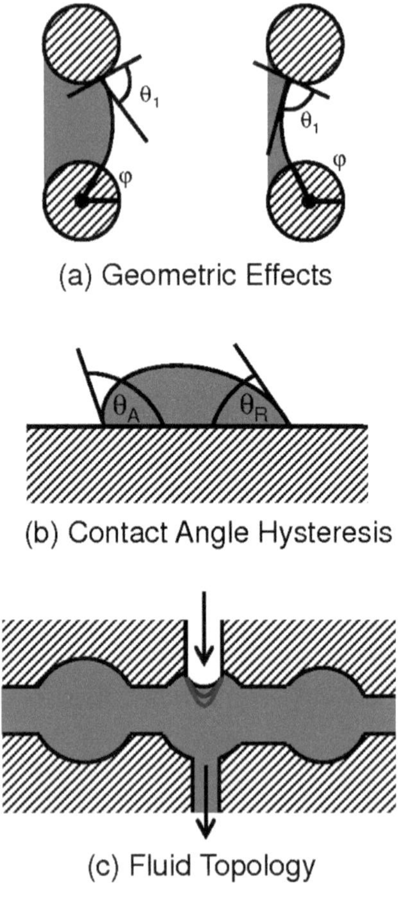

Figure 5. Fluid–solid interfacial configurations contributing to capillary pressure hysteresis. (a) Converging and diverging geometry causing mean curvature sign change. (b) Advancing and receding contact angles differing on rough surfaces. (c) Fluid topology leading to anticlastic (saddle-shaped) interfaces of negative mean curvature.

H, particularly when contact angles are in the range of intermediate wettability. Consequently, each mechanism should be expected to play a role in GDM capillary behavior. For instance, the attribution of capillary pressure hysteresis entirely to hysteresis of the contact angle in the sense of Eq. (3)[59] can be correct only if GDMs are composed of bundles of tubes of cylindrical cross section.[60]

A previously unrecognized manifestation of the hysteresis effect in GDMs can be seen in the experimental results of Benziger et al.[20] In their simple experiment, the height of a water column above a GDM sample was increased until water breakthrough occurred, after which the liquid head slowly decreased as water flowed through the GDM. It was observed that the flow rate of water through the sample decreased linearly with the height of water column, clearly indicating that the relative permeability and therefore the water saturation of the GDM remained constant at all positive capillary pressures ($P_L \geq P_G$). This is in agreement with direct capillary pressure measurements which clearly demonstrate negligible reduction in water saturation until negative capillary pressures are reached.[8,40,56] To further demonstrate this behavior, Fig. 6 shows internal withdrawal loops (i.e., withdrawal of water from $S_W < 1$)

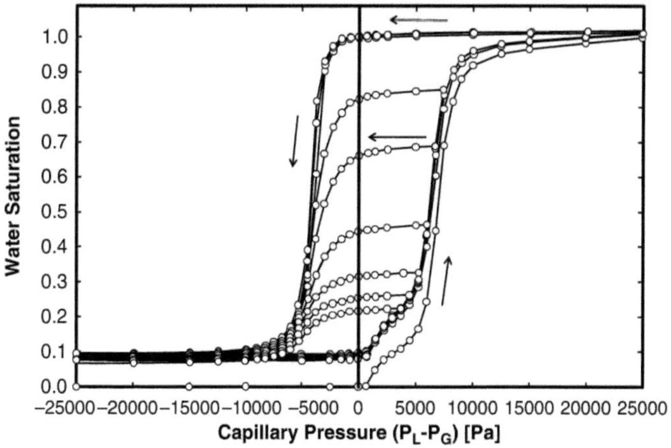

Figure 6. Air–water capillary pressure curves for treated Toray TGP-H-120 (10 wt%) showing internal withdrawal scanning loops obtained using the method of Gostick et al.[8]

obtained for a sample with PTFE treatment. Clearly, a negligible amount of water is removed until $P_C < 0$ for water withdrawal from any initial value of S_W. Similar observations have been made by Harkness et al.[56]

The observed hysteresis in GDMs has significant implications for fuel cell operation since it implies that water, once in the pores of the GDM, cannot be removed by normal processes operating in standard fuel cells. A number of innovative strategies for water removal have been devised which not surprisingly involve application of suction ($P_L < P_G$) to the liquid in the GDM.[61,62] Aside from such complex approaches, the only other means of dewatering a GDM in situ is by drying and evaporation. Since humidity cycles are damaging to the membrane and remedial action on an operating fuel cell may not be feasible, a preferred approach would be to prevent liquid water from accumulating on the GDM in the first place. It should be noted that capillary pressure hysteresis has not been considered in any PEMFC model published to date.

4.2. Effect of Hydrophobic Coating

Gostick et al.[21] performed a systematic study of the effect of PTFE on GDM capillary behavior materials treated by the standard PTFE emulsion technique.[63] Interestingly, although it was found that PTFE addition did increase hydrophobicity in terms of increased water injection pressures, no difference was seen between different amounts of PTFE (i.e., 10 wt% versus 20 wt%). This suggests that standard PTFE application techniques may not be fully effective and that increasing PTFE loading leads to thicker coatings rather than rendering more portions of the available surface hydrophobic as desired. Gallagher et al.[50] also evaluated treated and untreated GDMs, but they applied a treatment to render them hydrophilic, as evidenced by the amount of water imbibed at $P_C < 0$. This effect clearly demonstrates that native, untreated GDMs are intermediately wet, rather than hydrophilic. Harkness et al.[56] studied the effect of different types of coatings on GDM capillary properties. In order to quantify the wettability of the various coatings, contact angle measurements were made on coated flat surfaces. It was found that water injection pressures ranked according to the independently determined contact angles. Gostick et al.[21] assessed the effect of

PTFE-treatment on GDM wettability in terms of the US Bureau of Mines (USBM) wettability index.[26] The USBM wettability index compares the amount of work required for water injection to that for water withdrawal and is highly sensitive for intermediately wet systems. This approach proved to be particularly sensitive to slight changes in wettability caused by PTFE addition.

The effect of the hydrophobic coating on the capillary behavior of GDMs is not as pronounced as might be expected, but is still significant. Figure 7 shows a direct comparison of water–air capillary pressure curves for Toray TGP-H-120 with and without PTFE treatment.[21] The primary water injection curve for the treated sample is shifted to slightly higher capillary pressures, indicating that the addition of PTFE did indeed shift the wettability toward that of a more hydrophobic state. A more striking difference between the treated and untreated samples is evident when comparing water saturation at a given capillary pressure. For instance, at $P_C = 5000$ Pa during primary injection, the saturation in the untreated sample is twice that of the treated sample. The difference is a factor of four during the secondary injection. It is not clear whether the primary

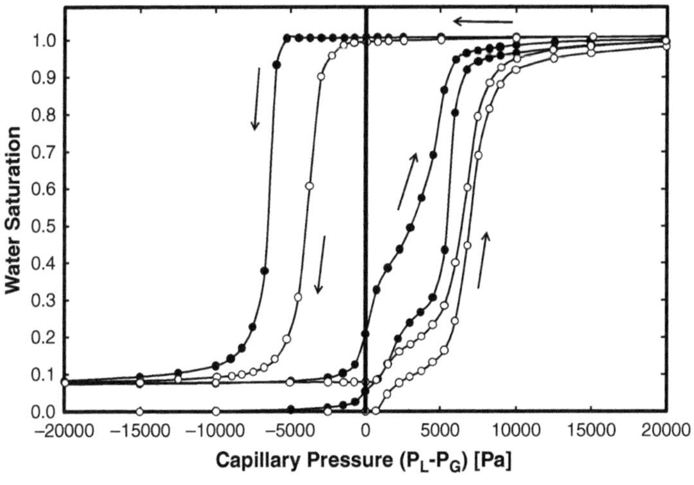

Figure 7. Air–water capillary pressure curves for untreated Toray TGP-H-120 (black markers) and treated Toray TGP-H-120 (10 wt%) (white markers) reported by Gostick et al.[21]

or secondary injection curve would be operative in a functioning fuel cell. Presumably, the GDM experiences periods of high and low saturation, but saturation reductions likely occur via evaporation and it is not known if evaporation restores primary injection behavior. This matter deserves further study. Another notable feature of the PTFE-treated GDM is the absence of spontaneous water imbibition (i.e., water uptake at $P_C < 0$) on both the primary and secondary injection curves, while the untreated material shows significant water uptake at $P_C < 0$. It has been observed that untreated GDMs can exhibit spontaneous imbibition of water to some extent during secondary injection, unlike PTFE-treated samples.[21] This general behavior may explain why untreated GDMs do not function well in fuel cells.

A final feature of interest in Fig. 7 is the absence of any spontaneous water ejection at $P_C > 0$ during the water withdrawal portion. Such a feature might be naively expected on grounds that water would be expelled from small PTFE-coated pores once the capillary pressure was sufficiently reduced. Early results[16] suggested that GDMs did indeed expel some water after being removed from a water bath to atmospheric pressure (i.e., $P_C = 0$), but this water loss was likely due to handling of the sample prior to weighing. Similar behavior was reported by Gallagher et al.[50] using a method which also involves removing excess surface water prior to weighing. The misinterpretation that some water was spontaneously expelled from PTFE-treated GDMs, combined with the fact that negative capillary pressures were required to remove the remaining water, has given support to the idea of mixed or fractional wettability for PTFE-treated GDMs. The idea that GDMs containing a mixture of hydrophobic PTFE-coated pores that expel water and hydrophilic graphite pores that wick water[17–19,64] is, however, refuted by the recent capillary pressure measurements discussed in this chapter.

4.3. Effect of Compression

Fuel cell assembly requires that considerable compression be applied to the stack to ensure tight gas seals and good electrical conductivity at the GDM–bipolar plate interface. This compression leads to reduced porosity, smaller average pore size, and possibly other microstructure changes in the GDM. Since capillary properties are influenced by pore size, it is expected that the capillary behavior of GDMs should differ depending on whether or not they are compressed. Harkness et al.[56] have devised a sample holder capable of applying a known compressive force to the GDM during testing. Gostick et al.[21] also applied compression during assembly and inferred the compressed thickness by the observed reduction in pore volume. Both approaches show compression to have similar effects on the capillary behavior of GDMs.

Figure 8 shows water–air capillary pressure curves measured on compressed and uncompressed samples of treated Toray TGP-H-120 (10 wt% PTFE). The pressure for water injection into the compressed samples is increased considerably as expected if pore sizes are reduced. This behavior is probably undesirable in fuel cell operation since the GDM is compressed almost exclusively under the rib, meaning that capillary forces might promote preferential accumulation of water in the uncompressed region over the channel. Such a localized water distribution might create much larger oxygen transport restrictions than expected based on the globally averaged water saturation alone. On the other hand, since heat is removed from the current collector ribs, areas of the GDM under the rib are likely colder, meaning that there is potential for increased

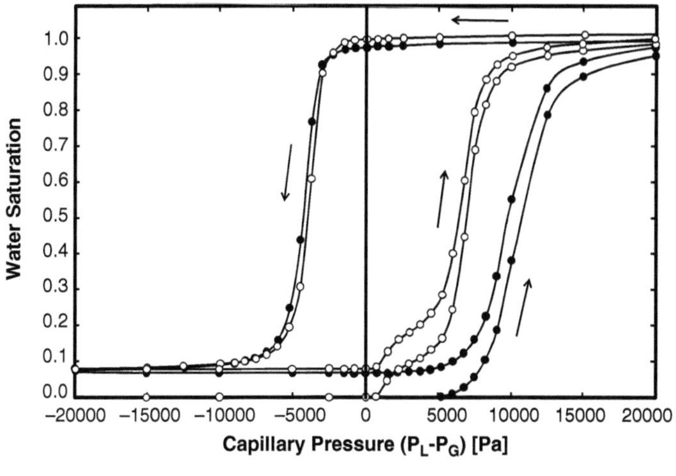

Figure 8. Air–water capillary pressure curves for treated Toray TGP-H-120 (10 wt%) under compression (black markers) and uncompressed (white markers).[21]

condensation and accumulation of liquid water there. In this scenario, GDM compression under the current collector ribs might be beneficial because water accumulation is likely to be limited by the onset of liquid flow from regions of high capillary pressure (under the rib) to regions of low capillary pressure (under the channel).

4.4. Water Breakthrough Condition

Water produced in the cathode catalyst layer must travel through the GDM to leave the cell. Water transport can occur by diffusion as a gas or by capillary flow of liquid through the pore space. In the latter case, it is of interest to determine the pressure necessary to establish a continuous percolating path through the GDM, as well as the resulting saturation. The saturation at breakthrough represents the minimum amount of liquid water that will exist in the GDM, which in turn places an upper limit on the effective diffusivity of oxygen. Surprisingly few attempts have been made to measure these quantities. Benziger et al.[20] have reported a simple experiment to measure breakthrough pressure, but saturation could only be roughly determined by weighing the sample after disassembly of the sample holder. This approach has been used by several others.[65,66] Bazylak et al.[67] injected water into the GDM using a syringe pump while monitoring the pressure in the water phase, but did not report the GDM saturation. Gostick et al.[68] have reported a method for simultaneously determining the pressure and saturation at breakthrough using an adaptation of their capillary pressure technique.[8] Breakthrough saturations in excess of 25% were found, even for PTFE-treated GDMs. Van Nguyen et al.[57] estimated breakthrough points by performing capillary pressure measurements without a hydrophobic barrier and visually determining breakthrough. The reported breakthrough pressures are in good agreement with other values,[20,68] but the saturations at breakthrough are unreasonably high. Buchi et al.[69] have recently reported X-ray micro-tomography images of water injection into GDMs. A rather high breakthrough saturation of 60% was observed, in part because the sample tested was very thin (~200 μm) and in part because the sample was very small (<1.8 mm). In such a small sample, even a single invading water cluster occupies a disproportionate fraction of the pore space – a fraction that would have been smaller had a larger sample area been used. Sample thinness has a similar effect on water saturation, but is not an experimental artifact, rather an intrinsic feature of GDMs, as explained next.

4.5. Finite-Size Effects

One of the defining features of GDMs as a porous media is their extreme thinness. With GDM thicknesses ranging from 200 to 400 μm and pore sizes ranging from 20 to 40 μm, it follows that only 10–20 pores span the thickness dimension. This leads to significant finite-size effects. Specifically, dead-end (non-percolating) clusters of water-filled pores formed during water injection in the GDM occupy a significant volume fraction of the pore space. Although these clusters do not contribute to liquid water flow, they would hinder gas access to the catalyst layer. Gostick et al.[68] have recently reported an experimental investigation of finite-size effects in GDMs. By placing an impermeable mask with a small hole between a GDM and the water source, the water saturation in the GDM at breakthrough was reduced to low values ($\sim$5%). In the absence of this impermeable mask, the water saturation in the GDM at breakthrough was in excess of 25%. Not only does this demonstrate that significant volume is occupied by dead-end clusters of water-invaded pores, it also suggests how the application of a microporous layer (MPL) to the GDM might improve PEMFC performance at high current density.

The effect of GDM thickness on the capillary pressure curves has also been investigated.[21] A pronounced shoulder was observed at low capillary pressures in the curves obtained for very thin samples. This feature was diminished for thicker samples and virtually non-existent in the capillary pressure curves obtained from multilayer stacks. The dependence of this feature on the sample surface area-to-volume ratio is a clear demonstration of finite-size effects.[70]

4.6. Effect of Microporous Layer

The addition of an MPL to one side of a GDM is known to improve cell performance.[71] The MPL consists of a mixture of carbon black and PTFE powders, which is applied as a thin coating to one side of the GDM, and subsequently sintered.[63] The result is a layer that has very small (<1 μm) pores. An unequivocal explanation for the actual function of the MPL is still lacking.[68,71–76]

Measurements of the capillary pressure characteristics of MPL-coated GDMs have been studied by Kumbur et al.[45–47] and Gostick et al.[16] using the MSP technique. As discussed in Section 3, the MSP method begins with an initially water saturated sample. Saturation of a sample with water involves evacuation, followed by immersion in water and, finally, establishment of ambient pressure conditions with the sample still submerged. This effectively forces water into the sample at a capillary pressure of 1 atm. In the presence of an MPL, complete saturation is not achieved by this procedure, demonstrating the hydrophobicity of the MPL. Using an alternative method, Gostick et al.[68] have studied the capillary behavior of MPL-coated GDMs. When the MPL faced the water outlet, the GDM substrate was nearly completely filled with water before the MPL was breached. When the MPL faced the water inlet, the MPL and GDM were breached simultaneously and very low GDM saturations were established. In either case, incomplete saturation was observed at the highest value of capillary pressure reached during the experiments (ca. 50 kPa) because water did not invade the MPL.

The capillary pressure and breakthrough behavior of MPL-coated GDMs suggest an explanation for the efficacy of MPLs in fuel cell operation.[68] If liquid water is generated at the catalyst layer, then it must first percolate through the MPL before it reaches the GDM. Liquid breaks through the MPL at a few isolated locations on the MPL-GDM interface in the same way that water droplets emerge from the GDM at the GDM–gas channel interface.[77] Water percolation through the GDM subsequently proceeds from few point-like sources. Many surface pores are thus inaccessible to water. In view of the preceding discussion regarding finite-size effects, it is clear that the number of dead-end (non-percolating) clusters of water-invaded pores (and therefore the GDM water saturation at breakthrough) is dramatically reduced by the presence of a MPL.

5. CONCLUSION AND OUTLOOK

Great progress has been made towards understanding the air–water capillary properties of GDMs as a result of recent efforts by a number of researchers. Originally motivated by the need to provide continuum PEMFC models with $P_C(S_W)$ constitutive relationships appropriate for GDMs, these studies have provided a wealth of information about the physics of immiscible displacement in fibrous media. Specifically, compelling evidence has been found against the widely held view that GDMs possess mixed wettability with regards to the air–water fluid pair. Instead, it has been demonstrated that GDMs will not spontaneously imbibe either water or air. Regardless of the PTFE treatment, GDMs possess all of the characteristics of intermediate wettability systems. The significance of this realization is that the use of the Washburn equation to interpret and model interfacial curvature along paths of increasing or decreasing saturation is a gross simplification and should be used with caution. Rather, interfacial curvatures controlling water injection and withdrawal depend non-trivially on pore structure, contact angle, and saturation history. Capillary pressure hysteresis is ubiquitously manifested in GDMs. The complexity of interfacial configurations controlling water injection and withdrawal in fibrous media of intermediate wettability presents a challenge to analytical and numerical approaches alike. Recent advances, however, make possible direct measurement of the $P_C(S_W)$ relationships for GDMs of varying thickness, PTFE content, and state of compression. Of the experimental techniques now available, some are reliable and accurate and should be particularly useful in the development and characterization of GDMs with novel structural and wetting properties. Thus far, only carbon paper GDMs have been extensively studied. Other types (carbon cloth, metal foam/mesh) need to be investigated.

In view of recent experimental air–water capillary pressure data which refute Eq. (2) (see Fig. 1), a re-examination of at least some of the results of continuum multiphase PEMFC models is in order. A reliable Leverett-type correlation appropriate for GDMs has not been established. Such a correlation is desirable, but inclusion of contact angle in Eq. (1) is not justified, potentially limiting the applicability of such a correlation. The detailed capillary pressure information now available for different GDM materials highlights the lack of equivalent information for other porous components of the fuel cell electrode, such as the catalyst layer and the microporous layer. Extending existing techniques to study these components will be a challenge due to their very low pore volume and small pore size. Rate-controlled methods are better suited to this task. A further challenge is the evaluation of capillary properties of GDMs at the higher temperatures encountered in fuel cell operation. Perhaps more importantly, correlations between fuel cell performance and GDM capillary characteristics need to be explored.

REFERENCES

[1] Z.H. Wang, C.Y. Wang, K.S. Chen, J. Power Sources **94**, 40 (2001)
[2] C.Y. Wang, Z.H. Wang, Y. Pan, ASME HTD **364**, 351 (1999)
[3] W. He, J.S. Yi, T.V. Nguyen, AlChE J. **46**, 2053 (2000)
[4] M.C. Leverett, AIME – Petro. Dev. Tech. **142**, 152 (1941)
[5] W. Rose, W.A. Bruce, Trans AIME **186**, 127 (1949)
[6] F.A.L. Dullien, *Porous Media: Fluid Transport and Pore Structure* (Academic Press, New York, 1992)
[7] K.S. Udell, Int. J. Heat Mass Transfer **28**, 485 (1985)
[8] J.T. Gostick, M.A. Ioannidis, M.W. Fowler, M.D. Pritzker, Electrochem. Commun. **10**, 1520 (2008)
[9] J.T. Gostick, M.A. Ioannidis, M.W. Fowler, M.D. Pritzker, J. Power Sources **173**, 277 (2007)
[10] D. Mattia, H.H. Ban, Y. Gogotsi, Langmuir **22**, 1789 (2006)
[11] A. Yan, X. Xiao, I. Kulaots, B.W. Sheldon, R.H. Hurt, Carbon **44**, 3116 (2006)
[12] A.W. Adamson, A.P. Gast, *Physical Chemistry of Surfaces* (Wiley, New York, 1997)
[13] N.R. Morrow, J. Can. Pet. Technol. **15**, 49 (1976)
[14] S. Goswami, S. Klaus, J. Benziger, Langmuir **24**, 8627 (2008)
[15] A.Z. Weber, R.M. Darling, J. Newman, J. Electrochem. Soc. **151**, A1715 (2004)
[16] J.T. Gostick, M.W. Fowler, M.A. Ioannidis, M.D. Pritzker, Y.M. Volfkovich, A. Sakars, J. Power Sources **156**, 375 (2006)
[17] R.P. Ramasamy, E.C. Kumbur, M.M. Mench, W. Liu, D. Moore, M. Murthy, Int. J. Hydrogen Energy **33**, 3351 (2008)
[18] S. Park, B.N. Popov, Electrochim. Acta **54**, 3473 (2009)
[19] P.K. Sinha, C.Y. Wang, Chem. Eng. Sci. **63**, 1081 (2008)
[20] J. Benziger, J. Nehlsen, D. Blackwell, T. Brennan, J. Itescu, J. Membrane Sci. **261**, 98 (2005)
[21] J.T. Gostick, M.A. Ioannidis, M.W. Fowler, M.D. Pritzker, J. Power Sources **194**, 433 (2009)
[22] A. Defay, I. Prigogine, A. Bellemans, *Surface Tension and Adsorption* (Wiley, New York, 1966)
[23] R. Finn, *Equilibrium Capillary Surfaces* (Springer, New York, 1986)
[24] J.C. Melrose, Ind. Eng. Chem. **60**, 53 (1968)
[25] W.G. Anderson, J. Petro. Tech. **39**, 1283 (1987)
[26] W.G. Anderson, J. Petro. Tech. **38**, 1246 (1986)
[27] N.J. Shirtcliffe, G. McHale, M.I. Newton, F.B. Pyatt, S.H. Doerr, Appl. Phys. Lett. **89**, 9 (2006)
[28] J.C. Melrose, SPE J **5**, 259 (1965)
[29] N.R. Morrow, Ind. Eng. Chem. **62**, 32 (1970)
[30] D. Wilkinson, J.F. Willemsen, J Phys A-Math Gen **16**, 3365 (1983)
[31] R. Lenormand, E. Touboul, C. Zarcone, J. Fluid Mech. **189**, 165 (1988)

[32] O. Chapuis, M. Prat, M. Quintard, E. Chane-Kane, O. Guillot, N. Mayer, J. Power Sources **178**, 258 (2008)
[33] M.A. Ioannidis, I. Chatzis, J. Colloid Interface Sci. **161**, 278 (1993)
[34] R. Lenormand, C. Zarcone, A. Sarr, J. Fluid Mech. **135**, 337 (1983)
[35] Y. Li, N.C. Wardlaw, J. Colloid Interface Sci. **109**, 473 (1986)
[36] M.A. Ioannidis, I. Chatzis, A.C. Payatakes, J. Colloid Interface Sci. **143**, 22 (1991)
[37] C.D. Tsakiroglou, A.C. Payatakes, Adv. Colloid Interface Sci. **75**, 215 (1998)
[38] V. Joekar, S.M. Hassanizadeh, L.J. Pyrak-Nolte, C. Berensten, Water Resour. Res. **45**, W02430 (2009)
[39] Y. Li, N.C. Wardlaw, J. Colloid Interface Sci. **109**, 461 (1986)
[40] J.D. Fairweather, P. Cheung, J. St Pierre, D.T. Schwartz, Electrochem. Commun. **9**, 2340 (2007)
[41] Y.M. Volfkovich, V.S. Bagotzky, J. Power Sources **48**, 327 (1994)
[42] Y.M. Volfkovich, V.S. Bagotzky, J. Power Sources **48**, 339 (1994)
[43] Y.M. Volfkovich, V.S. Bagotzky, V.E. Sosenkin, I.A. Blinov, Colloids Surf. A **187–188**, 349 (2001)
[44] I.N. Tsimpanogiannis, Y.C. Yortsos, S. Poulou, N. Kanellopoulos, A.K. Stubos, Phys. Rev. E **59**, 4353 (1999)
[45] E.C. Kumbur, K.V. Sharp, M.M. Mench, J. Electrochem. Soc. **154**, B1295 (2007)
[46] E.C. Kumbur, K.V. Sharp, M.M. Mench, J. Electrochem. Soc. **154**, B1305 (2007)
[47] E.C. Kumbur, K.V. Sharp, M.M. Mench, J. Electrochem. Soc. **154**, B1315 (2007)
[48] E.C. Kumbur, K.V. Sharp, M.M. Mench, J. Power Sources **176**, 191 (2008)
[49] Y.M. Volfkovich, V.E. Sosenkin, N.F. Nikol'skaya, T.L. Kulova, Russ. J. Electrochem. **44**, 278 (2008)
[50] K.G. Gallagher, R.M. Darling, T.W. Patterson, M.L. Perry, J. Electrochem. Soc. **155**, B1225 (2008)
[51] D. Rensink, S. Fell, J. Roth, in *Proceedings of the Sixth International Conference on Nanochannels, Microchannels and Minichannels*, Darmstadt, Germany, 23–25 June 2008, pp. 1–7
[52] H.H. Yuan, B.F. Swanson, SPE Formation Evaluation **4**, 17 (1989)
[53] P.G. Toledo, L.E. Scriven, H.T. Davis, SPE Formation Evaluation **9**, 46 (1994)
[54] M.A. Knackstedt, A.P. Sheppard, W.V. Pinczewski, Phys. Rev. E **58**, R6923 (1998)
[55] J. Sole, M.W. Ellis, in *Proceedings of the 6th International Conference on Fuel Cell Science, Engineering and Technology*, Denver, CO, 16–18 June 2008, pp. 829–840
[56] I.R. Harkness, N. Hussain, L. Smith, J.D.B. Sharman, J. Power Sources **193**, 122 (2009)
[57] T.V. Nguyen, G. Lin, H. Ohn, X. Wang, Electrochem. Solid-State Lett. **11**, B127 (2008)
[58] T. Koido, T. Furusawa, K. Moriyama, J. Power Sources **175**, 127 (2008)
[59] P. Cheung, J.D. Fairweather, D.T. Schwartz, J. Power Sources **187**, 487 (2009)
[60] M. Gladkikh, S. Bryant, J. Colloid Interface Sci. **288**, 526 (2005)
[61] A.Z. Weber, R.M. Darling, J. Power Sources **168**, 191 (2007)
[62] S. Litster, C.R. Buie, T. Fabian, J.K. Eaton, J.G. Santiago, J. Electrochem. Soc. **154**, B1049 (2007)
[63] M.F. Mathias, J. Roth, J. Fleming, W. Lehnert, in *Handbook of Fuel Cells – Fundamentals, Technology and Applications*, Vol. 3, Part 1, ed. by W. Vielstich, H.A. Gasteiger, A. Lamm (Wily, New York, 2003), pp. 517–537
[64] A.Z. Weber, M.A. Hickner, Electrochim. Acta **53**, 7668 (2008)

[65] D. Spernjak, A.K. Prasad, S.G. Advani, J. Power Sources **170**, 334 (2007)
[66] E. Kimball, T. Whitaker, Y.G. Kevrekidis, J.B. Benziger, AlChE J. **54**, 1313 (2008)
[67] A. Bazylak, D. Sinton, N. Djilali, J. Power Sources **176**, 240 (2008)
[68] J.T. Gostick, M.A. Ioannidis, M.W. Fowler, M.D. Pritzker, Electrochem. Commun. **11**, 576 (2009)
[69] F.N. Buchi, R. Fluckiger, D. Tehlar, F. Marnoe, M. Stampanoni, ECS Transactions **16**, 587 (2008)
[70] R.G. Larson, N.R. Morrow, Powder Technol. **30**, 123 (1981)
[71] H.K. Atiyeh, K. Karan, B. Peppley, A. Phoenix, E. Halliop, J. Pharoah, J. Power Sources **170**, 111 (2007)
[72] J.H. Nam, M. Kaviany, Int. J. Heat Mass Transfer **46**, 4595 (2003)
[73] U. Pasaogullari, C.Y. Wang, Electrochim. Acta **49**, 4359 (2004)
[74] U. Pasaogullari, C.Y. Wang, K.S. Chen, J. Electrochem. Soc. **152**, A1574 (2005)
[75] A.Z. Weber, J. Newman, J. Electrochem. Soc. **152**, A677 (2005)
[76] J.H. Nam, K.-J. Lee, G.-S. Hwang, C.-J. Kim, M. Kaviany, Int. J. Heat Mass Transfer **52**, 2779 (2009)
[77] X.G. Yang, F.Y. Zhang, A.L. Lubawy, C.Y. Wang, Electrochem. Solid-State Lett. **7**, A408 (2004)

8

Mesoscopic Modeling of Two-Phase Transport in Polymer Electrolyte Fuel Cells

Partha P. Mukherjee[1] and Chao-Yang Wang[2]

[1]Oak Ridge National Laboratory, One Bethel Valley Road, P.O. Box 2008, MS 6164, Oak Ridge, TN 37831-6164, USA
[2]Electrochemical Engine Center (ECEC), and Department of Mechanical and Nuclear Engineering, The Pennsylvania State University, University Park, PA 16802, USA

1. INTRODUCTION

Fuel cells, owing to their high energy efficiency, environmental friendliness and low noise, are widely considered as the twenty-first century energy-conversion devices for mobile, stationary and portable power. Among the different types of fuel cells, the polymer electrolyte fuel cell (PEFC) has emerged as a promising power source for a wide range of applications.

A typical PEFC, shown schematically in Fig. 1, consists of the anode and cathode compartments, separated by a proton conducting polymeric membrane. The anode and cathode sides each comprises of gas channel, gas diffusion layer (GDL) and catalyst layer (CL). Despite tremendous recent progress in enhancing the overall cell performance, a pivotal performance/durability limitation in PEFCs centers on liquid water transport and resulting flooding in the constituent components.[1,2] Liquid water blocks the porous pathways in the CL and GDL thus causing hindered oxygen transport to the

U. Pasaogullari, C.-Y. Wang (eds.), *Modeling and Diagnostics of Polymer Electrolyte Fuel Cells*, Modern Aspects of Electrochemistry 49,
DOI 10.1007/978-0-387-98068-3_8,

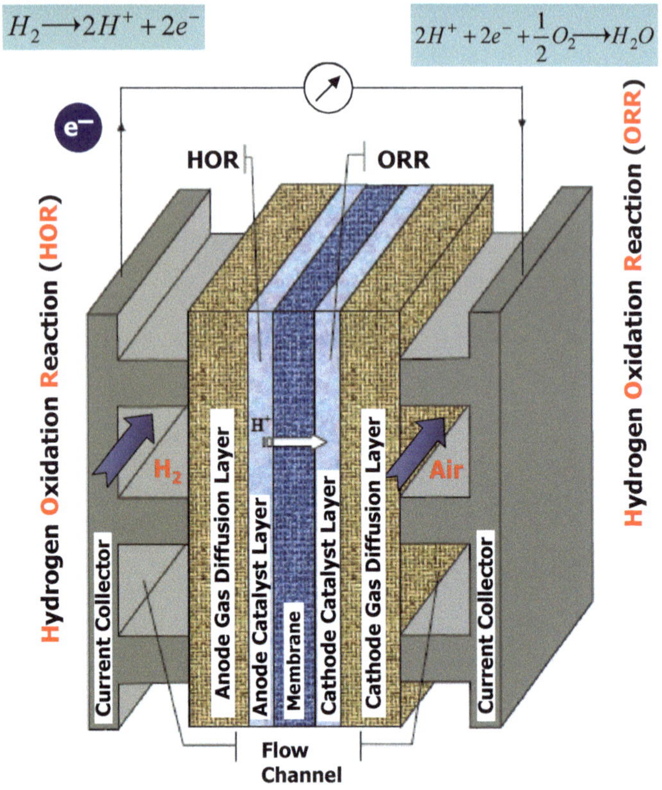

Figure 1. Schematic diagram of a polymer electrolyte fuel cell.

reaction sites as well as covers the electrochemically active sites in the CL thereby increasing surface overpotential. This phenomenon is known as "flooding" and is perceived as the primary mechanism leading to the limiting current behavior in the cell performance which is illustrated by the typical polarization curve in Fig. 2. The catalyst layer and gas diffusion layer, therefore, play a crucial role in the PEFC water management[1,2] aimed at maintaining a delicate balance between reactant transport from the gas channels and water removal from the electrochemically active sites.

Recent years have witnessed an explosion of interest in water management research, evidenced by the development of several

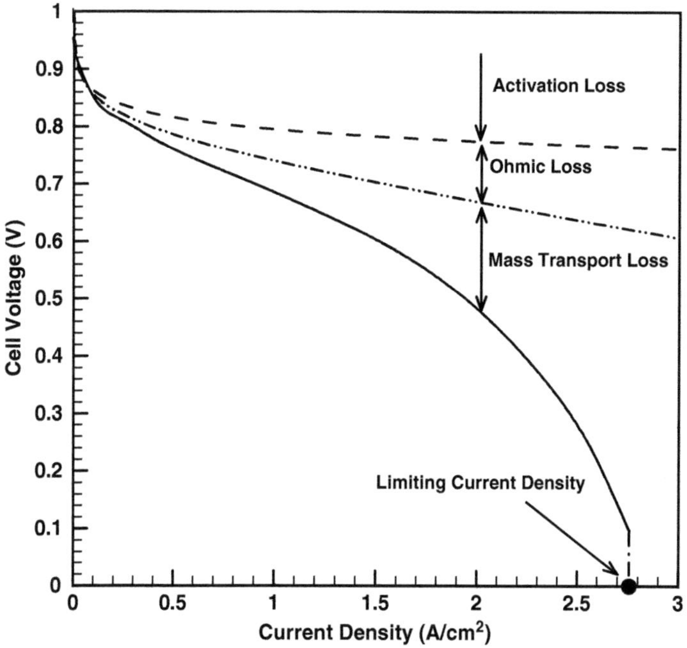

Figure 2. Typical polarization curve of a PEFC.

macroscopic computational models for liquid water transport in PEFCs.[3–17] Comprehensive overview of the various PEFC models is furnished by Wang[2] and Weber and Newman.[18] The macroscopic models, reported in the literature, are based on the theory of volume averaging and treat the catalyst layer and gas diffusion layer as macrohomogeneous porous layers. Due to the macroscopic nature, the current models fail to resolve the influence of the structural morphology of the CL and GDL on the underlying two-phase dynamics. Additionally, there is serious scarcity of two-phase correlations, namely capillary pressure and relative permeability as functions of liquid water saturation, as constitutive closure relations for the two-phase PEFC models tailored specifically for the CL and GDL. Current two-phase fuel cell models often employ a capillary pressure – saturation relation for modeling liquid water transport in hydrophobic gas diffusion media adapted by Pasaogullari and

Wang[4] and Nam and Kaviany[14] from Udell's work[19] in the form of Leverett-J function.[20] Very recently few attempts to experimentally evaluate the capillary pressure for the PEFC GDL have emerged in the literature,[21–24] while the CL still remains experimentally intractable. Furthermore, hardly any experimental efforts to measure relative permeability for both GDL and CL, more important for two-phase transport than capillary pressure, have been reported. In addition, in the macroscopic two-phase fuel cell models, the site coverage effect in the CL and the pore blockage effect in the CL and GDL owing to liquid water are taken into account through an electrochemically active area reduction model and the Bruggeman type correction for the effective oxygen diffusivity, respectively. These two empirical correlations cannot be separately discerned through experimental techniques. Despite substantial research, both theoretical and experimental, there is serious paucity of fundamental understanding regarding the overall structure-transport-performance interactions and underlying two-phase dynamics in the catalyst layer and the gas diffusion layer.

In this chapter, the development of a mesoscopic modeling formalism is presented in order to gain fundamental insight into the structure-wettability influence on the underlying liquid water transport and interfacial dynamics in the PEFC CL and GDL.

2. MODEL DESCRIPTION

The mesoscopic modeling approach consists of a stochastic reconstruction method for the generation of the CL and GDL microstructures, and a two-phase lattice Boltzmann method for studying liquid water transport and flooding phenomena in the reconstructed microstructures.

2.1. Stochastic Microstructure Reconstruction Model

Detailed description of a porous microstructure is an essential prerequisite for unveiling the influence of pore morphology on the underlying two-phase behavior. This can be achieved either by 3-D volume imaging or by constructing a digital microstructure based on stochastic reconstruction models. Non-invasive techniques, such as X-ray micro-tomography, are the popular methods for 3-D imaging of pore structure. Additionally, 3-D porous structure can be generated using stochastic simulation technique, which creates 3-D replicas of the random microstructure based on specified statistical information obtained from high-resolution 2-D micrographs of a porous sample. The low cost and high speed of data generation put forth the stochastic generation methods as the preferred choice over the experimental imaging techniques. In this work, CL and GDL microstructures are generated through the development of stochastic reconstruction techniques.

2.1.1. Catalyst Layer Structure Generation

For the electrochemical reaction to occur, the state-of-the-art catalyst layer of a PEFC is a three-phase composite comprising of: (1) ionomer, i.e. the ionic phase which is typically Nafion® to provide a passage for protons to be transported in or out, (2) Pt catalysts supported on carbon i.e. the electronic phase for electron conduction, and (3) pores for the reactant to be transferred in and product water out. In the present study, the catalyst layer is delineated as a two-phase (pore/solid) structure consisting of the gas phase (i.e. the void space) and a mixed electrolyte/electronic phase (i.e. the solid matrix). The assumption of the mixed electrolyte/electronic phase is well justified from the perspective of ion transport in the electrolyte phase as the limiting mechansim as compared to the electron conduction via the electronic (C/Pt) phase within the CL and henceforth is referred to as the "electrolyte" phase.[25–27] Figure 3 shows a high resolution transmission electron microscope (TEM) image of a conventional PEFC electrode[28] with the three-phase interface. This figure further illustrates a pseudo dual phase (pore/solid) scenario with the ionomer film inpregnated C/Pt as an effective solid phase and the adjacent pore space leading toward an electrochemically active interface.

The stochastic reconstruction method is based on the idea that an arbitrarily complex porous structure can be described by a binary phase function which assumes a value 0 in the pore space and 1 in the solid matrix.[29] The intrinsic randomness of the phase function can be adequately qualified by the low order statistical moments, namely porosity and two-point autocorrelation function.[29] The porosity is the probability that a voxel is in the pore space. The two-point autocorrelation function is the probability that two

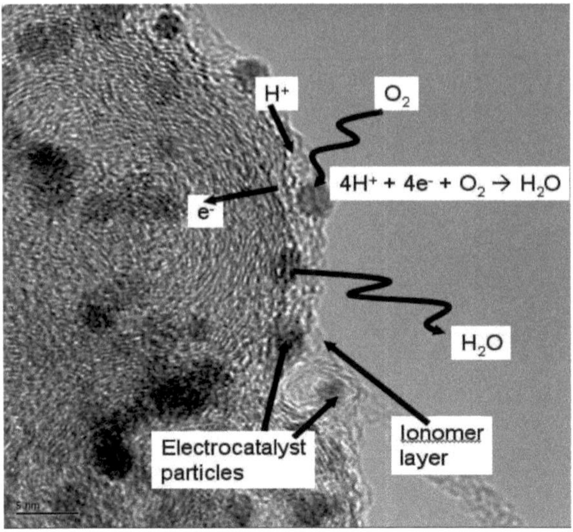

Figure 3. High resolution TEM image of a PEFC catalyst layer microstructure. Reproduced from Gostick et al.[21] with permission from Karren L. More, Oak Ridge National Laboratory, USA.

voxels at a specific distance are both in the pore space.[29] Details about the CL microstructure reconstruction along with the underlying assumptions are elaborated in our recent work,[25] which is based on the stochastic generation method originally reported by Adler et al.[30] and Bentz and Martys.[31] In brief, the stochastic reconstruction technique starts with a Gaussian distribution which is filtered with the two-point autocorrelation function and finally thresholded with the porosity, which creates the 3-D realization of the CL structure. The autocorrelation function is computed from a 2-D TEM image of an actual CL.[25] The porosity can be calculated by converting the mass loading data of the constituent components available from the CL fabrication process.[26] For example, the porosity of the CL can be evaluated using the following relation:[26]

$$\varepsilon_{CL} = 1 - \left[\frac{1}{\rho_{Pt}} + \frac{R_{C/Pt}}{\rho_c} + \frac{R_{C/Pt} \cdot R_{I/C}}{\rho_{\mathrm{Nafion}}} \right] \frac{L_{Pt}}{\Delta X_{CL}} \tag{1}$$

where ρ_{Nafion}, ρ_{Pt}, and ρ_c are the density of Nafion®, Pt and carbon, respectively. $R_{I/C}$ is the weight ratio of ionomer to carbon, $R_{C/Pt}$ is the weight ratio of carbon to Pt in 40% Pt/C catalyst,

L_{Pt} is the Pt loading and ΔX_{CL} is the CL thickness. Now, using Eq. (1), with $\Delta X_{CL} = 10\ \mu m$, $L_{Pt} = 0.35$ mgPt/cm^2, $R_{C/Pt} = 1.5$, $R_{I/C} = 0.42$, $\rho_{Nafion} = \rho_C = 2$ g/cc and $\rho_{Pt} = 21.5$ g/cc, the nominal porosity of the CL miscrostructure can be evaluated as $\varepsilon_{CL} \sim 60\%$.

The pore/solid phase is further distinguished as "transport" and "dead" phase. The basic idea is that a pore phase unit cell surrounded by solid phase-only cells does not take part in species transport and hence in the electrochemical reaction and can, therefore, be treated as a "dead" pore and similarly for the electrolyte phase.[25] The interface between the "transport" pore and the "transport" electrolyte phases is referred to as the electrochemically active area (ECA) and the ratio of ECA and the nominal CL cross-sectional area provides the "ECA-ratio". It is be noted that in this chapter, ECA is normalized with the apparent electrode area and therefore differs from the definition in terms of the electrochemically active area per Pt loading reported elsewhere in the literature.

Figure 4 shows the reconstructed microstructure of a typical catalyst coated membrane (CCM) CL with nominal porosity of 60% and thickness of 10 μm along with the input TEM image and the evaluated cross-section averaged pore and electrolyte phase volume fraction distributions across the CL thickness. The cross-section averaged pore/electrolyte volume fraction distribution illustrates the local tortuosity variation along the electrode thickness. Several 3-D structures are generated with varying number of unit cells. However, the reconstructed microstructure with the electrochemically active interfacial area ratio complying closely with the experimentally measured value using cyclic voltammerty (CV) is selected for the subsequent simulations. The reconstructed, 3-D microstructure with 100 elements in the thickness direction and 50 elements each in the span-wise directions produces an active interfacial area ratio of around 46, which matches reasonably well with the measured CV value (~45) and thereby reproduces the most important structural parameter responsible for electrochemical activity of the CL.

2.1.2. Gas Diffusion Layer Structure Generation

The multi-faceted functionality of a GDL includes reactant distribution, liquid water transport, electron transport, heat conduction and mechanical support to the membrane-electrode-assembly.

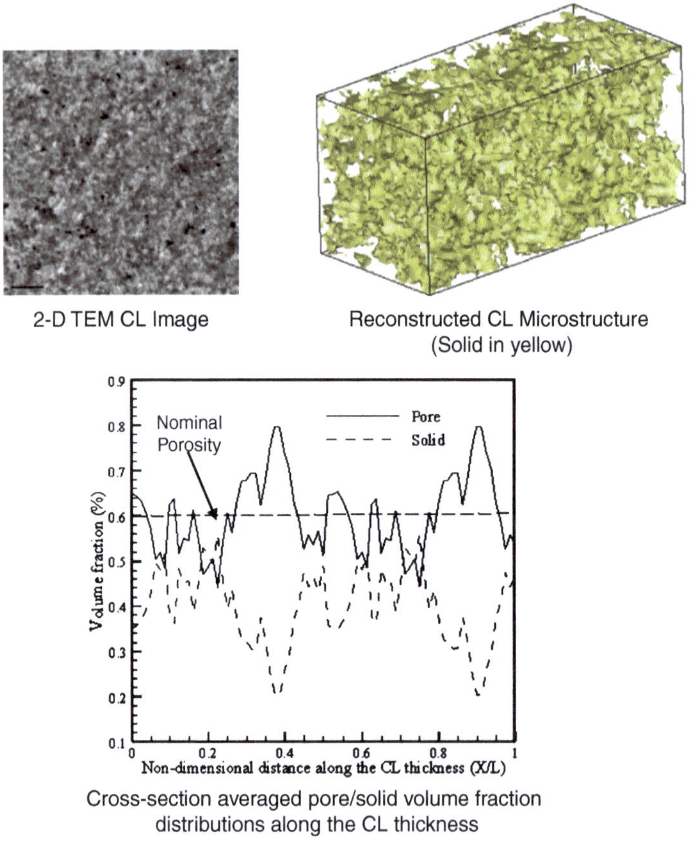

Figure 4. Reconstructed catalyst layer microstructure along with pore and electrolyte phase volume fractions distribution.

Carbon-fiber based porous materials, namely non-woven carbon paper and woven carbon cloth, shown in Fig. 5, have received wide acceptance as materials of choice for the PEFC GDL owing to high porosity (~70% or higher) and good electrical/thermal conductivity. Mathias et al.[32] provided a comprehensive overview of the GDL structure and functions. In this work, the reconstruction of non-woven carbon paper GDL is presented.

The stochastic simulation technique creates 3-D realization of the non-woven carbon paper GDL based on structural inputs,

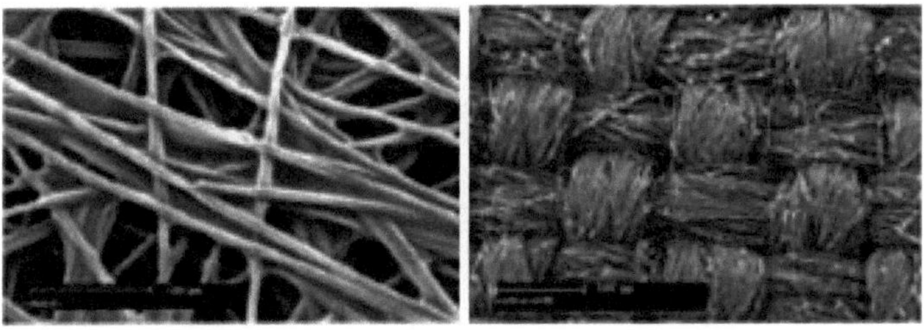

Carbon Paper Carbon Cloth

Figure 5. SEM micrographs of carbon paper and carbon cloth GDL.

namely fiber diameter, fiber orientation and porosity which can be obtained either directly from the fabrication specifications or indirectly from the SEM (scanning electron microscope) micrographs or by experimental techniques. Details about the carbon paper GDL microstructure reconstruction along with the underlying assumptions are elaborated in our recent work[33] which is based on the non-woven structure generation technique originally proposed by Schladitz et al.[34] The specific assumptions made in the reconstruction method, which can however be well justified by inspecting the corresponding carbon paper GDL SEM micrographs, include:[33] (1) the fibers are long compared to the sample size and their crimp is negligible; (2) the interaction between the fibers can be neglected, i.e., the fibers are allowed to overlap; and (3) the fiber system, owing to the fabrication process, is macroscopically homogeneous and isotropic in the material plane, defined as the *xy* plane, which implies that the distribution properties of the stochastic model are invariant with respect to translations as well as rotations about the *z* axis. With these assumptions, the stochastic reconstruction technique can be adequately described as a Poisson line process with one-parametric directional distribution where the fibers are realized as circular cylinders with a given diameter and the directional distribution provieds in-plane/through-plane anisotropy in the reconstructed GDL microstructure.[33]

Figure 6 shows the reconstructed microstructure of a typical non-woven, carbon paper GDL with porosity around 72% and thickness of 180 μm along with the structural parameters in terms of

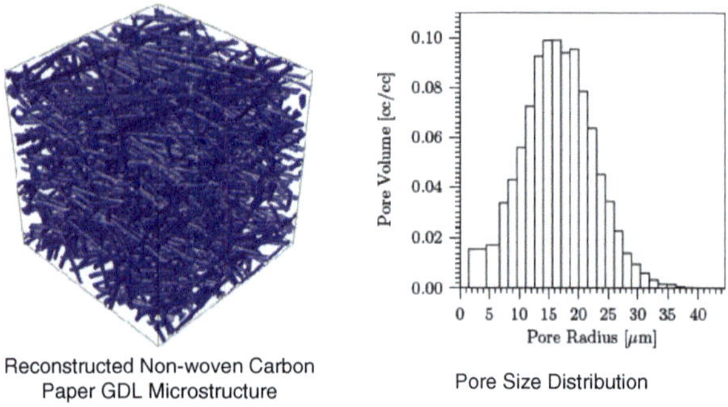

Permeability	In-plane	Through-plane	In-plane/Through-plane
Reconstructed GDL	30.9 darcy	21.1 darcy	1.46
Measured data	33 darcy	18 darcy	1.83

Figure 6. Reconstructed non-woven carbon paper GDL microstructure along with the evaluated structural properties.

the estimated pore size distribution (PSD) and the anisotropy in the in-plane vs. through-plane permeability values.

2.2. Lattice Boltzmann Model

In recent years, the lattice Boltzmann (LB) method, owing to its excellent numerical stability and constitutive versatility, has developed into a powerful technique for simulating fluid flows and is particularly successful in applications involving interfacial dynamics and complex geometries.[35] The LB method is a first-principle based numerical approach. Unlike the conventional Navier-Stokes solvers based on the discretization of the macroscopic continuum equations, lattice Boltzmann methods consider flows to be composed of a collection of pseudo-particles residing on the nodes of an underlying lattice structure which interact according to a velocity distribution function. The lattice Boltzmann method is also an ideal scale-bridging numerical scheme which incorporates simplified kinetic models to capture microscopic or mesoscopic flow physics and yet the macroscopic averaged quantities satisfy the desired macroscopic equations, e.g. Navier-Stokes equation.[35] Due to its underlying kinetic nature, the LB method has been found to be particularly useful in applications involving interfacial dynamics and complex boundaries, e.g. multiphase or multicomponent flows, and flow in porous medium. As opposed to the front-tracking and front-capturing multiphase models in traditional CFD (computational fluid dynamics) approach, due to its kinetic nature, the LB model incorporates phase segregation and surface tension in multiphase flow through interparticle force/interactions, which are difficult to implement in traditional methods. While the LB modeling approach better represents the pore morphology in terms of a realistic digital realization of the actual porous medium and incorporates rigorous physical description of the flow processes, it is computationally very demanding. However, because of the inherently parallel nature of the LB algorithm, it is also very efficient running on massively parallel computers.[35]

Several LB models have been presented in the literature to study multiphase/multicomponent flows. Gunstensen et al.[36] developed a multicomponent LB model based on a two-component lattice gas model. Shan and Chen[37] proposed a LB model with interparticle potential for multiphase and multicomponent fluid flows. Swift et al.[38] developed a LB multiphase and multicomponent model by using the free-energy approach. He et al.[39] proposed an LB multiphase model using the kinetic equation for multiphase flow. Among the afore-mentioned multiphase LB models, the interaction potential based approach is widely used due to its simplicity in implementing boundary conditions in complex porous structures and remarkable versatility in terms of handling fluid phases with different densities, viscosities and wettabilities, as well as the capability of incorporating different equations of state. In this work, we have developed the interaction potential based two-phase LB model to study the structure-wettability influence on the underlying two-phase dynamics in the CL and GDL of a PEFC.

2.2.1. Two-phase LB Model Description

The interaction potential model, originally proposed by Shan and Chen,[37] and henceforth referred to as the S-C model, introduces k distribution functions for a fluid mixture comprising of k components. Each distribution function represents a fluid component and satisfies the evolution equation. The non-local interaction between particles at neighboring lattice sites is included in the kinetics through a set of potentials. The LB equation for the kth component can be written as:

$$f_i^k(\mathbf{x} + \mathbf{e}_i\delta_t, t + \delta_t) - f_i^k(\mathbf{x}, t) = -\frac{f_i^k(\mathbf{x}, t) - f_i^{k(eq)}(\mathbf{x}, t)}{\tau_k} \qquad (2)$$

where $f_i^k(\mathbf{x}, t)$ is the number density distribution function for the kth component in the ith velocity direction at position $\mathbf{x}$ and time t, and δ_t is the time increment. In the term on the right-hand side, τ_k is the relaxation time of the kth component in lattice unit, and $f_i^{k(eq)}(\mathbf{x}, t)$ is the corresponding equilibrium distribution function. The right-hand-side of Eq. (2) represents the collision term, which is simplified to the equilibrium distribution function $f_i^{k(eq)}(\mathbf{x}, t)$ by the so-called BGK (Bhatnagar-Gross-Krook), or the single-time relaxation approximation.[40] The spatio-temporal discrete form of the LB evolution equation based on the BGK approximation, given by Eq. (2), is often referred to as the LBGK equation. For a three-dimensional 19-speed lattice (D3Q19, where D is the dimension and Q is the number of velocity directions), the schematic of which is shown in Fig. 7 with the velocity directions, the equilibrium distribution function, $f_i^{k(eq)}(\mathbf{x}, t)$, assumes the following form:[41,42]

$$\left.\begin{aligned}
f_0^{k(eq)} &= \beta_k n_k - \frac{1}{2} n_k \mathbf{u}_k^{eq} \cdot \mathbf{u}_k^{eq}, \\
f_i^{k(eq)} &= \frac{1-\beta_k}{12} n_k + \frac{1}{6} n_k (\mathbf{e}_i \cdot \mathbf{u}_k^{eq}) + \frac{1}{4} n_k (\mathbf{e}_i \cdot \mathbf{u}_k^{eq})^2 - \frac{1}{12} n_k \mathbf{u}_k^{eq} \cdot \mathbf{u}_k^{eq} \\
&\quad \text{for } i = 1, \ldots, 6, \\
f_i^{k(eq)} &= \frac{1-\beta_k}{24} n_k + \frac{1}{12} n_k (\mathbf{e}_i \cdot \mathbf{u}_k^{eq}) + \frac{1}{8} n_k (\mathbf{e}_i \cdot \mathbf{u}_k^{eq})^2 - \frac{1}{24} n_k \mathbf{u}_k^{eq} \cdot \mathbf{u}_k^{eq} \\
&\quad \text{for } i = 7, \ldots, 18.
\end{aligned}\right\} \qquad (3)$$

In the above equations, the discrete velocities, $\mathbf{e}_i$, are given by:

$$\mathbf{e}_i = \begin{cases} (0,0,0), & i = 0, \\ (\pm 1,0,0), (0,\pm 1,0), (0,0,\pm 1) & i = 1-6, \\ (\pm 1,\pm 1,0), (\pm 1,0,\pm 1), (0,\pm 1,\pm 1) & i = 7-18. \end{cases} \qquad (4)$$

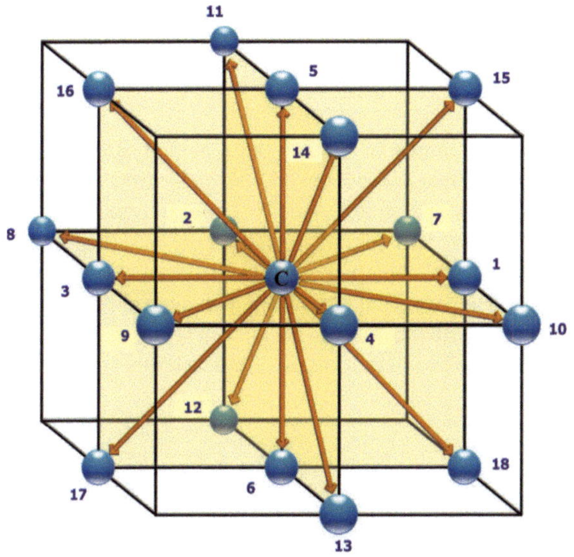

Figure 7. Schematic of D3Q19 lattice structure.

The free parameter, β_k, relates to the speed of sound of a region of pure kth component, as:

$$(c_s^k)^2 = \frac{1-\beta_k}{2} \tag{5}$$

The number density of the kth component, n_k, is defined as:

$$n_k = \sum_i f_i^k \tag{6}$$

The mass density of the kth component is defined as:

$$\rho_k = m_k n_k = m_k \sum_i f_i^k \tag{7}$$

The fluid velocity of the kth fluid is defined through:

$$\rho_k \mathbf{u}_k = m_k \sum_i \mathbf{e}_i f_i^k \tag{8}$$

where m_k is the molecular mass of the kth component.

The equilibrium velocity, u_k^{eq}, is determined by the expression:

$$\rho_k \mathbf{u}_k^{eq} = \rho_k \mathbf{u}' + \tau_k \mathbf{F}_k \tag{9}$$

where, $\mathbf{u}'$ is a common or base velocity, which can be perceived in the context of single component, single phase flow, on top of which an extra component-specific velocity due to interparticle interaction is added for each component. In this particle-based approach, interparticle interaction, which manifests in terms of surface tension, wall adhesion and external gravity force, is realized through the total force, $\mathbf{F}_{\mathrm{k}}$, acting on the kth component, including fluid/fluid interaction $\mathbf{F}_{1k}$, fluid/solid interaction, $\mathbf{F}_{2k}$, and external force $\mathbf{F}_{3k}$[42] and is expressed as:

$$\mathbf{F}_k = \mathbf{F}_{1k} + \mathbf{F}_{2k} + \mathbf{F}_{3k} \tag{10}$$

The conservation of momentum at each collision in the absence of interaction force (i.e. in the case of $\mathbf{F}_k = 0$), requires $\mathbf{u}'$ to satisfy the following relation,

$$\mathbf{u}' = \left(\sum_{k=1}^{s} \frac{\rho_k \mathbf{u}_k}{\tau_k} \right) \Big/ \left(\sum_{k=1}^{s} \frac{\rho_k}{\tau_k} \right) \tag{11}$$

A simple long-range interaction force between particles of the kth component at site $\mathbf{x}$ and the $\bar{k}$th component at site $\mathbf{x}'$ is introduced and the total fluid/fluid interaction force on the kth component at site $\mathbf{x}$ is given by:

$$\mathbf{F}_{1k}(\mathbf{x}) = -\psi_k(\mathbf{x}) \sum_{\mathbf{x}'} \sum_{\bar{k}=1}^{s} G_{k\bar{k}}(\mathbf{x}, \mathbf{x}') \psi_{\bar{k}}(\mathbf{x}')(\mathbf{x}' - \mathbf{x}) \tag{12}$$

where $G_{k\bar{k}}(\mathbf{x}, \mathbf{x}')$ is Green's function and satisfies $G_{k\bar{k}}(\mathbf{x}, \mathbf{x}') = G_{k\bar{k}}(\mathbf{x}', \mathbf{x})$. It reflects the intensity of interparticle interaction. $\psi_{k\bar{k}}(\mathbf{x})$ is called the "effective number density" and is defined as a function of $\mathbf{x}$ through its dependency on the local number density, $\psi_k = \psi_k(n_k)$. In the D3Q19 lattice model, the interaction potential couples nearest and next-nearest neighbors and the Green's function is given by:

$$G_{k\bar{k}}(\mathbf{x},\mathbf{x}') = \begin{cases} g_{k\bar{k}}, & |\mathbf{x}-\mathbf{x}'| = 1, \\ g_{k\bar{k}}/2, & |\mathbf{x}-\mathbf{x}'| = \sqrt{2}, \\ 0, & \text{otherwise.} \end{cases} \tag{13}$$

where $g_{k\bar{k}}$ represents the strength of interparticle interactions between component k and $\bar{k}$. The effective number density, $\psi_k(n_k)$ is taken as n_k in the present model and other choices will give a different equation of state (EOS). The interactive force between the fluid and wall is realized by considering the wall as a separate phase with constant number density and is given by:

$$\mathbf{F}_{2k}(\mathbf{x}) = -n_k(\mathbf{x}) \sum_{\mathbf{x}'} g_{kw} n_w(\mathbf{x}')(\mathbf{x}' - \mathbf{x}) \tag{14}$$

where n_w is the number density of the wall, which is a constant at the wall and zero elsewhere, g_{kw} is the interaction strength between kth component and the wall. The interactive strength, g_{kw}, defines the wall wettability, and is positive for a non-wetting fluid and negative for a wetting fluid. It is to be noted that $\mathbf{F}_{2k}$ is perpendicular to the wall and will not affect the no-slip boundary condition. The action of a constant body force can be incorporated as:

$$F_{3k} = \rho_k \mathbf{g} = m_k n_k \mathbf{g} \tag{15}$$

where $\mathbf{g}$ is the constant body force per unit mass.

The continuity and momentum equations can be obtained for the fluid mixture as a single fluid using Chapman-Enskog expansion procedure in the nearly incompressible limit:[43]

$$\left.\begin{aligned} &\frac{\partial \rho}{\partial t} + \nabla \cdot (\rho \mathbf{u}) = 0 \\ &\rho\left[\frac{\partial \mathbf{u}}{\partial t} + (\mathbf{u}\cdot\nabla)\mathbf{u}\right] = -\nabla p + \nabla \cdot [\rho\nu(\nabla \mathbf{u} + \mathbf{u}\nabla)] + \rho \mathbf{g} \end{aligned}\right\} \tag{16}$$

where the total density and velocity of the fluid mixture are given, respectively, by:[44]

$$\left.\begin{aligned} \rho &= \sum\nolimits_k \rho_k \\ \rho\mathbf{u} &= \sum\nolimits_k \rho_k \mathbf{u}_k + \frac{1}{2}\sum\nolimits_k \mathbf{F}_k \end{aligned}\right\} \tag{17}$$

The pressure, which is usually a non-ideal gas equation of state, is given by:[43]

$$p = \sum\nolimits_k \frac{(1-\beta_k)m_k n_k}{2} + 3\sum\nolimits_{k,\bar{k}} g_{k\bar{k}} \psi_k \psi_{\bar{k}} \tag{18}$$

In the present model, $m_k = 1$ and $\beta_k = 1/3$, which is commonly used in the literature. Then the equation of state can be written as:

$$p = \sum\nolimits_k \frac{n_k}{3} + 3\sum\nolimits_{k,\bar{k}} g_{k\bar{k}} \psi_k \psi_{\bar{k}} \tag{19}$$

The viscosity is given by:

$$\nu = \left(\sum\nolimits_k \alpha_k \tau_k - 1/2\right) \Big/ 3 \tag{20}$$

where α_k is the mass density concentration of the kth component and is defined as:[45]

$$\alpha_k = \rho_k \Big/ \sum\nolimits_k \rho_k \tag{21}$$

It should be noted that the introduction of fluid/solid interaction has no effect on the macroscopic equations since $\mathbf{F}_{2k}$ exists only at the fluid/solid interface. The relaxation time, τ_k, is estimated based on the viscosity and mass density representations given by Eqs. (20) and (21) of the kth component and is detailed in Ref. [37, 43, 44]. This model has been shown to satisfy Galilean invariance.[44] Furthermore, in this interparticle potential model, the separation of a two-phase fluid into its components is automatic.[37]

The primary physical parameters, such as the fluid/fluid and fluid/solid interaction parameters, need apriori evaluation through model calibration using numerical experiments. The fluid/fluid interaction gives rise to the surface tension force and the fluid/solid interaction manifests in the wall adhesion force. The fluid/fluid and fluid/solid interaction parameters are evaluated by designing two numerical experiments, *bubble test* in the absence of solid phase and *static droplet test* in the presence of solid wall, respectively. The details of these numerical experiments are detailed elsewhere.[27,45–48]

3. TWO-PHASE SIMULATION

Before discussing the details of the numerical experiments performed in this study, the primary mechanisms governing the two-phase transport in the PEFC *catalyst layer* and *gas diffusion layer* are discussed, which essentially build the foundations behind the specific assumptions and justifications pertaining to the subsequent two-phase simulations.

3.1. Two-phase Transport Mechanism

For two-phase flow through the porous catalyst layer and the fibrous gas diffusion layer in a PEFC, due to the complex structure as well as significantly small pore size, e.g. around 0.05–0.1 μm in the CL and 20–30 μm in the GDL, surface forces become dominant as compared to the gravity, viscous and inertia forces. In this regard, the representative values of the salient non-dimensional numbers governing the underlying transport in the CL are listed below. For a representative CL, the order of magnitude of the following non-dimensional numbers can be estimated as:

$$\text{Reynolds number: Re} = \frac{\rho_2 U_2 D}{\mu_2} \sim 10^{-4}$$

$$\text{Capillary number: Ca} = \frac{\mu_2 U_2}{\sigma} \sim 10^{-6}$$

$$\text{Bond number: Bo} = \frac{g(\rho_2 - \rho_1)D^2}{\sigma} \sim 10^{-10}$$

$$\text{Weber number: W}e = \frac{\rho_2 U_2^2 D}{\sigma} = \text{Re.Ca} \sim 10^{-10}$$

U_2 and μ_2 are the non-wetting phase velocity and dynamic viscosity respectively; σ is the surface tension and g is the gravitational acceleration. Similarly, the afore-mentioned non-dimensional parameters exhibit significantly low values in the GDL within comparable order of magnitude variations. It should be noted that for the hydrophobic CL and GDL representative of a typical PEFC, water is the non-wetting phase (NWP) and air the wetting phase (WP).

As mentioned earlier, in this chapter, we identify the non-wetting phase with subscript 2 and the wetting phase with 1. From the Bond number, defined as the ratio of gravitational force to the surface tension force, it is evident that the effect of gravity is negligible with respect to the surface tension force, thus indicating strong capillary force dominance. Also within PEFC electrodes, velocity, U_2 is significantly small. Now, from the Reynolds number, representing the ratio of inertia force to viscous force, it is obvious that the inertial effect is negligible, as compared to the viscous force. Combining the implications of low Reynolds and Bond numbers, it can be inferred that the density difference, which is $\sim$1000 for air-water two-phase flow for the PEFC operation, should have significantly small influence on the overall transport in the CL and GDL. Now, from the Capillary number, Ca, which represents the ratio of viscous force to the surface tension force, it can be observed that the effect of viscous force is also negligible as compared to the surface tension force. Apart from these non-dimensional numbers, the Weber number, (We), defining the ratio of inertial force to the surface tension force, which is also a product of Re and Ca, emphasizes the fact that the effects of inertia and viscous forces are truly insignificant as compared to the surface tension force representative of two-phase transport in the PEFC CL and GDL. It should be noted that the viscosity ratio for the non-wetting and wetting phases in a fuel cell operating at 80°C is estimated to be, $M = \mu_2/\mu_1 \sim 18$. Based on the calculated representative viscosity ratio and capillary number values, the typical operating two-phase regime for the CL and GDL belongs to the capillary fingering zone on the "phase diagram" proposed by Lenormand et al.,[49] and is shown in Fig. 8. The notion of the "phase diagram", proposed by Lenormand et al.,[49] is based on their experiment, involving immiscible displacement of a wetting phase by a non-wetting phase, in a flat and horizontal porous medium where gravity forces were neglected. This phase diagram further bolsters the validity of our assumption that for air-water two-phase flow in the CL and GDL of a PEFC, the viscous forces are truly negligible and the principal force is owing to the action of capillarity, which would consequently lead to capillary fingering type displacement pattern. Viscous fingering pattern is observed at low viscosity ratio (M), where the principal force is the viscous force of the resident fluid and the capillary force and pressure drop in the displacing fluid can be neglected. In the stable displacement

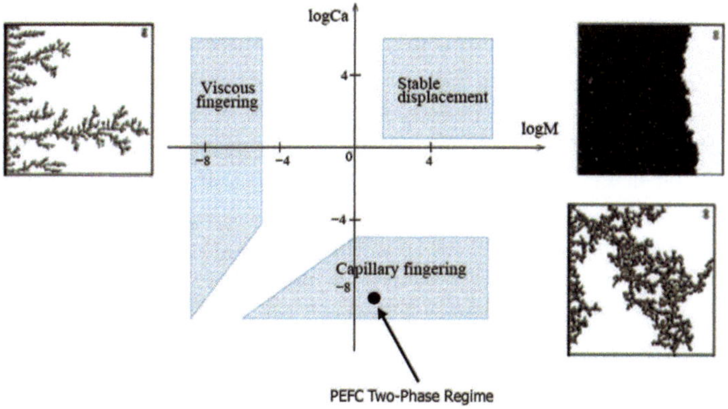

Figure 8. Phase diagram along with fluid displacement patterns based on the works by Lenormand et al.[49] and Ewing and Berkowitz.[50] Reproduced from Mukherjee and Wang[61] with permission from Elsevier.

flow regime, representative of high M and high Ca, the principal force is due to the viscosity of the displacing fluid and capillary effects in the resident fluid are negligible. Typical fluid displacement patterns pertaining to the three flow regimes are also shown in Fig. 8 along with the phase diagram and are adapted from the work by Ewing and Berkowitz,[50] who extended the two-dimensional phase diagram by Lenormand et al.[49] to a three-dimensional phase diagram by adding the effect of gravity force through Bond number (Bo) in the third direction. However, in general, the effect of gravity force in the overall fuel cell system, let alone in the porous catalyst layer and gas diffusion layer, has been shown to be truly insignificant.[51] Using this analysis, it can be safely adjudged that for modeling air-water two-phase transport in the PEFC CL and GDL the effects of high density ratio ($\sim$1000) and viscosity ratio ($\sim$18) variation can be assumed to be negligible represented by the significantly low capillary number. Apropos of this justification, in the present two-phase LB model, comparable density and viscosity values of the non-wetting and wetting phases are assumed.

3.2. Two-phase Numerical Experiments and Setup

In this study, two numerical experiments are designed for investigating liquid water transport and two-phase dynamics through the reconstructed CL and GDL microstructures. It should be noted that the primary objective of the numerical experiments is to study, in an "ex-situ setup", the influence of the microstructure and wetting characteristics on the underlying two-phase behavior and flooding dynamics in the PEFC porous catalyst layer and fibrous gas diffusion layer. Additionally, in the subsequent two-phase numerical simulations, isothermal condition is assumed as a first and reasonable approximation i.e. the thermal effects on the two-phase behavior are not considered. The development of non-isothermal, two-phase LB model with the phase change effects is left as a future exercise.

The first numerical setup is designed to simulate a *quasi-static displacement experiment*, typically devised in the petroleum/reservoir engineering applications and detailed elsewhere in the literature,[52–54,58] for simulating immiscible, two-phase transport through the CL and GDL structures. Figure 9 schematically shows the computational domain and setup of the displacement, i.e. *primary drainage* experiment. A NWP reservoir is added to the porous structure at the front end and a WP reservoir is added at the back end.[27,55,56] These two end reservoirs added to the CL/GDL domain in the through-plane (i.e. thickness) direction are composed of void space. It should be noted that for the primary drainage (PD) simulation in the hydrophobic CL and GDL, liquid water is the NWP and air is the WP.

Fixed pressure boundary conditions, which are equivalent to fixed densities within the LB framework, are imposed at the first layer of the NWP reservoir and the last layer of the WP reservoir.

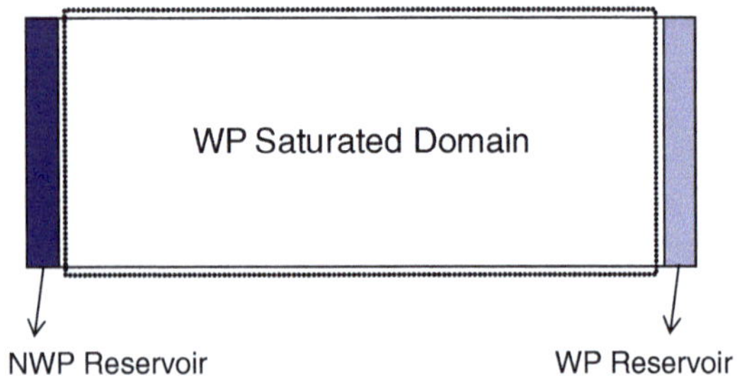

Figure 9. Schematic diagram of the immiscible two-phase displacement simulation setup.

Fixed pressure boundary conditions are implemented by assigning the equilibrium distribution functions, computed with zero velocity and specified density at the reservoirs, to the distribution functions. Periodic boundary conditions are applied in the span-wise directions. A particle distribution function bounce-back scheme[57] is used at the walls to obtain no-slip boundary condition.

The primary drainage process is simulated starting with zero capillary pressure, by fixing the NWP and WP reservoir pressures to be equal. Then the capillary pressure is increased incrementally by decreasing the WP reservoir pressure while maintaining the NWP reservoir pressure at the fixed initial value. The pressure gradient drives liquid water into the initially air-saturated CL and GDL by displacing it. A lattice site is assumed occupied by the NWP if the NWP density at that node is larger than half of the fixed NWP density; otherwise the node is assumed to be occupied by the WP. For each capillary pressure, the steady state is assumed achieved when, either the relative difference of the overall fluid velocity given by Eq. (17) over 1000 time steps is of the order of 10^{-6} or the NWP saturation shows negligible variation after 1000 time steps, whichever occurs first. It is worth mentioning that the current primary drainage simulation corresponds to an "ex-situ experimental procedure" where the CL or the GDL structure is not a part of an operational fuel cell. The primary objective of the quasi-static displacement simulation is to study liquid water behavior through the CL and GDL structures and the concurrent response to capillarity as a direct manifestation of the underlying pore-morphology.

In order to achieve enhanced fundamental insight into the underlying two-phase transport and interfacial dynamics, another numerical experiment is designed based on the *steady-state flow experiment*, typically devised in the petroleum/reservoir engineering applications and detailed elsewhere in the literature.[52–54] Briefly, in the steady-state flow experiment the two immiscible fluids are allowed to flow simultaneously until equilibrium is attained and the corresponding saturations, fluid flow rates and pressure gradients can be directly measured and correlated using Darcy's law. In the *steady-state flow simulation*, initially both the NWP and WP are randomly distributed throughout the CL and GDL porous structures such that the desired NWP saturation is achieved.[27,59] The initial random distribution of the liquid water phase (i.e. NWP) in the otherwise air (i.e. WP) occupied CL closely represents the physically perceived scenario of liquid water generation due to the electrochemical reaction at different catalytically active sites within the CL structure and subsequent transport by the action of capillarity. Furthermore, in the case of recent ultra-thin CL structures and also conventional catalyst layers characterized by very low hydrophobicity or mixed wettability characteristics,[9,27] liquid water generated in the CL fails to overcome the capillary pressure barrier imposed by the sufficiently hydrophobic microporous layer (MPL) situated between the CL and GDL. In this scenario, water vapor transports into the GDL and condenses into liquid water at several preferential sites throughout the GDL which subsequently transports due to the action of capillarity.[59] These two scenarios form the basis for the present dynamical equilibrium immiscible flow simulation. In this numerical experiment, the computational domain is assumed to be bounded by walls in the span-wise directions and periodic in the through-plane/thickness direction.[60] Counter-current flow is simulated by applying a body force for both the phases along the flow direction which mimics the redistribution of the phases under the capillary force corresponding to the typical capillary number ($Ca \sim 10^{-6}$). Steady state is considered attained when the saturation and flow rates of both phases exhibit infinitesimal change over 1000 time steps.

For efficient simulation of two-phase transport through the porous structures with sufficiently large number of lattice points along with the computational intensive nature of the particle-based LB modeling approach, the present two-phase LB model is parallelized based on full lattice representation (FLR) where both the fluid and solid nodes are stored and computed on a regular computational grid.[27] With the FLR approach, the parallelization is implemented based on 1-D slice decomposition along the thickness/through-plane direction using the standard Message Passing Interface (MPI) library. In order to keep the computational cost within a reasonable limit and without losing much on accuracy the simulations presented in this study are performed on a CL structure with $100 \times 50 \times 50$ lattice nodes and a GDL structure with 100^3 lattice nodes. A typical simulation using ten computer nodes on a typical Linux PC cluster takes approximately 100–120 h to reach steady state for a particular saturation level.

4. TWO-PHASE BEHAVIOR AND FLOODING DYNAMICS

The aforementioned numerical experiments, namely *quasi-static drainage* and *steady-state flow simulations*, are specifically designed to study the influence of microstructure and wetting characteristics on the underlying two-phase behavior and flooding dynamics in the PEFC *CL* and *GDL*.

4.1. Structure-Wettability Influence

Figure 10 displays the steady state invading liquid water fronts corresponding to increasing capillary pressures from the *primary drainage simulation* in the *reconstructed CL microstructure*

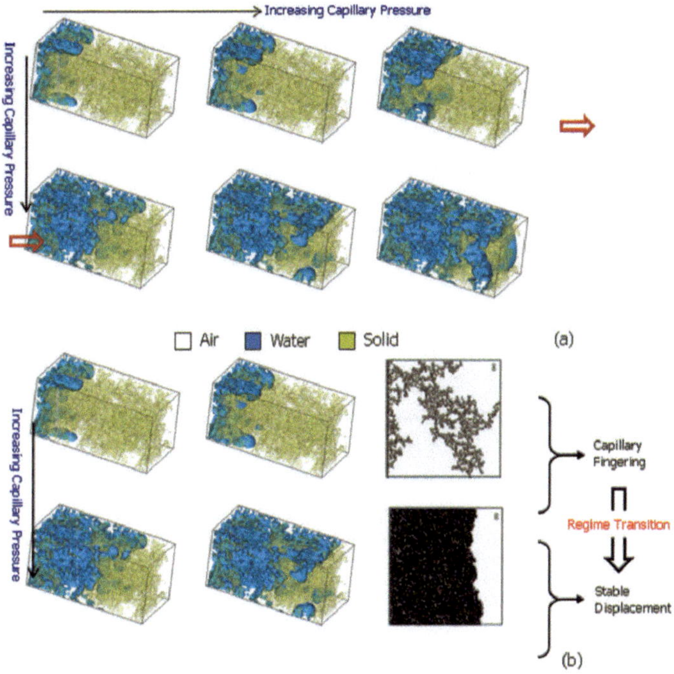

Figure 10. Advancing liquid water front with increasing capillary pressure through the initially air-saturated reconstructed CL microstructure from the primary drainage simulation.

characterized by mildly hydrophobic wetting characteristics with a static contact angle of 100°.[61] At lower capillary pressures, the liquid water saturation front exhibits finger like pattern, similar to the displacement pattern observed typically in the capillary fingering regime. The displacing liquid water phase penetrates into the body of the resident wetting phase (i.e. air) in the shape of fingers owing to the surface tension driven capillary force. However, at high saturation levels, the invading non-wetting phase tends to exhibit a somewhat flat advancing front. This observation, highlighted in Fig. 10b, indicates that with increasing capillary pressure, even at very low Capillary number (Ca), several penetrating saturation fronts tend to merge and form a stable front. The invasion pattern transitions from the capillary fingering regime to the stable displacement regime and potentially lies in the transition zone in between. Further investigations with varying CL microstructures and wetting characteristics (purely hydrophilic and mixed-wettability) are currently underway to understand the two-phase regime transition dynamics and capillary hysteresis in the PEFC electrode characterized by capillary dominated flow behavior. In an operating fuel cell, the resulting liquid water displacement pattern pertaining to the underlying pore-morphology and wetting characteristics would play a vital role in the transport of the liquid water and hence the overall flooding behavior.

Figure 11 shows the liquid water distribution as well as the invasion pattern from the *primary drainage simulation* with increasing capillary pressure in the initially air-saturated *reconstructed carbon paper GDL* characterized by hydrophobic wetting characteristics with a static contact angle of 140°.[61] It is to be noted that the reconstructed GDL structure used in the two-phase simulation consists of 100^3 lattice points as opposed to the high resolution structure with 512^3 lattice points shown in Fig. 6 in order to manage the computational overhead to a reasonable level. Physically, this scenario corresponds to the transport of liquid water generated due to the electrochemical reaction in a hydrophobic CL into the otherwise air-occupied GDL in an operating fuel cell. At the initially very low capillary pressure, the invading front overcomes the barrier pressure only at some preferential locations depending upon the pore size along with the emergence of droplets owing to strong hydrophobicity. As the capillary pressure increases, several liquid water fronts start to penetrate into the air occupied domain.

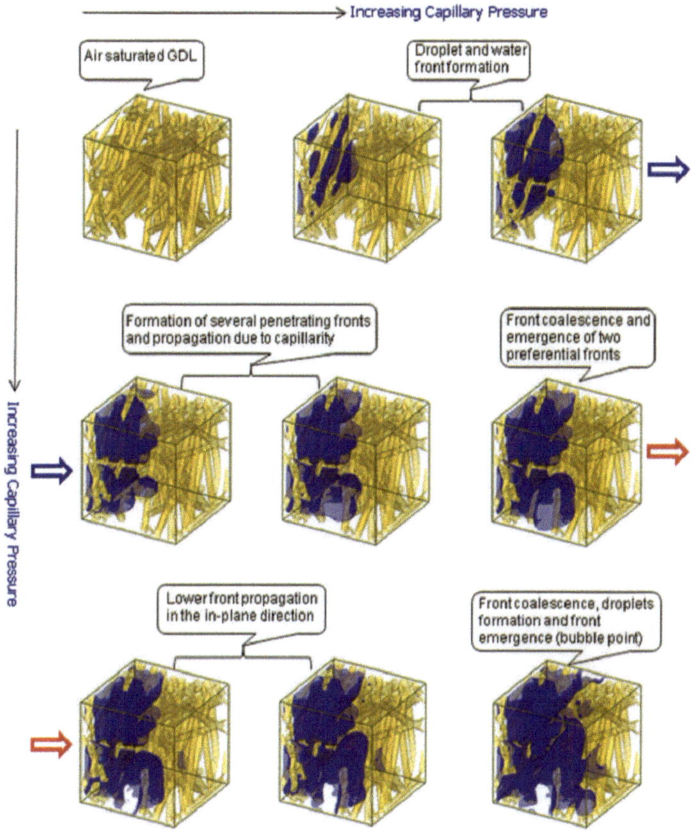

Figure 11. Advancing liquid water front with increasing capillary pressure through the initially air-saturated reconstructed GDL microstructure from the primary drainage simulation.

Further increase in capillary pressure exhibits growth of droplets at two invasion fronts, followed by the coalescence of the drops and collapsing into a single front. This newly formed front then invades into the less tortuous in-plane direction. Additionally, emergence of tiny droplets and subsequent growth can be observed in the constricted pores in the vicinity of the inlet region primarily due to strong wall adhesion forces from interactions with highly hydrophobic fibers with the increasing capillary pressure. One of the several invading fronts finally reaches the air reservoir, physically the GDL/channel interface, at a preferential location corresponding to the capillary pressure and is also referred to as the bubble point. Additionally, 2-D liquid water saturation maps on different cross-sections along the GDL through-plane direction corresponding to a representative liquid water saturation level are shown in Fig. 12 which demonstrates the porous pathways actually available for oxygen transport from the channel to the CL reaction sites. It is worth mentioning that the LB simulation is indeed able to capture the intricate liquid water dynamics including droplet interactions, flooding front formation and propagation through the hydrophobic fibrous GDL structure. Furthermore, it is important to note that the liquid water transport and flooding dyanmics through a woven carbon cloth GDL would lead to very different scenario owing to liquid water motion along individual fibers as well as between fiber bundles as opposed to that in the non-woven carbon paper GDL. Further

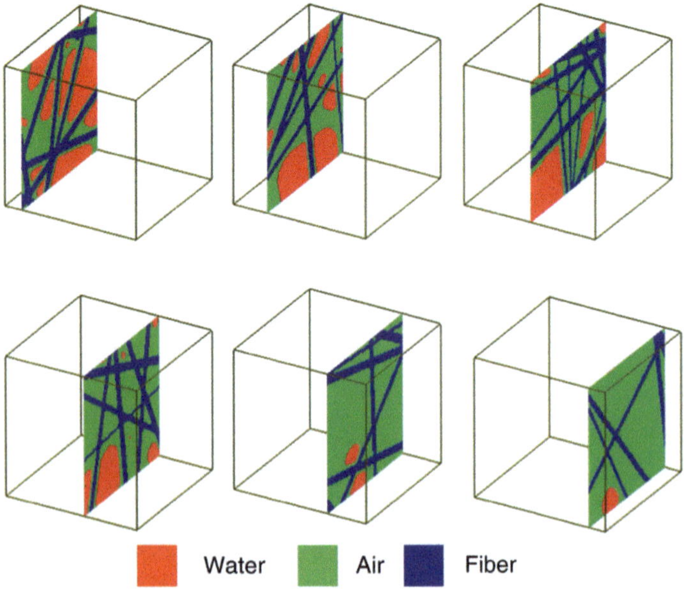

Figure 12. 2-D phase distribution maps on several cross-sections along the GDL through-plane direction from the primary drainage simulation.

investigations are currently underway to understand the influence of different pore morphology and wetting characteristics of GDL structures on the resulting flooding dynamics. It is important to note that the mesoscopic LB simulations provide fundamental insight into the pore-scale liquid water transport through different GDL structures and would likely enable novel GDL design with better water removal and flooding mitigation.

Figure 13 exhibits the 3-D liquid water distributions corresponding to several low saturation levels (below 15%) for the *CL* from the *steady state flow simulation* at equilibrium.[61] It can be observed that below 10% saturation level there is hardly any connected pathway for the liquid water phase to transport through the CL structure and hence the relative mobility of the liquid water phase with respect to the incumbent air phase is negligible. As the saturation level increases, the initially random liquid water phase redistributes owing to the action of capillarity and finds a connected pathway for transport through the CL structure. Figure 14 exhibits the 2-D liquid water saturation maps corresponding to a representative saturation level of 20% on several cross-sections along the thickness of the *reconstructed CL.* The implication of the 2-D saturation maps is that the reduction in electrochemically active interfacial area (ECA) owing to liquid water coverage in the CL can be estimated.[62] Furthermore, it is imperative to emphasize that this pore-scale two-phase model delineates the impact of the underlying structure and wetting characteristics of the CL on liquid water

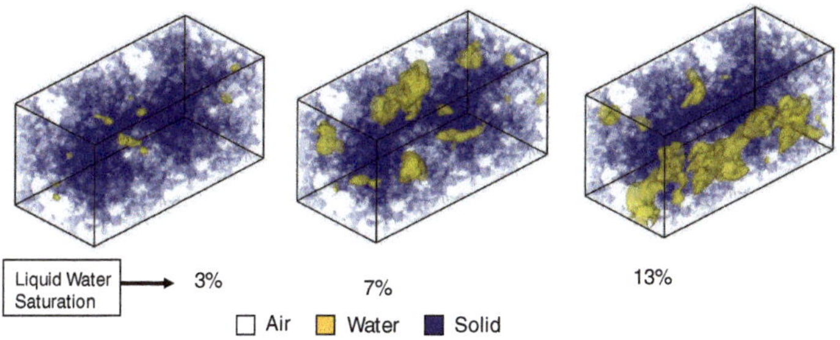

Figure 13. 3-D liquid water distributions in the reconstructed CL microstructure from the steady state flow simulation at equilibrium.

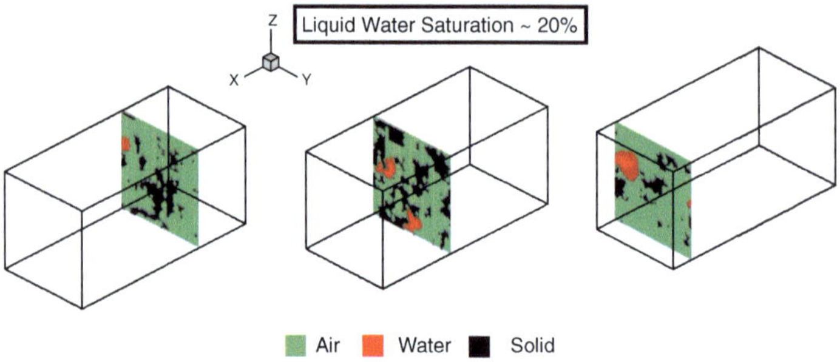

Figure 14. 2-D phase distribution maps in the CL structure from the steady state flow simulation at equilibrium.

distribution and dynamics which would subsequently influence the relative phase mobility and transport.

Figure 15 shows the equilibrium liquid water distribution after the randomly dispersed initial liquid water redistributes by the action of capillary force for a purely hydrophobic *GDL* with contact angle of around 140° and also for a mixed wettability GDL with hydrophilic and hydrophobic contact angles of 80 and 140°, respectively.[61] In the mixed wettability GDL, a hydrophilic pore fraction of 50% is considered and the hydrophilic pores are assumed to be randomly distributed through the GDL structure. It can be observed that at similar saturation level, the liquid water distribution is quite different for the two GDLs, underscoring the influence of the wetting characteristics and interfacial dynamics on transport processes. In a mixed wettability GDL, due to the hydrophilic pores, liquid water exhibits a somewhat stable front preferentially oriented along the through-plane direction. On the other hand, in a purely hydrophobic GDL, liquid water tends to distribute preferentially in the less tortuous in-plane direction and exhibits a narrow flooding front, which would facilitate effective oxygen transport along the through-plane direction from the gas channels to the CL active sites. The far reaching implication of this study is to understand the liquid water behavior and flooding dynamics in a beginning-of-life GDL which is predominantly hydrophobic as opposed to that in an ageing GDL after prolonged operation leading to partially hydrophilic fibers and mixed wetting characteristics thereby causing enhanced flooding and severe performance degradation. Further

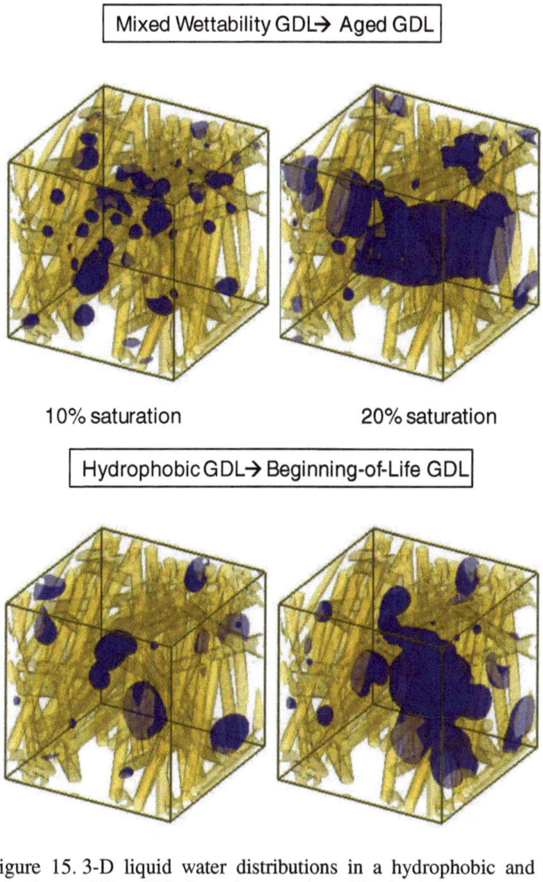

Figure 15. 3-D liquid water distributions in a hydrophobic and a mixed wettability GDL from the two-phase LB simulations.

investigations are presently underway related to GDL durability influence on two-phase flooding phenomena. Additionally, Fig. 16 exhibits the 2-D phase distributions on several cross-sections for a representative saturation level around 20% in the *GDL*. The liquid water distributions from such study could be further used to quantify the averaged saturation-dependent effective transport properties (e.g. effective species diffusivity). Finally, it is worth pointing out that the mesoscopic two-phase model not only captures the pore occupancy by liquid water due to pore size distribution but also

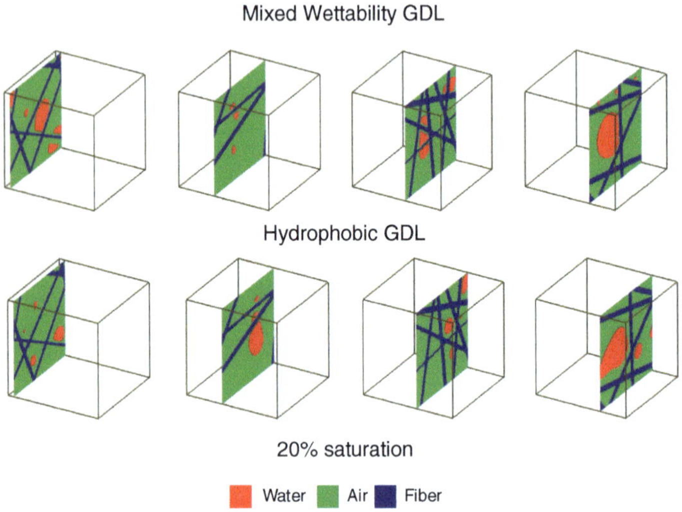

Figure 16. 2-D phase distribution maps on several cross-sections for a hydrophobic and a mixed wettability GDL from the two-phase LB simulations.

phase redistribution owing to the underlying interfacial dynamics and wetting characteristics.

4.2. Effect of GDL Compression

The importance of cell clamping pressure on fuel cell performance has been studied by several researchers. Notable works include Mathias et al.,[32] Wilde et al.[63] and Ihonen et al.[64] Mathias et al.[32] reported compression and flexural behavior of carbon paper and carbon cloth GDLs and indicated the effect of compressive characteristics on the channel flow-field pressure drop. Wilde et al.[63] studied the impact of compression force on the GDL properties, namely electrical resistivity, pore size, and permeability for different materials and briefly described the resulting influence on PEFC performance. Ihonen and co-workers,[64] on the other hand, tried to assess experimentally the influence of clamping pressure on the GDL flooding behavior. In the present study, we briefly report the influence of GDL compression on the underlying microstructure and flooding dynamics using a reduced compression model and the two-phase LB model.

Detailed modeling of a porous material under compression is a challenging task of applied structural mechanics. The reduced compression model employed in the current study is based on the unidirectional morphological displacement of solid voxels in the GDL structure under load and with the assumption of negligible transverse strain. The reduced compression model is detailed in our recent work.[33] However, with the reduced compression model, it is difficult to find a relation between the compression ratio and the external load. The compression ratio is defined as the ratio of the thickness of compressed sample to that of the uncompressed sample. Nevertheless, this approach leads to reliable 3-D morphology of the non-woven GDL structures under compression. Figure 17a shows compressed, reconstructed non-woven GDL

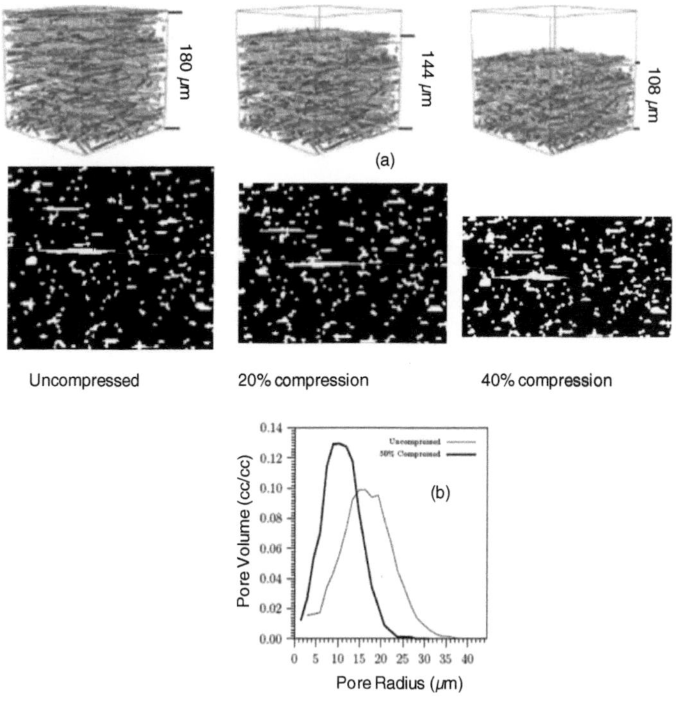

Figure 17. Representative uncompressed and compressed non-woven GDL structures along with pore size distribution.

microstructures with compression ratio of 0.8 and 0.6 along with the uncompressed structure and representative 2-D cross-sections.[33] Due to the compression of the structure, the pore size distribution is expected to shift toward smaller pores. Figure 17b shows the pore size distributions of the uncompressed and 50% compressed GDL microstructures. It can be observed that the mean pore size shifts from around 17 μm to approximately 10 μm under 50% compression, while the width of the pore size distribution shows negligible variation.

Figure 18 shows the liquid water distribution as well as the invasion pattern from the *primary drainage simulation* with increasing capillary pressure in the initially air-saturated *30% compressed carbon paper GDL* microstructure characterized by hydrophobic wetting characteristics with a static contact angle of 140°, similar to its uncompressed counterpart, described in Section 4.1. The reduced compression model was used to generate the compressed structure. Similar to the uncompressed GDL, at very low initial capillary pressures, the invading front overcomes the barrier pressure only at some preferential locations depending upon the pore

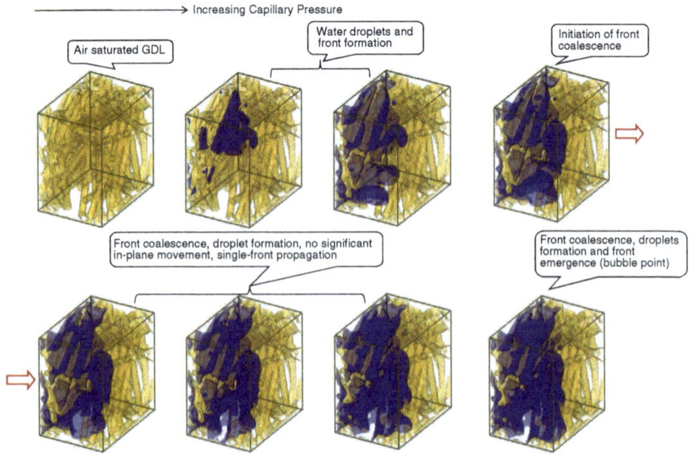

Figure 18. Advancing liquid water front with increasing capillary pressure through the initially air-saturated reconstructed GDL microstructure with 30% compression from the primary drainage simulation.

size. As the capillary pressure increases, several liquid water fronts start to penetrate into the air occupied domain. Owing to the compression induced microstructural change, further increase in capillary pressure exhibits coalescence of drops and concurrent merging of the fronts into a single front. This newly formed front propagates in the through-plane direction with liquid water build-up and subsequent action of capillarity. However, no preferential front migration toward the in-plane direction, as opposed to that in the uncompressed structure, is observed, which can be attributed to the increased tortuosity owing to the compression of the GDL. Emergence of droplets and subsequent growth can be also observed in the constricted pores primarily due to strong wall adhesion forces from interactions with highly hydrophobic fibers with the increasing capillary pressure. Finally, the predominant front finally reaches the air reservoir, physically the GDL/channel interface, at a preferential location and the corresponding capillary pressure is referred to as the bubble point. Concurrent to the bubble point, the microstructural change also reflects in the movement of water in the in-plane direction which could rather be attributed to the liquid water build-up at the corresponding saturation level, and differs significantly from the two-phase transport in the uncompressed GDL shown in Fig. 11. Additionally, the 2-D liquid water saturation maps on different cross-sections along the *GDL through-plane* direction corresponding to a representative liquid water saturation level are shown in Fig. 19 which demonstrates the porous pathways actually available for oxygen transport from the channel to the CL reaction sites in the compressed GDL. The roughened fiber cross-sections also suggest overlaying of fiber layers from compression leading to higher tortuosity. It is worth mentioning that the LB simulation is indeed able to capture the intricate liquid water dynamics and reflects the influence of compression induced GDL microstructural change on the flooding behavior. Detailed investigations are currently underway to understand the influence of different pore morphology and wetting characteristics of GDL structures under the influence of compression on the resulting flooding dynamics. Furthermore, a micro-FE (finite element) formalism is also under development to capture the realistic structural mechanics effect in order to directly relate the compression load to the microstructural change of the GDL.

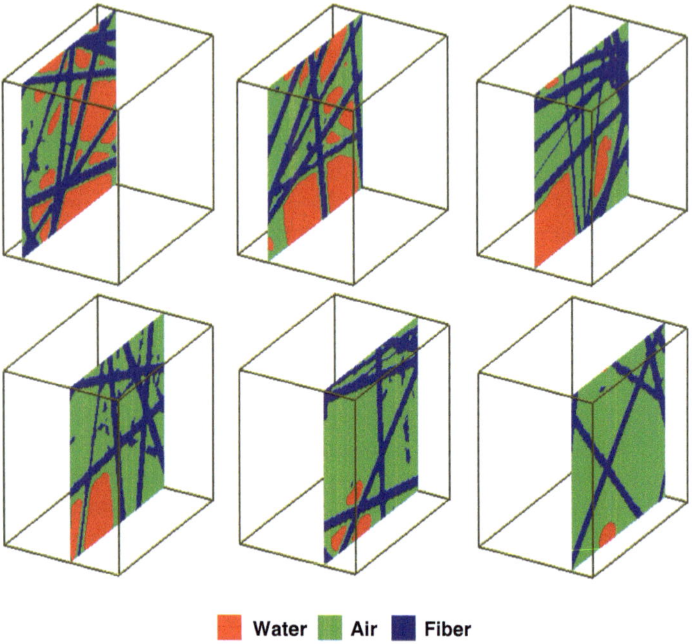

Figure 19. 2-D phase distribution maps on several cross-sections along the through-plane direction in the 30% compressed GDL structure from the primary drainage simulation.

4.3. Evaluation of Two-Phase Relations

The primary drainage and steady-state flow simulations, described in the earlier section and typically devised in the petroleum/reservoir engineering applications,[53,54,58,60] were deployed to evaluate the capillary pressure and relative permeability relations as functions of liquid water saturation, respectively.

The *capillary pressure* response, a direct manifestation of the underlying pore morphology, can be evaluated from the two-phase LB drainage simulation and the corresponding transport relation as function of liquid water saturation can be devised as shown in Fig. 20 for the reconstructed *CL microstructure*.[27] The overall shape of the capillary pressure curve agrees well with those reported in the literature for synthetic porous medium.[55] The capillary

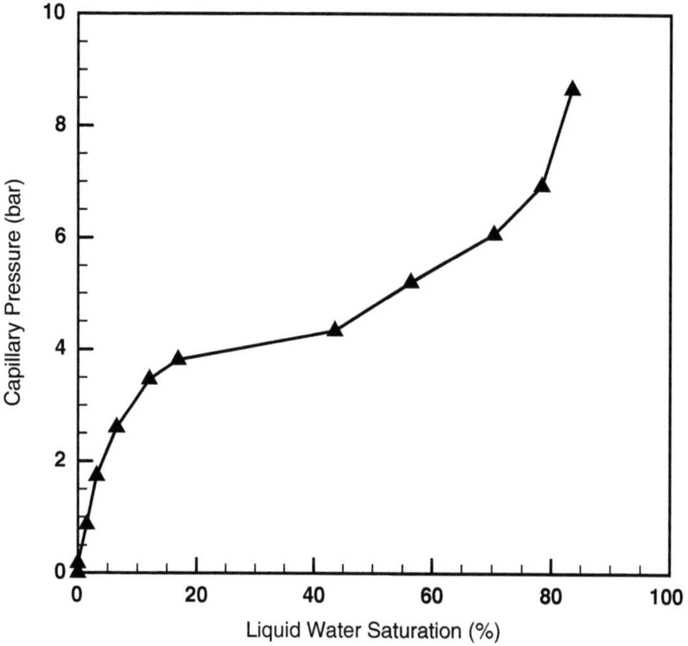

Figure 20. Capillary pressure – liquid water saturation relation for the reconstructed CL microstructure.

pressure – saturation curve exhibits a non-zero entry pressure for the NWP to initiate invading into the WP-saturated CL structure characterized by slightly hydrophobic wetting characteristics. Another point to be noted is that for the CL microstructure, the NWP saturation increases gradually with increase in capillary pressure till around 20%, after which NWP saturation jumps to around 45% with negligible increase in capillary pressure and finally exhibits a residual WP saturation of around 18%.

Similarly, the capillary pressure – saturation relations for the reconstructed GDL structures can be constructed from the two-phase LB drainage simulations. The capillary pressure response for reconstructed non-woven *GDL microstructures* was also evaluated using a full morphology (FM) approach, detailed in our recent work.[33] Briefly, the FM model relies on morphological decomposition of the 3-D digital image of the reconstructed GDL to determine the pore space accessible to the non-wetting phase using the pore radius as the ordering parameter via overlapping spherical elements corresponding to a specified pressure during drainage.[33] Figure 21 shows a typical capillary pressure - saturation relation for a hydrophobic *non-woven, reconstructed GDL microstructure* characterized by a static contact angle of 120° from the FM simulation. The effect of GDL compression on the capillary pressure response was also elucidated using the reduced compression model and the FM model.[33] It is to be noted that the quasi-static two-phase distribution in the FM model is based on purely morphological consideration of overlapping spherical elements and therefore cannot accurately capture the effect of the interfacial shape on the invasion process.[33] However, the FM model is a fast, direct simulation method based on the actual pore morphology and the two-phase distributions obtained thereof can be useful in finding saturation-dependent effective transport properties, e.g. effective gas diffusivity.

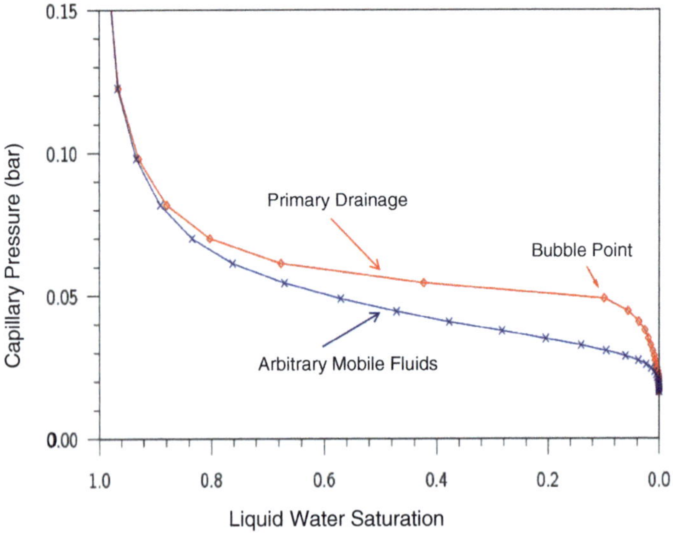

Figure 21. Capillary pressure – liquid water saturation relation for a non-woven GDL structure using the FM model.

The steady-state flow numerical experiment was primarily designed to evaluate the phasic *relative permeability* relations. The numerical experiment is devised within the two-phase lattice Boltzmann modeling framework for the *reconstructed CL microstructure*, generated using the stochastic reconstruction technique described earlier. Briefly, in the steady-state flow experiment two immiscible fluids are allowed to flow simultaneously until equilibrium is attained and the corresponding saturations, fluid flow rates and pressure gradients can be directly measured and correlated using Darcy's law, defined below.

$$v_i = -\frac{\kappa\kappa_{ri}}{\mu_i}(\nabla p_i - \rho_i g) \tag{22}$$

where v_i is the Darcy velocity for the wetting phase and non-wetting phase, p_i is the fluid phase pressure, $\rho_i g$ is the body force, μ_i is the dynamic viscosity of the fluid, κ is the intrinsic permeability determined by the pore structure of the porous medium alone, and κ_{ri} is the relative permeability of each phase that depends upon fluid saturation and the underlying two-phase dynamics. The term "steady-state", however needs to be duly qualified in the sense that the process is intrinsically a dynamical equilibrium of two moving fluids, although macroscopically stable.[58] The numerical experiment for the immiscible two-phase flow is described below in the context of evaluation of relative permeability relations:[27,60]

- Initially, both the NWP and WP are randomly distributed throughout the CL porous structure such that the desired NWP saturation is achieved. The initial random distribution of the liquid water phase (i.e. NWP) in the otherwise air (i.e. WP) occupied CL closely represents the physically perceived scenario of liquid water generation due to the electrochemical reaction at different catalytically active sites within the CL structure and subsequent transport by the action of capillarity. In the present numerical experiment, the CL domain is assumed to be bounded by walls in the span-wise directions and periodic in the thickness direction.
- Counter-current flow is simulated by applying a body force for both the phases along the flow direction which mimics the redistribution of the phases under the capillary force corresponding to the typical capillary number (Ca $\sim 10^{-6}$). The body force, implemented via Eq. (15), is adjusted such

that a representative low capillary number is achieved and the two phases interact via surface tension force at the fluid/fluid interface and wall adhesion force corresponding to the contact angle leading to phase distribution.

- Once steady state is achieved, the flux of each phase is calculated. Steady state is considered attained when the saturation and flow rates of both phases do not change any more. The corresponding absolute flux is calculated by modifying the two-phase LB model, where a body force is applied to one phase and the density of the other phase is rendered zero at all locations. Finally, the ratio of flux of each phase from the two-phase calculation to the one obtained from the single-phase calculation gives the relative permeability related to the saturation level.

The two-phase distribution from such simulations is already shown in Fig. 13 for a *reconstructed CL microstructure*. Figure 22 shows the NWP and WP relative permeability as functions of

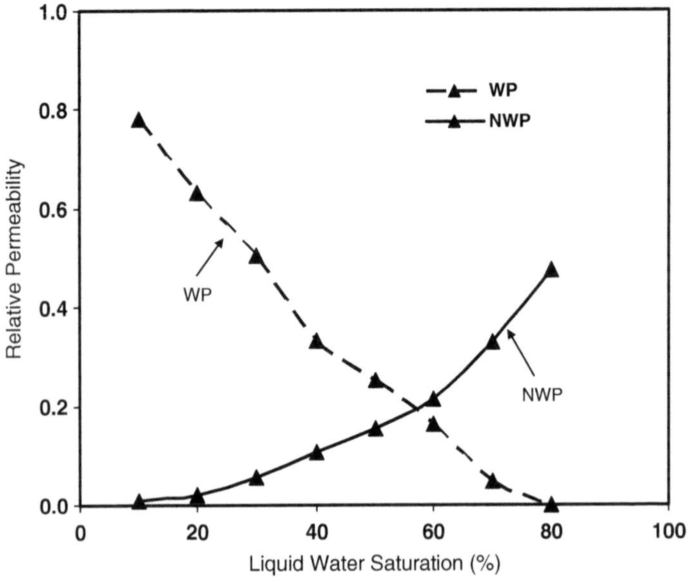

Figure 22. Relative permeability – liquid water saturation relation for the reconstructed CL microstructure.

liquid water (NWP) saturation for the reconstructed CL.[27] The general trend of the relative permeability curves agrees well with those reported elsewhere in the literature for geological flows.[60] It can be observed that the liquid water relative permeability values below 10% saturation is negligibly small due to the non-existence of a connected pathway for transport apropos of the 3-D liquid water distributions in Fig. 13. In general it is observed that at this low Ca the relative permeability curves exhibit non-linear relationships for both phases between the flow rate and the driving pressure gradient and this observation is in agreement with the results reported by Li et al. for geologic porous samples.[60] This observation further emphasizes that for the capillary number of interest for fuel cell operation, i.e. Ca $\sim 10^{-6}$, the relative permeability relations for both phases will exhibit strong non-linearity. Despite the linear relation between the pressure gradient and the flow-rate demonstrated by the Darcy's law, the current prediction from the numerical experiment further underscores the capability of the LB model in capturing the underlying interfacial dynamics inherent to the low capillary number flow regime of interest. This relative permeability – saturation relation can be further deployed in the two-phase computational fuel cell dynamics models for reliable transport predictions.

The above-mentioned approach for estimating the relative permeability – saturation relation applies exactly for the reconstructed GDL microstructure as well and work is currently under progress. It is important to note that for the evaluation of the relative permeability – saturation relation, detailed description of the underlying interfacial dynamics dependent on the pore morphology and wetting characetristics is of paramount importance. In this regard, it is also worth mentioning the recent works by Niu et al.[64] and Koido et al.[65] on capillary pressure and relative permeability evaluations as functions of liquid water in the carbon paper GDL using similar two-phase lattice Boltzmann formalisms reported in this chapter.

4.4. Effect of Liquid Water on Performance

The detrimental consequence of liquid water on the electrochemical performance manifests in terms of coverage of the electrochemically active area in the CL leading to reduced catalytic activity and blockage of the porous pathways in the CL and GDL rendering hindered oxygen transport to the active reaction sites. In the macroscopic fuel cell models, where the CL and GDL are treated as macrohomogeneous porous layers, the site coverage and pore blockage effects owing to liquid water are taken into account through an electrochemical area reduction relation and the Bruggeman type correction for the effective oxygen diffusivity, respectively. These two empirical correlations, however, cannot be separately discerned through experimental techniques. A combination of the predictions of liquid water distributions from the LB two-phase simulations and species transport using a direct numerical simulation (DNS) model can be deployed to quantify the site coverage and pore blockage effects.

The computational approach couples the two-phase LB model for the liquid water transport and the DNS model for the species and charge transport for the CL.[25–27,68] The two-phase simulation using the LB model is designed based on the ex-situ, steady-state flow experiment for porous media, detailed earlier in the section 4.3, in order to obtain the liquid water distributions within the CL microstructure for different saturation levels resulting from the dynamic interactions between the two phases and the underlying pore morphology. The details of the simulation setup are provided in our work.[27,61,62] Once steady state is achieved, 3-D liquid water distributions can be obtained within the CL, as shown in Fig. 13. From the liquid water distributions within the CL structure, the information about the catalytic site coverage effect can be extracted directly.

The DNS model can be deployed subsequently on the liquid water blocked CL structure pertaining to a saturation level for the evaluation of the hindered oxygen transport. In brief, the DNS model is a "top-down" numerical approach based on a fine-scale CFD framework which solves point-wise accurate conservation equations for species and charge transport in the CL with appropriate source terms due to the oxygen reduction reaction (ORR) directly on the CL microstructures.[25–27,67] The conservation equations for proton, oxygen and water vapor transport, respectively, are given by:[25–27,68]

$$\nabla \cdot (\kappa_e \nabla \phi_e) + S_{\phi_e} = 0 \tag{23}$$

$$\nabla \cdot (D^g_{O_2} \nabla c_{O_2}) + S_{O_2} = 0 \tag{24}$$

$$\nabla \cdot (D^g_{H_2O} \nabla c_{H_2O}) + S_{H_2O} = 0 \tag{25}$$

$S_{\phi e}$, S_{O_2} and S_{H_2O} refer to the respective source terms owing to the ORR, ϕ_e is the electrolyte phase potential, c_{O_2} is the oxygen concentration and c_{H_2O} is the water vapor concentration, κ_e is the proton conductivity duly modified w.r.t. to the actual electrolyte volume fraction, $D^g_{O_2}$ is the oxygen diffusivity and $D^g_{H_2O}$ is the vapor diffusivity. The details about the DNS model for pore-scale description of species and charge transport in the CL microstructure along with its capability of discerning the compositional influence on the CL performance as well as local overpotential and reaction current distributions are furnished in our work.[25–27,67]

With the liquid water distribution available from the two-phase LB simulation corresponding to a saturation level, the reduction in electrochemically active interfacial area (ECA) owing to liquid water coverage can be estimated from the 2-D saturation maps and subsequently a correlation between the effective ECA and the liquid water saturation can be established as the following:[27,62]

$$ECA^{eff} = ECA(1 - S_r)^c \tag{26}$$

where S_r is the liquid water saturation and c the site coverage factor.

Using the 2-D saturation maps from the two-phase LB simulation, shown in Fig. 14, the effective ECA can be evaluated and correlated according to Eq. (26). Based on several liquid water saturation levels, the catalytic surface coverage factor for the CL microstructure is estimated and the following correlation can be constructed, which can be used as valuable input to macroscopic two-phase fuel cell models.[27,62]

$$ECA^{eff} = ECA(1 - S_r)^{1.05} \tag{27}$$

Figure 23 shows the variation of the effective ECA with liquid water saturation from the evaluated correlation given in Eq. (27) along with the typical correlations with ad-hoc fitting of the coverage parameter otherwise used in the macroscopic fuel cell models.[2] It is to be noted that the effect of liquid water is manifested via a reduction of the active area available for electrochemical

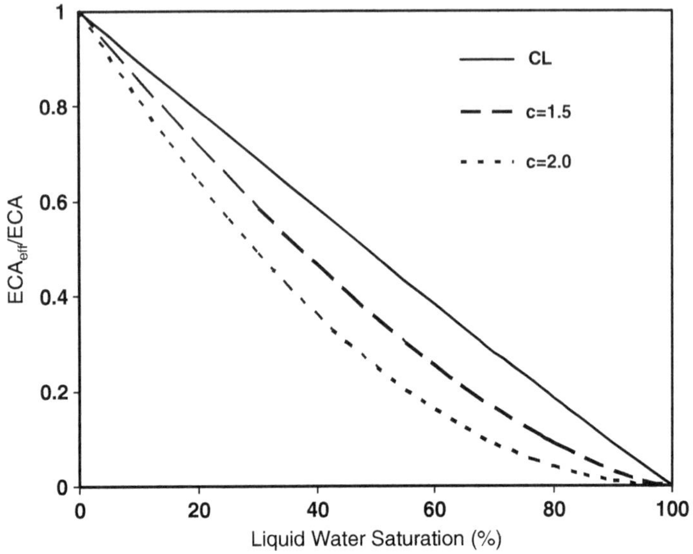

Figure 23. Catalytic site coverage relation as function of liquid water saturation for the CL structure.

reaction through the coverage parameter. In the mass transport control regime, the hindered oxygen transport owing to liquid water blockage, described in the following section, becomes the limiting factor.

In order to evaluate the effect of pore volume blockage in the presence of liquid water causing hindered oxygen transport to the active reaction sites, the direct numerical simulation (DNS) model, mentioned earlier and detailed in our work,[25–27,68] is deployed for the pore-scale description of species and charge transport through the reconstructed CL microstructure.

From the 3-D liquid water signature obtained from the two-phase LB simulation in the *CL* and *representatively shown in* Fig. 13, the pores occupied by liquid water are identified corresponding to a particular saturation level and these pores are rendered as "dead" pores. The idea is to generate a modified CL structure where the blocked pores do not take part in the oxygen reduction reaction (ORR) as well as produce extra resistance by reducing the effective porosity of the structure. With this virtual morphology of the liquid water blocked CL, the point-wise accurate species and charge conservation equations are solved within the DNS modeling framework. It should be noted that unlike in the GDL, the reactive transport in the CL requires the solution of the electrochemistry coupled oxygen transport for the true oxygen concentration field to evolve. The pore blockage effect is finally evaluated from the oxygen concentration field and can subsequently be correlated in terms of the effective oxygen diffusion coefficient based on the oxygen flux as the following:[27,62]

$$D_{O_2}^{eff} = D_{O_2,0} f(\varepsilon_{CL})\, g(S_r) = D_{O_2,0}(\varepsilon_{CL})^m (1 - S_r)^b \qquad (28)$$

ε_{CL} is the CL porosity, S_r the liquid water saturation, m the Bruggeman factor for the oxygen transport through the unblocked CL microstructure and b is the volume blockage factor representing the extra resistance to oxygen transport in the presence of liquid water in the CL. The effect of the resistance due to the tortuous pore pathways to oxygen transport in the absence of liquid water is evaluated using the DNS model in terms of a Bruggeman factor, m and is detailed in our work in.[25,27,61] Similarly, the resistance to ion transport due to the tortuous electrolyte phase network in terms of Bruggeman correction factor can also be evaluated using the DNS model.[25,27] With $\varepsilon_{CL} = 0.6$ and the estimated Bruggeman factor, $m = 3.5$,[25,27] the pore blockage factor for the CCM CL is evaluated and the following correlation is constructed.[27,62]

$$D_{O_2}^{eff} = D_{O_2,0}(\varepsilon_{CL})^{3.5} (1 - S_r)^{1.97} \qquad (29)$$

This estimate could prove to be valuable input for more accurate representation of the pore blockage effect in the macroscopic two-phase fuel cell models. Figure 24 shows the variation of the effective oxygen diffusivity with liquid water saturation from the correlation in Eq. (29) along with the typical macrohomogeneous correlation with $m = 1.5$ and $b = 1.5$ otherwise used arbitrarily in the macroscopic fuel cell modeling literature.

Similarly, with the 3-D liquid water distribution in the *GDL microstructure* available from the two-phase LB model and *representatively shown in* Fig. 15, the DNS model[25,27] can be applied to solve the Laplace equation for oxygen transport given below.[68]

$$\nabla \cdot \left(D_{O_2}^{g} \nabla c_{O_2}\right) = 0 \qquad (30)$$

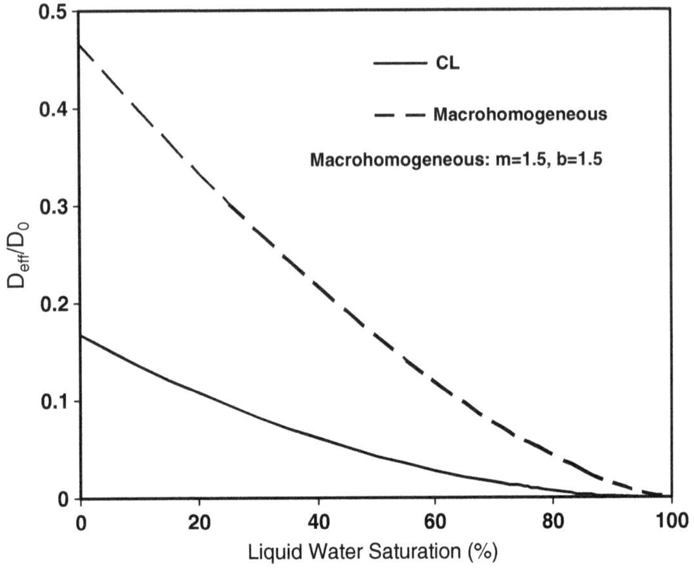

Figure 24. Pore blockage relation as function of liquid water saturation for the CL structure.

Similar to the approach outlined for the CL, the liquid water blocked pores are considered as "dead pores" which act as solid obstacles leading to extra resistance to oxygen diffusion. The pore blockage effect is evaluated from the oxygen concentration field pertaining to different liquid water satuartion levels and can be subsequently correlated in terms of the effective oxygen diffusion coefficient based on the oxygen flux.

$$D_{O_2}^{eff} = D_{O_2,0} f(\varepsilon_{GDL})\, g(S_r) = D_{O_2,0}(\varepsilon_{GDL})^m (1 - S_r)^b \qquad (31)$$

ε_{GDL} is the GDL porosity, S_r the liquid water saturation, m the Bruggeman factor for the oxygen transport through the unblocked GDL microstructure and b is the volume blockage factor representing the extra resistance to oxygen transport in the presence of liquid water in the GDL. The effect of the reistance due to the tortuous pore pathways to oxygen transport in the absence of liquid water can be evaluated using the DNS model in terms of a Bruggeman factor, m. It should be noted that the anisotropy resulting from the fiber orientations in the GDL structure manifests in terms of disparate resistance behavior to oxygen transport in the through-plane and in-plane directions. With GDL porosity of 0.72, the pore blockage effect in terms of the effective oxygen diffusivity is evaluated and the following correlations are constructed for the through-plane and in-plane variations, respectively.

$$D_{O_2,th}^{eff} = D_{O_2,0}(\varepsilon_{GDL})^{5.2}(1 - S_r)^{2.3} \tag{32}$$

$$D_{O_2,in}^{eff} = D_{O_2,0}(\varepsilon_{GDL})^{2.3}(1 - S_r)^{2.3} \tag{33}$$

Figure 25 displays the anisotropic effective oxygen diffusivity variations with liquid water saturation in the *GDL* based on the evaluated pore blockage correlations. Furthermore, the impact of GDL compression on the pore blockage effect was also investigated.[67]

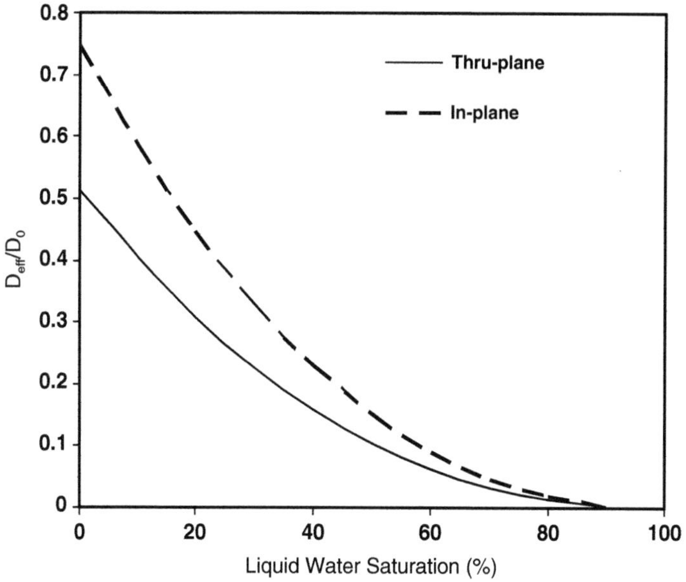

Figure 25. Pore blockage correlations for the uncompressed GDL structure.

This estimate could prove to be a valuable input for more accurate representation of the GDL pore blockage effect in the macroscopic two-phase fuel cell models.

The detrimental consequence of liquid water on the *CL voltage loss* primarily comes from the impeded oxygen transport and reduced electrochemically active area as explained above which can be described by the electrochemical kinetics in terms of the reaction current density, *j*, through the Tafel equation.[27]

$$j = i_0 a^{eff} \frac{c_{O_2}}{c_{O_2}^{ref}} \exp\left(-\frac{\alpha_c F}{RT}\eta\right) \tag{34}$$

i_0 is the exchange current density, c_{O_2} and $c_{O_2}^{ref}$ refer to local oxygen concentration and reference oxygen concentration respectively, α_c is the cathode transfer co-efficient for ORR, F is the Faraday's constant, R is the universal gas constant, and T is the cell operating temperature. In the above expression, η is the cathode overpotential and a^{eff} represents the effective electrochemically active area (ECA) due to the catalytic site coverage effect. Details about the above equation along with the pertinent assumptions are furnished in our work.[25,27]

The pore blockage effect comes into play through the oxygen concentration, c_{O_2} distribution given by the following equation:

$$\nabla \cdot \left(D_{O_2}^{eff} \nabla c_{O_2}\right) = \frac{j}{4F} \tag{35}$$

With the evaluated site coverage and pore blockage correlations for the effective ECA and oxygen diffusivity, respectively, and the intrinsic active area available from the reconstructed CL microstructure, the electrochemistry coupled species and charge transport equations can be solved with different liquid water saturation levels within the 1-D macrohomogeneous modeling framework,[25,27] and the cathode overpotential, η can be estimated.

Figure 26 exhibits the polarization curves in terms of the cathode overpotential variation with current density *for the CL*[27] obtained from the 3-D, single-phase DNS model prediction,[25,27] the experimental observation[25,27] and the liquid water transport corrected 1-D macrohomogeneous model.[27] The "polarization curve" refers to the cathode overpotential vs. current density curve in the

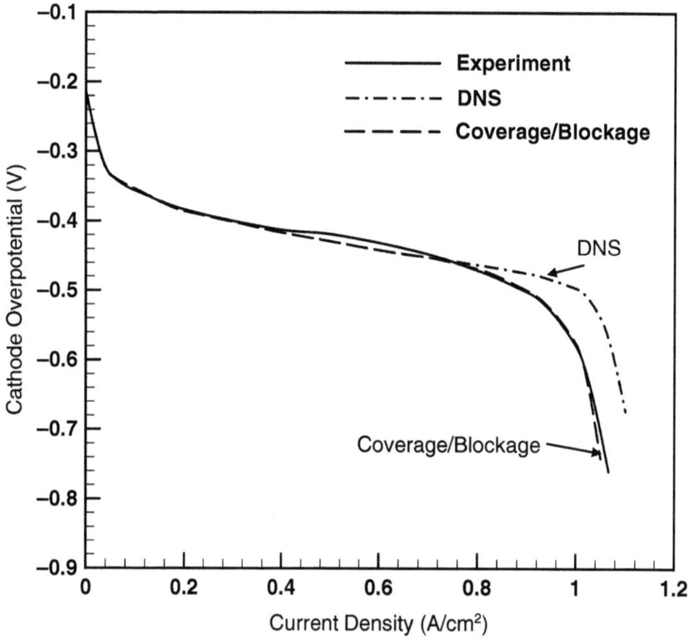

Figure 26. Electrochemical performance in terms of polarization curve for the reconstructed CL structure.

present study and hence differs from the standard performance curve in terms of the cell voltage vs. current density variation shown in Fig. 2 as a representative cell performance behavior otherwise used popularly in the literature. The experimental polarization curves were obtained from electrochemical performance evaluations conducted in a 5-cm^2 graphite cell fixture with identical anode and cathode single-pass, serpentine flow fields with computer-controllable test parameters, such as temperature, pressure, fuel/oxidant flow rates, current and cell voltage using a fuel cell test stand (Arbin®, TX). The cell was operated at 70°C, 200 kPa and 100% RH conditions at both anode and cathode sides with fixed flow rates of hydrogen and air. The experimentally obtained cell voltage (V_{cell}) vs. current density (I) data was further processed to extract the variation of cathode overpotential (η_c) with current density according to the following equation:

$$\eta_c = V_{\text{cell}} + I \times HFR - U_0 \tag{36}$$

where *HFR* refers to the high frequency resistance measured experimentally and U_0 is the thermodynamic equilibrium potential corresponding to the fuel cell operating temperature. It is important to note that in the above equation, the anodic overpotential for hydrogen oxidation and protonic resistance in the anode catalyst layer are assumed to be negligible. However, the cathode overpotential defined in Eq. (36) contains the protonic resistance or ohmic loss in the cathode catalyst layer.

In Figure 26, "DNS" refers to the polarization curve predicted by the single-phase, electrochemistry coupled DNS model for the reconstructed CL microstructure,[25,27] whereas "Coverage/Blockage" refers to the performance curve predicted by the 1-D macrohomogeneous model with correction for liquid water transport taken into account via the correlations for pore blockage and surface coverage effects evaluated from the two-phase simulations detailed in our work.[27] It should be noted that the disagreement between the DNS calculations and experimental data in the transport-control regime is due to the fact that the resistance from liquid water transport was not considered in the DNS model. The details about the DNS calculations, the corresponding operating and boundary conditions along with the experimental data are furnished in our work.[25,27] It can be observed that the estimated catalytic site coverage and pore blockage correlations for the CL from the combined two-phase LB and the DNS modeling framework can indeed capture the transport limiting regime and agrees well with the experimental data.

Finally, a key highlight of this investigation is that the systematic estimation of the effective transport parameters for the porous CL and GDL from the mesoscopic modeling can quantitatively predict the fuel cell performance from the macroscopic fuel cell models.

5. SUMMARY AND OUTLOOK

The catalyst layer and gas diffusion layer play a crucial role in the overall PEFC performance due to the transport limitation in the presence of liquid water and flooding phenomena. The macroscopic two-phase fuel cell models cannot address the effects of the complex pore morphology and wetting characteristics on the liquid water transport. In this chapter, we discuss the development of a comprehensive mesoscopic modeling framework comprising a stochastic microstructure reconstruction model and a two-phase LB model in order to reveal the underlying structure-wettability-performance relationship and to predict two-phase closure relations. The stochastic reconstruction model generates 3-D, statistically meaningful CL and GDL microstructures. The mesoscopic two-phase LB model simulates liquid water transport through the CL and GDL microstructures in order to gain insight into the influence of structure and wettability on the pore-scale two-phase dynamics as well as to evaluate the two-phase constitutive relations in terms of capillary pressure and relative permeability as functions of liquid water saturation. The liquid water flooding front in the CL apparently tends to exhibit a transition from capillary fingering to somewhat stable displacement pattern even in a strictly capillary dominated flow regime and with benign wetting characteristics. The liquid water transport in the fibrous GDL characterized by highly hydrophobic fibers shows intricate interfacial dynamics including droplet interactions, flooding front formation and propagation. The influence of partially wetting pore structure on the flooding behavior is elucidated for a mixed wettability GDL observed in an aged GDL. The effect of GDL compression on the flooding dynamics is also illustrated using a reduced compression model and the two-phase LB model. A quantitative estimate of the detrimental consequence of liquid water transport on the CL and GDL performance in terms of the pore blockage and catalytic site coverage effects is predicted by combining the two-phase LB model and the pore-scale multicomponent, electrochemically reactive DNS model. The DNS model not only provides pore-level description of charge and species transport in the CL as well as oxygen transport in the GDL but also acts as a macroscopic upscaling formalism to predict performance based on the pore-scale physics. In the dearth of two-phase correlations for the PEFC CL and GDL, the two-phase transport parameters evaluated from such pore-scale study could be adapted into two-phase computational fuel cell dynamics (CFCD) models for more reliable performance predictions.

Finally, the overriding implications of the mesoscopic modeling formalism coupled with realistic microstructure reconstruction techniques lie in quantifying the structure-wettability-performance interlinks in the CL and GDL and fostering novel water management strategies for PEFCs.

ACKNOWLEDGEMENTS

PPM would like to thank V. P. Schulz, A. Wiegmann and J. Becker from Fraunhofer ITWM, Germany for collaboration with GDL microstructure generation. Financial support from NSF through grant No. 0609727, ECEC industrial sponsors and the Director's Fellowship to PPM from Los Alamos National Laboratory LDRD Program is gratefully acknowledged.

REFERENCES

[1] C.Y. Wang, in *Handbook of Fuel Cells – Fundamentals, Technology and Applications*, vol. 3, ed. by W. Vielstich, A. Lamm, H.A. Gasteiger (Wiley, Chichester, 2003), p. 337
[2] C.Y. Wang, Chem. Rev. **104**, 4727 (2004)
[3] Z.H. Wang, C.Y. Wang, K.S. Chen, J. Power Sources **94**, 40 (2001)
[4] U. Pasaogullari, C.Y. Wang, J. Electrochem. Soc. **151**, 399 (2004)
[5] U. Pasaogullari, C.Y. Wang, J. Electrochem. Soc. **152**, A380 (2005)
[6] U. Pasaogullari, P.P. Mukherjee, C.Y. Wang, K.S. Chen, J. Electrochem. Soc. **154**, B823 (2007)
[7] Y. Wang, C.Y. Wang, J. Electrochem. Soc. **153**, A1193 (2006)
[8] Y. Wang, C.Y. Wang, J. Electrochem. Soc. **154**, B636 (2007)
[9] A.Z. Weber, J. Newman, J. Electrochem. Soc. **152**, A677 (2005)
[10] T. Berning, N. Djilali, J. Electrochem. Soc. **150**, A1589 (2003)
[11] M. Noponen, E. Birgersson, J. Ihonen, M. Vynnycky, A. Lundblad, G. Lindbergh, Fuel Cells **4**, 365 (2004)
[12] S. Dutta, S. Shimpalee, J.W. Van Zee, Int. J. Heat Mass Transfer **44**, 2029 (2001)
[13] S. Mazumder, J.V. Cole, J. Electrochem. Soc. **150**, A1510 (2003)
[14] J. Nam, M. Kaviany, Int. J. Heat Mass Transfer **46**, 4595 (2003)
[15] W. He, J.S. Yi, T.V. Nguyen, AIChE J. **46**, 2053 (2000)
[16] L. You, H. Liu, Int. J. Heat Mass Transfer **45**, 2277 (2002)
[17] J.J. Baschuk, X. Li, J. Power Sources **86**, 181 (2000)
[18] A.Z. Weber, J. Newman, Chem. Rev. **104**, 4679 (2004)
[19] K.S. Udell, Int. J. Heat Mass Transfer **28**, 485 (1985)
[20] M.C. Leverett, AIME Trans. **142**, 152 (1941).
[21] J.T. Gostick, M.W. Fowler, M.A. Ioannidis, M.D. Pritzker, Y.M. Volfkovich, A. Sakars, J. Power Sources **156**, 375 (2006)
[22] J.D. Fairweather, P. Cheung, J. St-Pierre, D.T. Schwartz, Electrochem. Commun. **9**, 2340 (2007)
[23] T.V. Nguyen, G. Lin, H. Ohn, X. Wang, Electrochem. Solid-State Lett. **11**, B127 (2008)

[24] K.G. Gallagher, R.M. Darling, T.W. Patterson, M.L. Perry, J. Electrochem. Soc. **155**, B1225 (2008).
[25] P.P. Mukherjee, C.Y. Wang, J. Electrochem. Soc. **153**, A840 (2006)
[26] P.P. Mukherjee, C.Y. Wang, J. Electrochem. Soc. **154**, B1121 (2007)
[27] P.P. Mukherjee, PhD Dissertation, Pennsylvania State University (2007)
[28] K.L. Moore, K.S. Reeves, DOE Hydrogen Program Annual Merit Review Proceedings, Arlington, VA, USA, May 23–26 (2005)
[29] Adler, P.M., 1992, "Porous Media: Geometry and Transports," Butterworth-Heinemann, Stoneham
[30] P.M. Adler, C.G. Jacquin, J.A. Quiblier, Int. J. Multiphase Flow **16**, 691 (1990)
[31] D.P. Bentz, N.S. Martys, Transport in Porous Media **17**, 221 (1995)
[32] M.F. Mathias, J. Roth, J. Fleming, W. Lehnert, in *Handbook of Fuel Cells – Fundamentals, Technology and Applications*, vol.3, Ch. 42, p. 517, ed. by W. Lietsich, A. Lamm, H.A. Gasteiger (Wiley, Chicester, 2003), p. 517
[33] V.P. Schulz, J. Becker, A. Wiegmann, P.P. Mukherjee, C.Y. Wang, J. Electrochem. Soc. **154**, B419 (2007)
[34] K. Schladitz, S. Peters, D. Reinel-Bitzer, A. Wiegmann, J. Ohser, Comput. Mater. Sci. **38**, 56 (2006)
[35] S. Chen, G. Doolen, Ann. Rev. Fluid Mech. **30**, 329 (1998)
[36] A.K. Gunstensen, D.H. Rothman, S. Zaleski, G. Zanetti, Phys. Rev. A **43**, 4320 (1991)
[37] X. Shan, H. Chen, Phys. Rev. E **47**, 1815 (1993)
[38] M.R. Swift, W.R. Osborn, J.M. Yeomans, Phys. Rev. Lett. **75**, 830 (1995)
[39] X.Y. He, S.Y. Chen, R.Y. Zhang, J. of Comp. Phys. **152**, 642 (1999)
[40] P. Bhatnagar, E. Gross, M. Krook, Phys. Rev. **94,** 511 (1954)
[41] H. Chen, S. Chen, W.H. Matthaeus, Phys. Rev. A **45**, R5339 (1992)
[42] N. Martys, H. Chen, Phys. Rev. E **53**, 743 (1996)
[43] X. Shan, G.D. Doolen, Phys. Rev. E **54**, 3614 (1996)
[44] X. Shan, G.D. Doolen, J. Stat. Phys. **81**, 379 (1995)
[45] S. Hou, X. Shan, Q. Zou, G. Doolen, W. Soll, J. Comput. Phys. **138**, 695 (1997)
[46] Z.L. Yang, T.N. Dinh, R.R. Nourgaliev, B.R. Sehgal, Int. J. Heat Mass Transf. **44**, 195 (2001)
[47] Q. Kang, D. Zhang, S. Chen, Phys. Fluids **14**, 3203 (2002)
[48] Q. Kang, D. Zhang, S. Chen, J. Fluid Mech. **545**, 41 (2005)
[49] R. Lenormand, E. Touboul, C. Zarcone, J. Fluid Mech. **189**, 165 (1988)
[50] R.P. Ewing, B. Berkowitz, Adv. Water Res. **24**, 309 (2001)
[51] X.G. Yang, F.Y. Zhang, A. Lubawy, C.Y. Wang, Electrochem. & Solid-State Letters **7**, A408 (2004)
[52] F.A.L. Dullien, *Porous Media: Fluid Transport and Pore Structure* (Academic, San Diego, CA 1992)
[53] J. Bear, *Dynamics of Fluids in Porous Media* (Dover, New York 1972)
[54] D. Tiab, E. Donaldson, *Petrophysics: Theory and Practice of Measuring Reservoir Rock and Transport Properties* (Gulf Publishing Company, Houston 1996)
[55] C. Pan, M. Hilpert, C.T. Miller, Water Resour. Res. **40**, 1 (2004)
[56] C. Pan, Ph. D. Dissertation, Univ. of North Carolina, Chapel Hill, North Carolina, USA (2003)
[57] P. Lavallee, J.P. Boon, A. Noullez, Physica D **47**, 233, 1991.
[58] D.G. Avraam, A.C. Payatakes, Transp. Porous Media **20**, 135 (1995)
[59] P.K. Sinha, P.P. Mukherjee, C.Y. Wang, J. Mater. Chem. **17**, 3089 (2007)

[60] H. Li, C. Pan, C.T. Miller, Phys. Rev. E **72**, 026705 (2005)
[61] P.P. Mukherjee, C.Y. Wang, Q. Kang, Electrochimica Acta **54**, 6861 (2009)
[62] P.P. Mukherjee, C.Y. Wang, ECS Trans. **3**, 1085 (2006)
[63] P.M. Wilde, M. Mandel, M. Murata, N. Berg, Fuel Cells **4**, 180 (2004)
[64] J. Ihonen, M. Mikkola, G. Lindbergh, J. Electrochem. Soc. **151**, A1152 (2004)
[65] X.D. Niu, T. Munekata, S.A. Hyodo, K. Suga, J. Power Sources **172**, 542 (2007)
[66] T. Koido, T. Furusawa, K. Moriyama, J. Power Sources **175**, 127 (2008)
[67] P.P. Mukherjee, G. Wang, C.Y. Wang, in *Modern Aspects of Electrochemistry*, vol. 40, ed. by R.E. White, C.G. Vayenas, M.E. Gamboa-Aldeco (Springer, New York, 2007), p. 285
[68] V.P. Schulz, P.P. Mukherjee, J. Becker, A. Wiegmann, C.Y. Wang, ECS Trans. **3**, 1069 (2006)

9

Atomistic Modeling in Study of Polymer Electrolyte Fuel Cells – A Review

Xiangyang Zhou, Juanjuan Zhou, and Yijin Yin

Department of Mechanical and Aerospace Engineering, University of Miami, P.O. Box 248294, Coral Gables, FL 33157, USA

1. INTRODUCTION

Polymer electrolyte fuel cell (PEFC) is considered as one of the most promising power sources for futurist's hydrogen economy. As shown in Fig. 1, operation of a Nafion-based PEFC is dictated by transport processes and electrochemical reactions at catalyst/polymer electrolyte interfaces and transport processes in the polymer electrolyte membrane (PEM), in the catalyst layers consisting of precious metal (Pt or Ru) catalysts on porous carbon support and polymer electrolyte clusters, in gas diffusion layers (GDLs), and in flow channels. Specifically, oxidants, fuel, and reaction products flow in channels of millimeter scale and diffuse in GDL with a structure of micrometer scale. Nafion, a sulfonic acid tetrafluorethylene copolymer and the most commonly used polymer electrolyte, consists of nanoscale hydrophobic domains and proton conducting hydrophilic domains with a scale of 2–5 nm. The diffusivities of the reactants (O_2, H_2, and methanol) and reaction products (water and CO_2) in Nafion and proton conductivity of Nafion strongly depend on the nanostructures and their responses to the presence of water. Polymer electrolyte clusters in the catalyst layers also play a critical

U. Pasaogullari, C.-Y. Wang (eds.), *Modeling and Diagnostics of Polymer Electrolyte Fuel Cells*, Modern Aspects of Electrochemistry 49,
DOI 10.1007/978-0-387-98068-3_9,

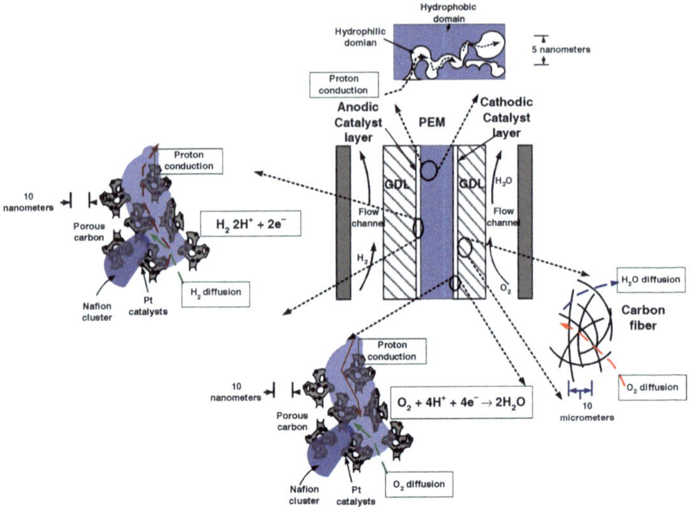

Figure 1. Schematic illustration of electrochemical and transport processes in a PEFC.

role in the catalysis of the nano-sized Pt catalysts. Electrochemical reactions occur at the interfaces between catalysts (Pt or Pt/Ru) and Nafion. The catalytic activity of the Pt catalysts is believed to be dictated by transport processes, adsorption/desorption, and charge transfer in the interfacial area. While transport processes may occur in an area of a few nanometers, adsorption/desorption and charge transfer occur within a region of a few angstroms from the surface of a nano-particulate catalyst. Thus, modeling or simulation of PEFC is a multiscale problem.

Two primary goals of modeling and simulation of PEFC seem to be, first, to supplement an experimental study performed in laboratory to investigate what have not been or cannot be investigated with the experimental approach; and second, to conduct a computer design and to evaluate the design with virtual tests that are difficult or impossible to be performed in laboratory under the current status of technology. First of all, all experimental methods have limitations on their spatial and temporal resolutions and sensitivities. Although there are a number of claims that high-resolution transmission electron microscopy (HRTEM), atomic force microscopy

(AFM), and scanning tunneling microscopy (STM) can achieve an atomic resolution, these technologies neither allow an investigator to actually "see" electrons, the actors of chemical reactions, nor permit recording vibrations of molecules in the temporal resolution of a femtosecond (10^{-15} s). In many cases, empirical theories are created on the basis of a few snapshots of a part of a molecular system. The limited information is far from the whole story or sometimes may even result in misleading impressions. Computer experiments especially atomistic simulations provide virtually continuous cartoons of motion of all atoms and molecules in a system or even changes of bond and electronic orbitals in the case of ab initio atomistic simulations. Secondly, real experiments are limited within a realistic range of parameters. An atomistic computer design of a new material can be conducted in a virtual space and tested in a virtual experiment.

With intensive competitions from the other clean and renewable energy technologies, the PEFC technology needs breakthroughs or so-called "quantum jumps" in development of polymer electrolytes and catalysts, or effective combinations of polymer electrolytes and catalysts in near future. Although there are spectra of alternative polymer electrolytes and catalysts, the most reliable combination is still Nafion- and Pt-based nanoparticles. Designing materials by computer atomistic modeling and virtual experiment has been a relatively new development in the study of polymer electrolytes and composite catalysts in contrast to "trial-and-error" approaches that have long been the main stream of PEFC research. The number of publications on atomistic modeling aiming to understand the physicochemical behaviors of materials in the atomic level and to realize designing materials in the atomic level for applications in PEFCs increases rapidly.

Figure 2 illustrates major modeling methods, i.e., ab initio molecular dynamic (AIMD), molecular dynamic (MD), kinetic Monte Carlo (KMC), and continuum methods in terms of their spatial and temporal scales. Models for microscopic and macroscopic components of a PEFC are placed in the figure in terms of their characteristic dimensions for comparison. While continuum models are successful in rationalizing the macroscopic behaviors based on a few assumptions and the average properties of the materials, molecular or atomistic modeling can evaluate the nanostructures or molecular structures and microscopic properties. In computational

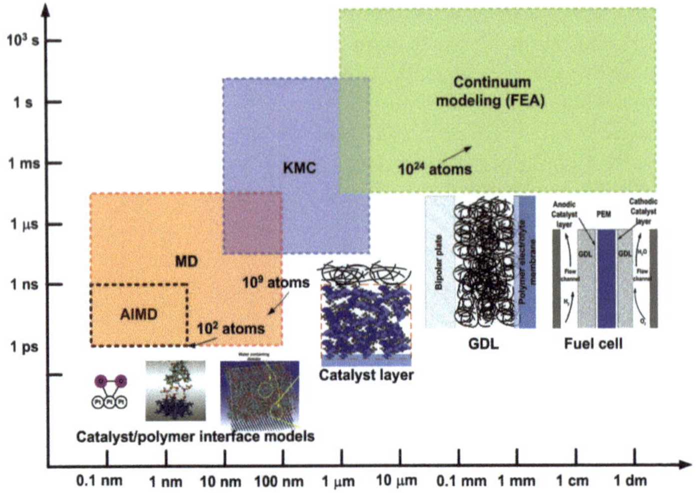

Figure 2. Multiscale modeling hierarchy. AIMD: ab initio molecular dynamics, MD: molecular dynamics, KMC: kinetic Monte Carlo modeling, and FEA: finite element analysis.

chemistry and computational materials science, the first principle or ab initio atomistic methods based on Schrödinger equation are virtually exact in the sense that they can be as accurate as possible, subject only to computational time or computational resource limitations. Thus, in theory, the processes in a PEFC can be precisely modeled if a sufficient computational capability is available. In this regard, AIMD methods can evaluate the forces acting on the nuclei from the electronic structure calculations that are performed as the trajectory of the molecules is generated.[1–6] Due to the high demand for the computational resources, AIMD methods are limited to analyses of small clusters (a few nanometers) and within a very short period of time (a few picoseconds) that are unable to represent real-world macroscopic phenomena. Classical molecular dynamic (MD) methods are based on the Newton's second law and inter-atom force fields (FFs) that are evaluated using quantum mechanics.[7] In comparison to ab initio methods, a MD method requires much less computational resources and are able to analyze a model with a larger size but still can achieve a sufficient precision for evaluating some transport properties. MD methods are particularly suitable for analyzing phenomena in the systems with a size up to 100 nm and characteristic timescale up to a few nanoseconds with a desktop workstation. However, while the continuous modeling is too coarse to precisely describe the motions of atoms and molecules, the atomistic modeling can reveal the motion trajectories of atoms and molecules in a microscopic system but cannot display collective behaviors of all atoms in a macroscopic system in a realistic time period (e.g., 1 s) due to the limited capacity of existing computational resources.

A great effort has been made by ultra-computer developers to magnify the capacity of the hardware. Ab initio modeling/simulation of a cluster of multi-million atoms is now being attempted with one of the most powerful computers.[8] Even if similar attempts are successful, they must be very limited in terms of number of simulations and in terms of ranges of parameters. It is predicted that a Petaflop computer (with a processing speed of 10^{15} float point operations per second) will be developed in 2012. This computer will be able to execute MD simulations for a cluster of 10^{12} atoms (about a micrometer in size) or perhaps ab initio simulations for a cluster of 108 atoms. The increase of computer capacity roughly follows Moore's law that predicts that the capacity of computer doubled in every 24 months. However, according to a recent analysis by Thompson and Parthasarathy, there will be a limit for the increase of computer capacity if future computers are based on architectures similar to currently existing ones.[9] This limit is related to the minimum size (10 nm) of a complementary metal-oxide semiconductor (CMOS) device and is called CMOS limit. This means that roughly speaking, the capacity can be further increased by 100–1000 times. They are not optimistic with the realization of a computer based on molecular electronic storage devices because wiring of such devices would be a formidable task. It seems that modeling a molecular system with a macroscopic size (10^{24} atoms) in a timescale of 10^3 s cannot be achieved in a foreseeable future.

Instead of enhancing the performance of computers, many theoreticians have tried to enhance the efficiency of computation by improving computational codes. One of the approaches is to reduce the dependence of the computation for the present step on the computational results for previous steps along the system trajectory or increase the parallelism of the computation. The other effective approach is the use of concept of hierarchical coupling of paradigms on different scales.[10] This approach is so-called multiscale modeling. In an optimal situation, as shown in Fig. 2, a modeling method for specific spatial scale and timescale ranges should share an overlapping range with a method at the higher scale end and another at the lower scale end. However, all of these multiscale approaches inevitably result in loss of certain atomic details and are not general or system specific. The goal of this review paper is to summarize recent developments in the area of atomistic modeling of PEFCs with an emphasis on the comparison between methods. This may help to enable a progress in atomistic modeling of PEFCs.

2. FUNDAMENTALS OF ATOMISTIC MODELING

2.1. Ab Initio Modeling of Materials

2.1.1. Adiabatic Approximation

The basis of ab initio modeling of materials is the time-independent Schrödinger equation in which the state of a molecular system is described with a wavefunction:

$$H(\{\vec{r}\},\{\vec{R}\})\Psi(\{\vec{r}\},\{\vec{R}\}) = E(\{\vec{R}\})\Psi(\{\vec{r}\},\{\vec{R}\}) \tag{1}$$

where $H(\{\vec{r}\},\{\vec{R}\})$ is a Hamailtonian of a many-ion-many-electron system with n electrons and N nuclei, $\Psi(\{\vec{r}\},\{\vec{R}\})$ is the wavefunction, $E(\{\vec{R}\})$ is energy, $\{\vec{r}\}$ are the position vectors of electrons, and $\{\vec{R}\}$ are the position vectors of ions.[11] The first step toward simplification and solutions of the Schrödinger equation is the adiabatic or Born–Oppenheimer (BO) approximation that separates slow nuclear motion from fast electronic motion. In order to do this, one has to assume that irrespective of the instantaneous configuration of the ions, the electrons are at every moment in their ground state. The system is described by two effective Hamiltonians (Eqs. 2 and 3):

$$H_{elec} = \sum_{i=1}^{n}\left[-\frac{\hbar}{2m}\frac{\partial^2}{\partial \vec{r}_i^2} + \sum_{l=1}^{N} V_{\text{el-ion}}(\vec{r}_i,\vec{R}_l)\right] + \frac{1}{2}\sum_{i}^{n}\sum_{j\neq i}^{n}\frac{e^2}{|\vec{r}_i - \vec{r}_j|} \tag{2}$$

for electrons and

$$H_{ion} = \sum_{I=1}^{N}\left[-\frac{\hbar}{2M_I}\frac{\partial^2}{\partial \vec{R}_I^2}\right] + \frac{1}{2}\sum_{I=1}^{N}\sum_{J\neq I}^{N} V_{\text{ion-ion}}\left(\left|\vec{R}_I - \vec{R}_J\right|\right) + E_{\text{tot}}\left(\left\{\vec{R}_I\right\}\right) \tag{3}$$

for ions, where $\vec{r}_i$ is the position vector for the ith electron, $\vec{R}_I$ is the position vector for the ith ion, m is the mass of electron, M_I is the mass of the ith ion, e is electronic charge, $V_{\text{el-ion}}$ is potential for the direct electron–ion interaction, $V_{\text{ion-ion}}$ is potential for the direct ion–ion interaction, and E_{tot} is the electronic total energy as a function of the ionic coordinates $\{R_I\}$.

2.1.2. Hatree–Fock Approximation and Single Electron Hamiltonian

Under the Hartree–Fock (i.e., HF) approximation, the function of $3n$ variables for the solutions of the electronic Hamiltonian is reduced to n functions, which are referenced as molecular orbitals (MOs), each dependent on only three variables. Each MO describes the probability distribution of a single electron moving in the average field of all other electrons. Because of the requirements of the Pauli principle or antisymmetry with respect to the interchange of any two electrons, and indistinguishability of electrons, the HF theory is to approximate the many-electron wavefunction by an antisymmetrized product of one-electron wavefunctions and to determine these wavefunctions by a variational condition applied to the expected value of the Hamiltonian in the resulting one-electron equations,

$$\left(\sum_{i=1}^{n}\left[-\frac{\hbar}{2m}\frac{\partial^2}{\partial \vec{r}_i^2} + \sum_{I=1}^{N} V_{\text{el-ion}}(\vec{r}_i, \vec{R}_I)\right] + \frac{1}{2}\sum_{i}^{n}\sum_{j\neq i}^{n}\frac{e^2}{\left|\vec{r}_i - \vec{r}_j\right|}\right)\psi_i(\vec{r})$$
$$-\sum_j \delta_{s_i s_j}\int \frac{\psi_j^*(\vec{r}')\psi_i(\vec{r}')}{|\vec{r} - \vec{r}'|}d\vec{r}'\,\psi_j(\vec{r}) = \varepsilon_i\psi_i(\vec{r}) \tag{4}$$

where the pair-wise electron–electron repulsion is replaced by the interaction of the ith electron with the average electrostatic field created by the charge distribution of all other electrons and an additional exchange term keeping electrons of like spin ($S_i = S_j$) away from each other to account for the Pauli principle. Note that in this functional equation, correlations arising from short-distance Coulomb interactions are neglected.

2.1.3. Density Function Theory

Directly solving the HF equations is difficult and time consuming. For analyzing many quantum chemical problems, a density function theory (DFT) is often employed. DFT methods are based on the famous theorem by Hohenberg and Kohn who demonstrated that the total energy of a many-electron system in an external potential is a unique functional of the electron density and that this functional has its minimum at the ground-state density.[12] Expressing the electron density *n(r)* as a sum over one-electron densities $|\psi(\vec{r}_i)|^2$ and using the one-electron wave functions as the variational parameters leads to the Kohn–Sham one-electron equations as follows:

$$\left[-\frac{\hbar}{2m}\frac{\partial^2}{\partial\vec{r}^2} + \sum_{l=1}^{N} V_{\text{el-ion}}(\vec{r}_i, \vec{R}_l) + e^2\int\frac{n(\vec{r}')}{|\vec{r}-\vec{r}'|}d\vec{r}' + V_{XC}(\vec{r})\right]\psi_i(\vec{r}) = \varepsilon_i\psi_i(\vec{r}) \tag{5}$$

$$n(\vec{r}) = \sum_i^n |\psi_i(\vec{r})|^2, \; V_{XC}(\vec{r}) = \frac{\delta E_{XC}\left[n(\vec{r})\right]}{\delta n(\vec{r})} \tag{6}$$

where the exchange-correlation potential has been expressed as the functional derivative of the exchange-correlation energy. While Eq. (5) is exact within DFT, the energy functional $E_{XC}\left[n(\vec{r})\right]$ is usually unknown.[13]

2.1.3.1. Local Density Approximation Local density approximation (LDA) is the first practical approximation based on the assumption that the exchange-correlation energy of a single electron at the position $\vec{r}$ is equal to the exchange-correlation energy, $E_{XC}\left[n(\vec{r})\right]$, of an electron in a homogeneous electron gas of a density $n(\vec{r})$ equal to the local electron density $n(\vec{r})$. The most widely used exchange-correlation energy functionals within the framework of a LDA were developed by von Barth, Hedin, Perdew and Zunger.[14,15]

2.1.3.2. Generalized-Gradient Corrections In order to improve the LDA that neglects the dependence of the exchange-correlation functional on the local variations of the electron density, generalized-gradient approximations (GGA) have been introduced. The GGA has an explicit dependence of the exchange-correlation functional on the gradient of the electron density. The GGAs proposed by Perdew and co-workers have been widely used for solid-state systems and enjoy successes in correcting the LDA.[16,17] However, there are also cases where the GGA over corrects the deficiencies of the LDA and leads to an underbinding. One of the most noticeable deficiencies of GGAs is the incorrect estimation of the strength of hydrogen bonds. Hydrogen bonds seem to be better described on the basis of hybrid functionals including an exact description of exchange.

2.1.3.3. Basis Functions In an executable computational method, the wavefunctions $\psi(\vec{r}_i)$ must be represented as a linear combination of a finite number of basis functions. The choice of the basis determines the achievable accuracy and computational efficiency of the computational method. In addition, a basis set may be more or less convenient for evaluating a given material's property than another basis set. Three types of basis sets are commonly used: (i) linear combinations of atomic orbitals (LCAOs), (ii) linearly augmented plane waves (LAPWs), or (iii) plane waves (PWs) in combination with pseudopotentials for describing the electron–ion interaction. The choice of a basis usually should consider that near to an atom, the wavefunctions and potential are atom-like (i.e., almost spherical-symmetric and strongly varying with radial distance), while in the interstitial regions the potential and wavefunctions are quite smooth.

2.1.4. Ab Initio Quantum Chemistry Computation

The most important feature of the ab initio quantum chemistry computation is the capability of describing the breaking and formation of chemical bonds, chemical reactions, or electron transfer. Although a few modern quantum mechanics methods including generalized valence bond (GVB) and Gaussian dual space density function theory (GDS-DFT) are available, the calculation is often too slow for studying polymers and catalysts.[1,4,18] Pseudo spectral-generalized valence bond (PS-GVB) combining pseudospectral grid with the many-body wavefunctions can enable a better scaling between effort and system size (N) of N^2. It is better than other methods for which the scaling is from N^3 to N^6.[8]

2.1.5. *Ab Initio Molecular Dynamics*

Another challenge for quantum mechanics theories is to compute the dynamic properties or to model the physicochemical processes of a system by solving the equations of motion based on the adiabatic Hamiltonian of the ionic subsystem (Eq. 3). The variational property of the ground-state energy has a very important consequence: for a given configuration $\{R_I\}$, the force $\vec{F}_I$ acting on the atom at site R_I is given by the expected value of the derivative of the Hamiltonian H with respect to R_I in the electronic ground state $|\Phi_0\rangle$

$$\vec{F}_I(\{\vec{R}_I\}) = -\left\langle \Phi_0 \left| \frac{\partial H}{\partial \vec{R}_I} \right| \Phi_0 \right\rangle \tag{7}$$

Equation (7) is the famous Hellmann–Feynman theorem which allows the full set of quantum-many-body forces $\vec{F}_I$ to be calculated which can then be used to optimize the atomic geometry or to study the dynamics of the atoms by integrating the Newtonian equations of motion,

$$M_I \ddot{\vec{R}}_I(t) = \vec{F}_I(\{\vec{R}_I(t)\}) \tag{8}$$

The challenge for this method is to perform a very accurate determination of the electronic ground state after each move of the ions.

In 1985, Car and Parrinello introduced a very useful concept that the optimization with respect to the ionic and electronic degrees of freedom need not be done separately. In this concept or theory, one can use a coupled set of pseudo-Newtonian equations of motion for the ionic coordinates $\{R_I\}$ and the electronic orbitals $\{\psi_i\}$ as follows to determine the motion and electronic states:

$$M_I \ddot{\vec{R}}_I(t) = -\frac{\partial E(\{\vec{R}_I\}, \{\psi_i\})}{\partial \vec{R}_I(t)} \tag{9}$$

and

$$m\ddot{\psi}_i(t) = -\frac{\partial E(\{\vec{R}_I\}, \{\psi_i\})}{\partial \psi_i^*(t)} + \sum_{i,j} \Lambda_{ij}(\langle \psi_j \mid \psi_i \rangle - \delta_{ij}) \tag{10}$$

where the Lagrange multipliers Λ_{ij} are used to ensure conservation of orthonormality of the $\{\psi_i\}$ during the dynamical evolution of the system.[19] Thus, the dynamics of nuclear centers and the electronic orbitals can be solved using two Newtonian equations (9) and (10). This method requires much less computational efforts than the Hellmann–Feynman method does. Unlike in the Hellmann–Feynman simulations, in the Car–Parrinello approach the system is not exactly in its adiabatic ground state, but moves at a "distance" from the Born–Oppenheimer surface. Thus, in Car–Parrinello molecular dynamics simulations it is essential to adjust to a fast dynamics of the electronic degrees. The time-step for the integration of motion in Car–Parrinello simulations is typically at least one order of magnitude shorter than for comparable Hellmann–Feynman simulations.

2.2. Classical Molecular Dynamic Modeling

Quantum molecular dynamics (MD) simulations can usually treat 10^3–10^5 atoms in a real time of a few nanoseconds. In a MD method, a molecular system is described with the Newton's laws. However, although the total energy of the system is not a function of electronic wavefunctions and is expressed explicitly as a function of the nuclear or ionic positions as $U(\vec{R}_1, \vec{R}_2, ..., \vec{R}_N)$, it is usually evaluated using the ab initio methods. Thus, the atoms are treated as classical particles moving on the Born–Oppenheimer potential energy surface (PES) where $U(\vec{R}_1, \vec{R}_2, ..., \vec{R}_N)$ equals to a constant, and the quantum mechanical motion governed by the Schrödinger equation is replaced by Newton's equation of classical mechanics:

$$\vec{F}_i = m_i a_i \tag{11}$$

where m_i is the mass of the ith atom, a_i is the acceleration of the ith atom, and $\vec{F}_i$ is the resultant force acting on the ith atom. The force is calculated from the interatomic potential energy according to Eq. (12),

$$\vec{F}_i = -\frac{\partial U}{\partial \vec{R}_i} = -\nabla_i U \tag{12}$$

Two primary aspects to the practical implementation of molecular dynamics are (i) the numerical integration of the equations of motion along with the boundary conditions and any constraints on the system and (ii) the choice of the interatomic potential. For a single-component system, the potential energy can be written as an expansion in terms of n-body potentials:

$$U(\{\vec{R}_i\}) = \frac{1}{2!}\sum_{i \neq j} U_2(\{\vec{R}_{ij}\}) + \frac{1}{3!}\sum_{i \neq j \neq k} U_3(\{\vec{R}_{ij}\}, \{\vec{R}_{ik}\}, \{\vec{R}_{jk}\}) + \ldots \tag{13}$$

where $\vec{R}_{ij} = \vec{R}_i - \vec{R}_j$ are inter-atomic separations, U_2 represents interactions between pairs of atoms, U_3 depends on the relative orientations of triplets of atoms, etc. The choice of potential is determined by the bond type, the desired accuracy, transferability, and the computational resources.[20] Quantum mechanical calculations can give a guide to obtain an ideal molecular force field. Of course, validation of simulation results with experimental measurements on thermophysical properties and vibration frequencies is very important for developing MD force field. Thus, force fields are material specific. For example, Goddard et al. developed DREIDING force field.[21] They claimed that the force field can be used to predict structures and dynamics of organic, biological, and several main-group inorganic molecules. For the DREIDING force field the total potential energy is given by

$$U_{total} = U(r_{ij}) + U(\theta_{ijk}) + U(\varphi_{ijkl}) + U_{LJ} + U_{\text{elec}} \tag{14}$$

where the first three terms are intramolecular interactions, representing bond stretching, angle bending, and torsion potentials, respectively:

$$U(r_{ij}) = \frac{1}{2}k_b(r_{ij} - r_0)^2 \tag{15}$$

$$U(\theta_{ijk}) = \frac{1}{2}k_\theta(\theta_{ijk} - \theta_0)^2 \tag{16}$$

$$U(\varphi_{ijkl}) = \frac{1}{2}v_{jk}\{1 - \cos[n_{jk}(\varphi - \varphi^0_{jk})]\} \tag{17}$$

and the last two terms are intermolecular interactions of the Lennard–Jones (LJ) and Coulombic type:

$$U_{LJ} = 4\varepsilon_{ij}\left[\left(\frac{\sigma}{r_{ij}}\right)^{12} - \left(\frac{\sigma}{r_{ij}}\right)^{6}\right] \tag{18}$$

$$U_{\text{Coul}} = 322.0637\sum_{i>j}\frac{q_i q_j}{\varepsilon_{DC} r_{ij}}\text{kcal/mol} \tag{19}$$

The parameters are defined by fitting the results of ab initio computations for various polymer/catalyst systems.

For all MD force fields, the parameters are determined by quantum chemical calculations or by fitting to thermophysical and phase coexistence data.

There are two major limitations for MD simulation methods. Although the computational effort required for MD simulations is much less than that for ab initio simulations, the processes of interest, such as atomic diffusion and conduction, are multiscale phenomena that span a spatial scale range from nanometers to millimeters and time scale range from 10^{-15}–10^{-13} s of molecular vibration frequency to 10^{-6}–10^{-3} s a minimum time required for conventional diffusivity measurements in studying PEFCs. Unless the results of a MD simulation can be comparable with the experimental results measured in a macroscopic system and in a timescale during which the phenomenon of interest occurs, it is not convinced that the MD method is useful for studying PEFCs or for atomistic computer design of new materials for PEFCs. The second major limitation is that classic MD methods cannot describe bond breaking and formation and chemical reactions.

2.3. Monte Carlo Modeling

Monte Carlo (MC) methods can address the "time gap" problem of MD. The basis of MC methods is that the deterministic equations of the MD method are replaced by stochastic transitions for the slow processes in the system.[3] MC methods are stochastic algorithms for exploring the system phase space although their implementation for equilibrium and non-equilibrium calculations presents some differences.

2.3.1. *The Metropolis Algorithm*

The thermodynamic average $\langle A \rangle$ of a variable with values A_i in state i that have energy E_i is given by:

$$\langle A \rangle = \sum_i A_i p_i / \sum_i p_i \tag{20}$$

in which the probabilities $p_i = \exp(-E_i/k_B T)$, where k_B is Boltzmann's constant and T is the absolute temperature. If the system is initially in a state i, detailed balance requires that the rate of transitions TR_{ij} from state i to state j satisfies the relation:

$$\frac{TR_{ij}}{TR_{ji}} = \frac{p_j}{p_i} = \exp[-(E_j - E_i)/k_B T] \tag{21}$$

Because the energies and temperature are known, to generate a set of states with the distribution p_i, the TR_{ij} are given by

$$TR_{ij} = \begin{cases} 1 & \text{if} \quad p_j > p_i (E_j < E_i) \\ \exp[-(E_j - E_i)/k_B T] & \text{if} \quad p_j \leq p_i (E_j \geq E_i) \end{cases} \tag{22}$$

A random number $\boldsymbol{r}\varepsilon(0,1)$ is then selected and the system is moved to state j only if $r < \exp[-(E_j - E_i)/k_B T]$.

2.3.2. *Kinetic Monte Carlo Modeling*

Suppose that the probability of finding a system in state σ at time t is $p(\sigma, t)$ and that the rate of transitions per unit time from σ

to σ' is $W(\sigma,\sigma')$. Then the evolution of p is the so-called Markov master equation[3]:

$$\frac{\partial p}{\partial t} = \sum_{\sigma'} p(\sigma',t)W(\sigma',\sigma) - \sum_{\sigma'} p(\sigma,t)W(\sigma,\sigma') \qquad (23)$$

The transition rates for an event i is assigned a rate $r_i = v_i \exp(-E_i/k_BT)$, where v_i is a frequency factor, typically of the order of a vibrational frequency ($10^{13}\,\mathrm{s}^{-1}$) for surface processes, E_i is the free-energy barrier for the process, and T is the temperature. KMC algorithms solve this equation by accepting and rejecting transitions with probabilities that yield the correct evolution of a non-equilibrium system. The factor v_i represents the timescale of the fastest process and is computed by MD simulations. The exponential factor, $\exp(-E_i/k_BT)$, extends the timescale of the KMC methods to that of the actual transition processes in the order of even several minutes. In MC or KMC simulations, each configuration in an MC or KMC simulation is generated stochastically in such a way that the molecular configuration is dependent only on the previous configuration at last step unlike in MD simulations where the current configuration is the result of an integration of all previous configurations. Thus, a MC method can simulate processes in a millisecond and even minutes but may lose the information with respect of the trajectory of the system and not particularly suitable for elucidating microscopic mechanisms.

2.4. Advancement of MD Methods

2.4.1. Empirical Valence Bond Models

Valence bond (VB) theories or empirical valence bond (EVB) methods have been developed in order to solve this problem with bond potential functions that (i) allow the change of the valence bond network over time and (ii) are simple enough to be used efficiently in an otherwise classical MD simulation code. In an EVB scheme, the chemical bond in a dissociating molecule is described as the superposition of two states: a less-polar bonded state and an ionic dissociated state. One of the descriptions is given by Walbran and Kornyshev in modeling of the water dissociation process.[4,5] As

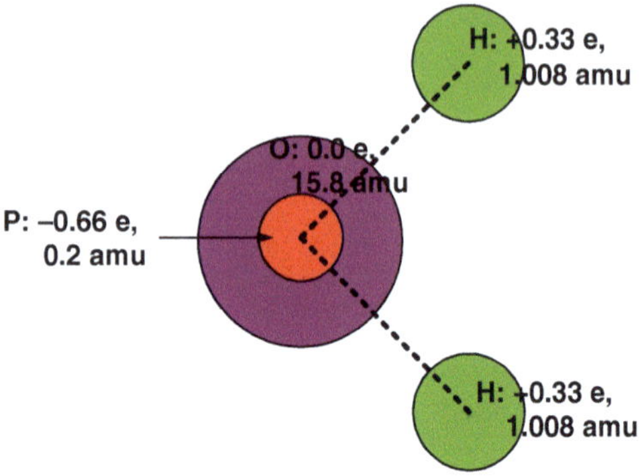

Figure 3. The four sites of the water molecule are shown schematically, along with their respective charges and masses.

shown in Fig. 3, the water molecule has four interaction sites, each of which undergoes Newtonian dynamics in the MD simulation. Two of these sites correspond to the protons in the water molecule, each having charge $q_{\rm H} = 0.33$ e and mass $m_{\rm H} = 1.008$ amu. Oxygen is a combination of two objects or sites: chargeless "oxygen" $m_{\rm O}$ and $m_{\rm P}$. The protons are bound to each other and to a chargeless "oxygen" site via the intramolecular interactions,

$$E_{\rm bond} = \frac{1}{2}a[(\Delta r_1)^2 + (\Delta r_2)^2] + \frac{1}{2}b(\Delta r_3)^2 + c(\Delta r_1 + \Delta r_2)\Delta r_3 + d(\Delta r_2 \Delta r_1) \tag{24}$$

where Δr_1 and Δr_2 are the stretches of O–H bond lengths and Δr_3 is the stretch of the H–H bond. The chargeless oxygen site is bound to a fourth site, with mass $m_{\rm P} = 0.20$ amu and charge $q_{\rm P} = q_{\rm H}$, by an interaction of the form,

$$E(r_{OP}) = \frac{1}{2}k_2 r_{OP}^2 + \frac{1}{4}k_4 r_{OP}^4 \tag{25}$$

The dynamics of this low-mass charged site provides an "inertialess" polarizability, intended to mimic the electronic polarizability of the frequency domain higher than the O–H stretch frequency. This feature is very similar to that of ab initio methods.

2.4.2. MD Modeling with Reactive Force Field

Goddard et al. developed and validated the reactive force field (ReaxFF) to describe complex reactions (including catalysis) nearly as accurately as QM in some cases but at computational effort comparable to classical molecular dynamics (MD).[10,18] Similar to empirical non-reactive force fields, the reactive force field divides the system energy up into various partial energy contributions,

$$\begin{aligned} E_{\text{system}} = {} & E_{\text{bond}} + E_{\text{over}} + E_{\text{under}} + E_{\text{val}} + E_{\text{pen}} + E_{\text{tors}} \\ & + E_{\text{conj}} + E_{\text{vdWaals}} + E_{\text{Coulomb}} \end{aligned} \tag{26}$$

A fundamental assumption of ReaxFF is that the bond order BO'_{ij} between a pair of atoms can be obtained directly from the interatomic distance r_{ij} as given in Eq. (27).

$$\begin{aligned} BO'_{ij} = {} & \exp\left[p_{\text{bo},1}\left(\frac{r_{ij}}{r_o}\right)^{p\text{bo},2}\right] + \exp\left[p_{\text{bo},3}\left(\frac{r^{\pi}_{ij}}{r_o}\right)^{p\text{bo},4}\right] \\ & + \exp\left[p_{\text{bo},5}\left(\frac{r^{\pi\pi}_{ij}}{r_o}\right)^{p\text{bo},6}\right] \end{aligned} \tag{27}$$

The bond order is the exponent of a summation of three terms. The first term is the contribution of σ bond characterized with $p$bo,1 and $p$bo,2. It is unity when r_{ij} is below 1.5 Å but negligible when r_{ij} is above 2.5 Å. The second term is the contribution of the first π bond characterized with $p$bo,3 and $p$bo,4. It is unity when r_{ij} is below 1.2 Å and negligible when r_{ij} is above 1.75 Å. The third term is the contribution of the second π bond characterized with $p$bo,5 and $p$bo,6. It is unity when r_{ij} is below 1.0 Å and negligible when r_{ij} is above 1.4 Å. The partial energy contributions related to bonds, E_{bond}, E_{over}, E_{under}, E_{val}, E_{pen}, E_{tors}, and E_{conj} are all functions of bond order BO'_{ij}. All valence terms (bonds, angles, torsions) go to zero, as the bonds are broken ($BO'_{ij} = 0$). Thus, in the ReaxFF modeling methods the description of bond formation/breaking is smooth as a function of bond order. The bond orders are determined only by the interatomic distances and are updated for every iteration. This allows automatically recognizing and handling changes in connectivity as reactions proceed. It was claimed that ReaxFF-based MD simulations can be carried out for non-equilibrium processes. Modeling the electrochemical processes at the interface of catalyst/polymer electrolyte in PEFCs may require simulation models on the scale of 10 nm or 1,000,000 atoms. This is now feasible with the ReaxFF-based MD simulation.

2.4.3. Methods for Accelerating Molecular Dynamics Simulations

Voter et al. summarized the methods that can be used to accelerate MD simulation.[6] These methods are underpinned by the existence of infrequent-event processes and derived from transition state theory. These methods are hyperdynamics, parallel replica dynamics, temperature accelerated dynamics, and on-the-fly kinetic Monte Carlo methods that can achieve a simulation speed several orders of magnitude greater than conventional MD methods while most of atomistic details can be retained. An infrequent-event system is such a system that the dynamical evolution consists of vibrational excursions within a potential basin, punctuated by occasional transitions between basins. These transition events are infrequent in the sense that the average time between events is many vibrational periods. Transition state theory (TST) is the formalism that underpins all of the accelerated dynamics methods, directly or indirectly. In the TST approximation, the classical rate constant for escape from state A (basin) to an adjacent state B (basin) is taken to be the equilibrium flux through the dividing surface between A and B. Thus, for TST, because this flux is an equilibrium property of the system, the TST rate can be computed without knowing an actual trajectory connecting A and B. The transition rate between A and B can be evaluated using the Monte Carlo method. Thus, the actual kinetic steps between A and B are neglected from the computational process.

Among the methods utilizing the TST, the parallel replica method is the simplest and most accurate dynamics techniques, because the only assumption for the method is that of infrequent events obeying first-order kinetics (exponential decay). In practice, replicated MD simulations are started in a particular basin state in search for a transition to an adjacent basin state on M non-correlated processors. Whenever a transition is detected on any processor, the exploring processes are stopped to start a new exploring process in a new state. The simulation clock is advanced by the accumulated trajectory time summed over all the total time spent exploring phase space within the previous basin.

The other methods rely on assumptions that are artificial. In the hyperdynamics approach, the potential surface is artificially modified to reduce the energy barrier for transition. In the temperature-accelerated dynamics (TAD) method, the idea is to speed up the transitions by increasing the temperature, while filtering out the transitions that should not have occurred at the original temperature. These methods are viable on the condition that the system trajectory is in the basin state for a long time and long enough to lose the memory on how it gets there.

3. MODELING OF OXYGEN ELECTROREDUCTION REACTION CATALYSTS

Seeking a breakthrough in electrocatalysis of oxygen electroreduction reaction (OER) in cathodic catalyst layer is one of the major issues in the current PEFC research. The exchange current density for oxygen reduction is about 6 orders of magnitude lower than that for hydrogen oxidation. The slow rate and associated high overpotential at the cathode of PEFC limits their specific power output (kW kg^{-1}) and overall efficiency. Generally speaking, if the voltage of a single cell is maintained at 0.8 V, a significant efficiency loss (30%) is due to the cathode overpotential although a high platinum loading (e.g., 0.5 mg cm^{-2}) that may contribute a high device cost is used. If the exchange current density for oxygen reduction is increased significantly by 2–3 orders of magnitude, the specific power can be increased by many times at the same output voltage and efficiency. Thus, the ratio of performance to the cost is increased by that many times. It is commonly accepted that the transport processes, the adsorption/desorption, and the reaction path in the specific interface structure between the polymer electrolyte and catalysts dictate the electrochemical activity of the catalyst layer. This underlies the fact that the electrochemical activity of the oxygen reduction at cathode for the Pt/Nafion electrolyte interface is 3 orders of magnitude lower than that of a Pt/H_2SO_4

interface.[22,23] Specifically, the exchange current density on Pt/C catalyst in conjunction with Nafion electrolyte is reported 8.0×10^{-7} A cm^{-2} for oxygen reduction at the cathode (ca. 10^{-3} A cm^{-2} for hydrogen oxidation at the anode) on the basis of the area of the catalyst layer.[24] This is equivalent to 2.0×10^{-9} A cm^{-2} on the basis of true active area of Pt crystallities. However, in 0.5 M sulfuric acid and with specific adsorption, Pt crystallites show an exchange current density between 5.0×10^{-7} and 6.0×10^{-6} A cm^{-2} on the basis of true active area of Pt.[25] Thus, the exchange current density at the Pt/Nafion interface is 2–3 orders of magnitude lower than that at the Pt/liquid electrolyte interface. Therefore, if the Pt/polymer electrolyte interface is as effective as the Pt/H_2SO_4 interface, the specific power of PEFC may be increased by many times. However, lack of understanding of the catalyst layer may underlie the fact that 30-year's great efforts on development of the catalysts and polymer electrolytes have not resulted in a desired progress. Better understanding on the catalyst/polymer electrolyte interface structures and its implication for electroreduction of oxygen may be the key for a breakthrough. In the last 10 years, a growing number of ab initio studies of electroreduction of oxygen on transition metals and alloys have been reported. These theoretical studies shed light on the understanding of microscopic mechanisms. However, at the present time, ab initio modeling is unable to provide convincing descriptions for the interface structure and detailed reaction steps due to the limitation with over-simplified models and extremely short timescale. Although MD methods can be used to establish an interface between the polymer electrolyte and catalysts with a realistic size, the results are not convincing because classical MD methods are not supposed to describe interactions between polymer electrolyte and catalysts involving bond breaking and formation. ReaxFF-based MD studies may enable us to determine the fundamental processes at realistic electrode–electrolyte interfaces. However, no result has been published yet.[10]

3.1. The Interface Structure

It is believed that the greatest challenge to understanding and optimizing PEM-FC performance is to comprehend the processes taking place at the interface between the hydrated polymer electrolyte

(Nafion) and the cathode (carbon-supported Pt nanoparticles). It is essential to understand how the protons and electrons are transferred across the membrane–cathode interface, particularly, how this is affected by the distribution of water at this interface. However, in earlier simulations on the electrochemical reactions no electrolyte molecules were involved.[26–30] It is amazing that these earlier simulation results based on the so-called "vapor-phase" models have been used as basis for development of catalysts in PEFCs. As shown in Fig. 4, the apparent oxygen reduction reaction rate is governed by a series of physicochemical processes that occurs at the buried electrode interface. First of all, diffusion of oxygen supplies reactants via two paths: (1) through the polymer electrolyte and (2) through the exposed region on the catalyst. Although most authors believe that the first path dominates, the possibility of the second path cannot be eliminated. Other issues regarding diffusion of oxygen include the effect of the support and catalysts on the diffusion of oxygen, the roles of the water layer, polymer clusters, and specific adsorbates at the interfacial area, and impact of the electrical field at the interface. Adsorption/desorption is often considered as the rate-determining step (rds) of electroreduction of oxygen. Adsorption can dramatically alter the structure and electrical field distribution and hence impact the entire OER process. Electron and proton transfer at the interface is directly related to the electrical field or electrical double layer (EDL) and must be heavily affected

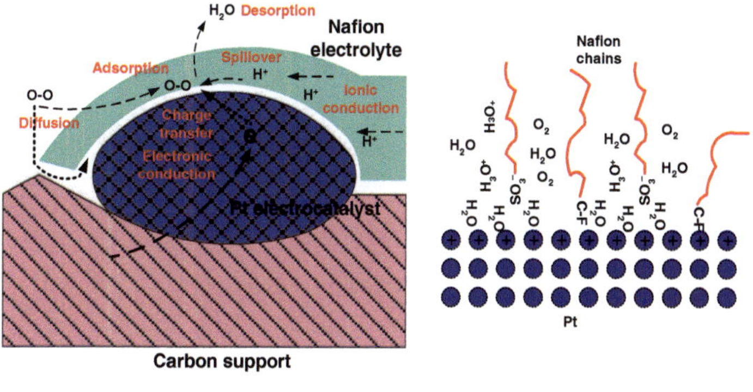

Figure 4. Reaction steps in oxygen reduction at the electrocatalyst in a PEFC (left) and a postulated molecular scale structure of the interface (right).

by the interface structure or specifically the adsorption of polymer functional groups, distribution and orientation of water, and adsorption of oxygen. Finally, formation and rejection of water are the last step of the reaction. Modeling of the electroreduction of oxygen first requires a precise description of the interface structure.

Indeed, the EDL structure has been one of the major topics of electrochemistry in liquid electrolytes. Numerous studies on atomistic modeling of the electrode/water interface have been reflected in a bulk of literatures. In earlier years, empirical force field models were created.[31–35] Latterly, ab initio simulation derived metal–water and metal–ion pair potential based MD models appeared.[36–40] Quantum mechanics models were created with periodic boundary conditions and with empirical electrical field of a continuum positive charge "jellium."[41–45] Pseudopotentials for the electrode were used to define the interface structure.[46–49] Ab initio valence electron cluster models are often specified as interface model but only include a few atoms.[50–57] Many theoretical analyses obtained a picture that the metal–water interface is highly organized and there is a peak of water density near the electrode at room temperature.[31–40,46–49] This ordered structure is often referred as "ice-like" or a prevalent hydrogen-bond ring structure. This has also been detected experimentally using X-ray reflectivity by Toney.[58] It has also been demonstrated, however, that although the time averaged structure is ice-like, at any individual moment the water arrangements at the surface are completely random. Understanding of the structure of water layer is relevant for modeling PEFC because presence of water on catalysts does greatly impact the performance of PEFC. However, it is also important to understand the impact of the polymer clusters on the EDL and OER process. Unfortunately, the number of publications on the EDL structure of polymer electrolyte/catalysts is very limited. To date, understanding of the EDL in PEFC largely relies on ab initio modeling for the EDL at the catalyst/water interface.

3.1.1. Ab Initio Modeling of Interface Structure in Aqueous Solutions

Unfortunately, there has hardly been any ab initio work on the structure of catalyst/polymer interface. On the other hand, there is a bulk of literature on the structures of catalyst/water interfaces.

These studies are considered as being relevant for PEFCs because it is commonly accepted that the water layers between the polymer and catalysts play a major role in electroreduction of oxygen. These studies were probably stimulated by the famous experimental findings by Toney et al. who investigated the interfacial structure of water molecules near charged metal surfaces using in situ X-ray scattering techniques.[58] The major findings are that water is ordered in three molecular layers above the metal surface and the orientation of interfacial water molecules depends on the sign of the charge density of the surfaces. In addition, they concluded that the area density of water next to a charged surface is very high compared to its bulk density. For example, the surface density of water molecules next to a positively charged silver electrode with a potential of 0.52 V relative to the potential of zero charge is about two times as great as that of bulk water.[58] Because of the ordering and high density of water, it was suggested that the dynamics of water molecules near charged surfaces would be very different from that of the bulk molecules.

Establishment of a correct electrical field as a function of electrode potential at the interface is a crucial and difficult issue for modeling the interface structure. If this issue cannot be addressed correctly, the real interface structure and its impact on the equilibrium and kinetic behaviors of the electrochemical reactions in PEFCs cannot be fully revealed. Lozovoi et al. pioneered the use of a potential reference and a homogeneous countercharge and modeled field-evaporation effects.[59] Filhol and Neurock applied this idea to the electrode/water systems via the creation of two potential references – a vacuum state and an aqueous state.[60] Briefly, a net surface charge density is introduced to the electrode via the variation of the number of electrons available to the self-consistent calculation, and overall cell neutrality condition is maintained by applying a homogeneous background charge for mimicking the presence of a diffuse ionic distribution close to the interface. This so-called double-reference method not only allows to freely vary the electrode potential versus a virtual reference electrode via varying the charge on the electrode and counter charges in the electrolyte, but also provides a description of electrode/water interface rather than electrode/vapor interface.

Taylor et al. further advance the method by introducing explicitly a plane of ions with the countercharge rather than a homogeneous countercharge in the liquid phase.[61] The electrochemical potential of the interface is established by adding a charge on the Pt slab while adding a compensating charge in the electrolyte side of the interface (Fig. 5). The potential at an arbitrary position *z* and Fermi potential of the electrode are given in reference to the potential at the center of the vacuum. The second reference is in the middle of a portion of electrolyte that is fixed far from the electrode

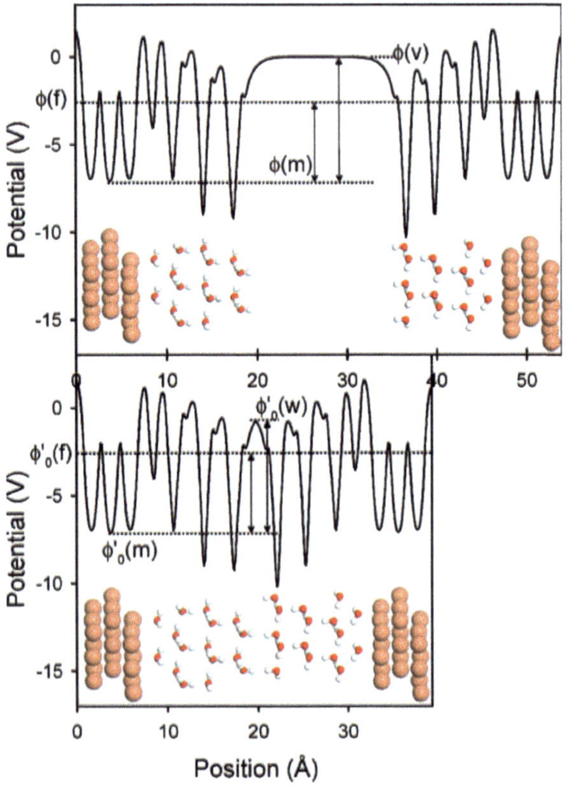

Figure 5. A schematic diagram illustrating the electrostatic potential profile as a function of position across the normal axis of the unit cell. The system shown here contains two symmetric Cu(1 1 1)/H_2O slab faces contained within the periodic simulation cell.[61] Reprinted from Physical Review B, 61C. D. Taylor, S. A. Wasileski, J.-S. Filhol, M. Neurock, Phys. Rev. B 73 (2006) 165402. Copyright (2006), with permission from American Physical Society.

while the rest of system is relaxed in response to the applied charge. The potential at all other position, z, including Fermi potential of the electrode is shifted according to the second reference point.

$$\phi_q(z) = \phi_q'(z) - \phi_q'(R) + \phi_0(R) \tag{28}$$

$$\phi_q(F) = \phi_q'(F) - \phi_q'(R) + \phi_0(R) \tag{29}$$

Where the subscript q represents the amount of charge on the Pt electrode, $\phi_q(z)$ and $\phi_q(F)$ are the referenced potential at z and the referenced Fermi potential, respectively, $\phi_q'(z)$ and $\phi_q'(F)$ are the un-referenced potential at z and the un-referenced Fermi potential respectively, and $\phi_0(R)$ is the potential at the reference point in the portion of electrolyte in reference to the vacuum reference point (the first reference). The electrode potential versus a SHE, U_q, is given by

$$U_q = -4.8 - \phi_q(F) \tag{30}$$

At a given potential, the AIMD was run up to 30 ps for structure optimization. Electrochemical reactions were analyzed using DFT methods. Other authors reported ab initio modeling of the interface involving only a few water molecules. They examined the charge distribution at the metal interface as a function of potential over Cu(1 1 1) and Pt(1 1 1) and the paths and potential-dependent reaction energetic for the electrocatalytical activation of methanol over Pt. Their results demonstrate that use of a homogeneous countercharge in aqueous solution is a good approximation but it is not in vapor phase. Although via other experimental and theoretical studies water is in a so-called ice-like ordered state, they did observe that the repulsion and rotation of water dipoles at the interface in a random way. This is one of the many cases showing that atomistic simulation in "liquid phase" can reveal some of very important behaviors that cannot be directly obtained with the "vapor phase" models.

Following this approach, Janik et al. studied the electrochemical reactions using 20–30 water molecules for mimicking the water layer on a Pt slab under periodic boundary conditions but without accounting for the role of polymer. They evaluated the interface structure considering the adsorbed oxygen (Fig. 6).[62] Their findings

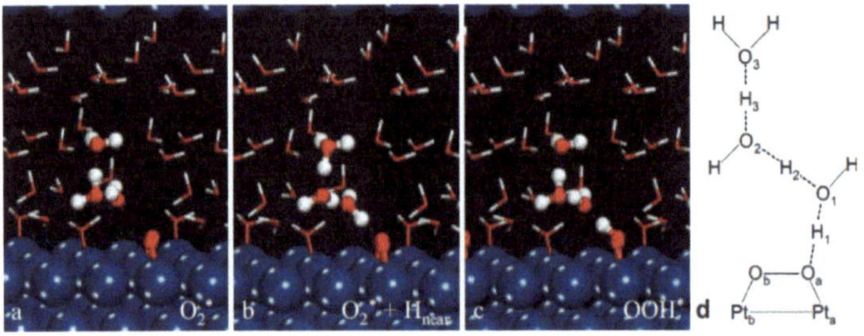

Figure 6. Optimized structures of the initial adsorbed molecular O_2 and the OOH* intermediates over aqueous solvated Pt(1 1 1) surface as determined by DFT.[62] Reprinted from Journal of Electrochemical Society, M.J. Janik, C.D. Taylor, M. Neurock, J. Electrochem. Soc. **156**, B126 (2009)– Reproduced by permission of the Electrochemical Society.

include that the O_2 molecule optimally binds to the Pt(1 1 1) surface in solution in a di-σ configuration, where the center of the molecule lies over a bridge site and both O atoms pointed toward atop sites. Interesting findings also include that O_2 may tends to have an η_1-like structure with a tilt to the Pt surface; however, previous vapor-phase calculations found that O_2 only adopts this η_1 mode at full surface coverage. They also evaluated the adsorption energetics as a function of potential in the standard hydrogen electrode (SHE). These energetics normally are not monotonic functions of potential.

Ab initio molecular dynamics (AIMD) simulation is a useful method in determining the interface structure without relying on any assumption. Izvekov et al. investigated adsorption of water on Cu(1 1 0).[48] They claimed that the simulation results were in good agreement with the experimental results. It was found that there is a strong coupling of the water overlayer with the metal crystal electronic states. However, the surface-state charge density is undisturbed by the presence of water. The empty surface states are seen to not be quenched by the presence of water, which is also in qualitative agreement with existing experiments.

3.1.2. MD Modeling of Interface Structure on Catalysts in Aqueous Solution

Raghavan et al. described a form interaction potential between rigid water and a rigid platinum metal surface.[63] Using this potential they performed a MD simulation on water lamina restricted by two Pt(1 1 1) surfaces. At 300 K the contact water layer to the metal surface displays a solid-like behavior that was evidenced by patches of ice-like structure observed in the simulation. The next two layers of water display ordering similar to ice-I. Beyond these three layers the structure and dynamics of water are bulk-like. The lateral diffusivity of water increases with the normal coordinates above the metal surface. The structure of water coordinates with the structure of the metal surface.

Spohr investigated the effects of the empirical electrostatic boundary conditions on the interface properties of water and on the adsorption of ions.[64] Spohr also conducted molecular dynamics simulations of water layers confined by two Pt(1 0 0) surfaces. The structure is revealed mainly with one-particle density profiles and solvent pair correlation functions. He found that the surface-induced structural heterogeneity ranged up to distances of 10 Å and that the central water layer is bulk-like. Hydrogen bonding in the vicinity of the interface is only slightly reduced in comparison to that of bulk water. The orientation of water is strongly influenced by water–water interactions and is considerably different from the preferential orientation according to the water–platinum interaction potential.

The structure of water near charged surfaces was studied by Berkowitz et al., Kiseley et al., and Snnapati and Chandra.[65–69] These authors carried out MD simulations of both non-polarizable SPC/E and polarizable PPC models of water near charged metal surfaces and found no dramatic increase of the water density and disruption of hydrogen bonding near the charged surfaces.

The dramatic increase of water density at a charged surface was observed by Toney et al. in their in situ X-ray scattering experiments, which has not yet been confirmed by simulation results.[58,70] In another MD simulation work, Kiselev et al. found that self-diffusion coefficient strongly decreases with increasing electric field.[27] However, no difference between the self-diffusion coefficients for motion parallel and perpendicular to the external field was observed.

Siepmanna and Sprik used a classical MD method to simulate the ordering of a water film adsorbed on an atomic model of a tip of scanning tunneling microscope (STM) probe approaching a planar metal surface.[71] They investigated the structural rearrangements of the water bilayer between the pair of scanning tunneling microscope (STM) electrodes in response to an applied external voltage difference. They found significant asymmetry in the dependence on the sign of the applied voltage. They also found that for the conditions typical for operation of a STM probe, the change in the work function is comparable to the applied voltage (a few hundred millivolts).

Halley et al. employed a MD method for the simulation of metal/water interfaces.[72] They found that the occupancy of on-top binding sites for water in this model as applied to a (1 0 0) surface of "copper" was very sensitive to potential. They suggested that this may provide an explanation for some previously unexplained features of X-ray data on water structure and noble metal/water interfaces. They also noticed that the strong bonding of water on a metal surface may result in metastable charging of the interface in molecular dynamics timescales.

3.1.3. MD Modeling of Interface Structure of Polymer Electrolyte/Catalysts Interface

To date, it is difficult to model the interface between catalyst and polymer electrolyte. MD methods allow simulation of a system with 10^3–10^5 atoms. Thus, MD simulations are suitable for describing a system with the realistic size of a metallic catalyst on carbon support in contact with a cluster of polymer electrolyte (Fig. 2). The problem is, however, whether these simulations are correct.

Balbuena et al. involve polymer molecules in the MD modeling of the metal catalyst/Nafion interface.[73] Figure 7 shows all the sulfonic acid groups surrounding a metal cluster and their distances to the catalyst particle. They used this method to estimate the number of Pt atoms that are directly accessible by protons migrating through the water clusters formed in the hydrophilic sites of a Nafion cluster. The diffusion of protons to the reaction centers on top of Pt surface is a crucial step of OER. They found that within simulation timescales (maximum production time of 400 ps), once the polymer is attached to the carbon support or catalyst, its constituent atoms generally vibrate around their equilibrium positions and little further configuration changes are observed in the side chain. In addition, the polymer greatly reduces self-diffusion rate of the catalyst over the graphite surface, thus reducing catalyst sintering.

Figure 7. Interactions of Nafion's side chains and a Pt nanoparticle. A light brown color indicates Pt atoms at distances closer than 6.79A° from an S atom. Atoms that are in regions around the sulfonic groups have been enlarged to improve visualization. Color codes are white and red for H and O in water, respectively, green for F, gray for C, and yellow for S.[73] Reprinted from Electrochim Acta, Vol. 50, Balbuena et al., Molecular modeling studies of polymer electrolytes for power sources, p. 3788, Copyright (2005), with permission from Elsevier.

Balbuena et al. also conducted simulations at various water concentrations for various water contents (Fig. 8). At low water contents ($\lambda = 5$), small water clusters are almost not connected with each other (Fig. 8a). At a very high water concentration ($\lambda = 45$), water forms a continuous phase (Fig. 8c). When λ is about 24, close to the amount in fully hydrated Nafion membranes at room temperature, the interface is defined by a semi-continuum water film (Fig. 8b) where some water clusters with diameters of about 1 nm are interconnected by multiple water bridges. The average water density in this phase is estimated to be about 0.682 g cm^{-3}, a much lower value than that of the bulk water phase at 353 K. These observations provide very valuable information for further investigating the OER and really highlight the power of atomistic simulations on the research topics for which currently existing experimental tools are lack of the resolutions in spatial and temporal scales.

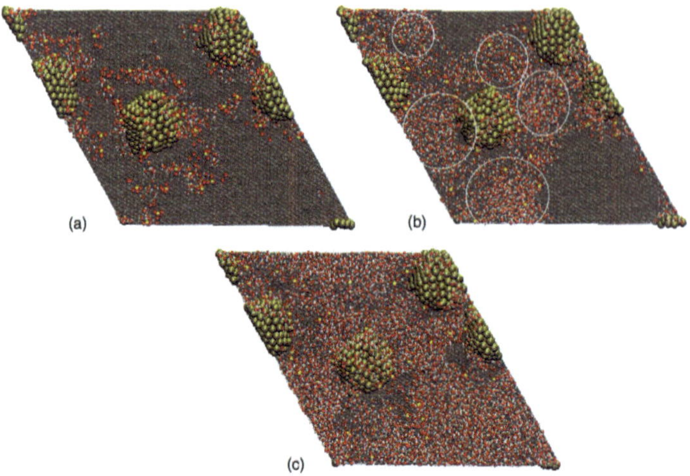

Figure 8. Distribution of water near the catalyst at different λ (water/S ratio) values: (a) $\lambda = 5$, (b) $\lambda = 24$, and (c) $\lambda = 45$. Nafion is not shown in order to clearly illustrate the water clusters. At $\lambda = 24$ the clusters (highlighted for clarity) are clearly defined showing connectivity among them.[73] Reprinted from Electrochim Acta, Vol. 50, Balbuena et al., Molecular modeling studies of polymer electrolytes for power sources, p. 3788, Copyright (2005), with permission from Elsevier.

Lamas and Balbuena studied the interfaces between nano-sized Pt catalysts on carbon and Nafion electrolyte using a classic MD method.[74] In this work, they carefully selected the interaction potentials. For example, Pt–Pt interactions are described with the Sutton–Chen many-body potential that has proven to successfully describe various properties in metallic systems.[75] Water and hydronium are represented with the rigid SPC/E model.[76,77] A two site model is employed for O_2. The O–O bond is described by a Morse potential and the intermolecular interactions by a 6–12 LJ potential, centered on each oxygen atom. The carbon atoms representing the graphite layers are fixed and the carbon interactions with other atoms in the system are represented by LJ potentials.[21] Nafion is represented by an oligomer model interacting through the Dreiding force field. Interactions between the species were calculated using Lorentz–Berthelot mixing rules.[78] The distribution of charges were obtained from a charge equilibration algorithm.[79,80] By this work, they demonstrate that different force fields can be selected for describing interactions within specific components and between different components.

In the computer simulations, water clustering is observed near the membrane hydrophilic sites constituted by sulfonic acid groups, which due to their affinity with platinum are located near the metallic surface. Not surprisingly they observed that the diffusion of water through the model hydrated polymer membrane depends strongly on the level of membrane hydration due to contribution from various mechanisms whose relative weights change with the degree of hydration.

Calculated water configurational and dynamics properties revealed how water organization in the vicinity of the catalyst surface changes as a function of Nafion contents. The diffusivity of water varies according to the contribution from three mechanisms whose relative weights were found to be a function of the water content. O_2 solubility in water may be a limiting factor in this acidic media with water clusters surrounded with Nafion. They claim that reported O_2 diffusion coefficients in the different phases cannot be used to prove or disprove one mechanism for O_2 transport to the catalyst site over the others because any observed property in experiment is an average of a large number of many different local structures but in the case of this modeling work only a few of these structures are investigated. This is one of the main issues of the MD modeling.

Goddard et al. have established an interconnected water cluster that exists throughout the Nafion membrane.[18] They have also proposed interface models to explain the distribution of water and O_2 at the Pt–Nafion interface. As the first step, they proposed to predict the structures at this interface using the ReaxFF. In addition to the structural features of the interface, they evaluated the processes of protons, O_2 molecules, and water entering the interface and reacting with the electrode. They anticipated that the interface will be heterogeneous in the way that O_2 molecules tend to concentrate in non-polar hydrophobic regions while protons (H_3O^+) and water molecules prefer the hydrophilic regions. Thus the important chemical reactions may occur at the boundaries between these regions on the catalyst.

To provide a basis for answering such fundamental primary questions, they carried out preliminary studies on a model of such an interface, including multiple phases (a hydrophobic Nafion backbone, water, O_2, and carbon support) in contact with each other. To describe the interaction of the Pt nanoparticle with other components (Nafion, water, proton, and oxygen) in this MD simulation, they optimized the force field to have a binding energy close to that from DFT calculations at the B3LYP/6-31G** level. The MD simulation showed that the hydrated Nafion had a well-developed and phase-segregated morphology with percolated water distributions reaching close to the Pt nanoparticle at the cathode.

Unlike the observations of Balbuena et al. Goddard et al. found that the Pt nanoparticle was entirely wetted by water, so that the proton (hydronium) particle can access the entire Pt nanoparticle surface. The O_2 molecules diffuse reasonably faster through the hydrophobic Nafion backbone than through water clusters. The author claimed that because this was a highly non-equilibrium system, it is required to use ReaxFF that can describe bond breaking/formation to investigate the interface in such a non-equilibrium state.

Recently, Goddard et al.[10,81] have developed ReaxFF potentials covering the following aspects of PEM-FC chemistry in the form shown in:

(1) Reactions at metal surfaces. They have developed ReaxFF for Pt/C/H/O, Ru/H/O, and Ni/C/H/O interactions in order to enable a large-scale dynamical description of the chemical events at the fuel cell metal anode and cathode.
(2) Oxidation/reduction reactions. They have fitted ReaxFF against a wide range of metal oxidation states to ensure that ReaxFF describes redox reactions accurately.
(3) Hydrogen transport. The approach that they have used to predict proton transport through complex membranes such as Nafion is to use ab initio methods to determine the barriers for migration of hydrogen as a function of the donor–acceptor separation and then to employ a statistical method that is based on the ab initio results. This method allows a proton jump among water clusters when the configuration around the proton is appropriate.

Although the interface models by the MD methods provide a picture of the atomic distribution of the interface, because of the limitation of the MD method, the details of the interface structure in the vicinity of the electrode is not accurate. The effect of polymer side groups and backbone on the interface structure, specifically the electrical double layer (EDL), is not addressed with sufficient confidence. The AIMD simulations have already demonstrated striking differences with those performed using classical pair potentials (MD force field). For example, the diminished density of water close to the interface and a sparsity of occupation of the atop sites observed by both Price and Halley and Izvekov et al. contrasts clearly with the close packing observed in the MD simulations by Spohr and Heinzinger.[33,44,46,48]) The complete exploration of dynamic effects as a function of metal surface has yet to be explored in a consistent ab initio modeling. On the other hand, existing ab initio interface models hardly include the side-chain functional groups and backbone of Nafion as can be seen in the next section.

3.2. Chemsorption on Catalysts

Chemisorption of oxygen on the electrocatalysts is the first step in the oxygen electroreduction reaction (OER). The adsorbate's structure, bonding type, and energy are key elements in understanding the effects of adsorption on the reaction kinetics. Because of the complex nature of adsorption at electrocatalysts, the computational modeling has been focused on a particular aspect of the problem. Usually the atomistic models include the adsorbate and metal but not the solvent. Theoretical studies of chemisorption can provide information about the nature of bonding at the surface, bond strength, geometry, and site preferences of adsorbates.

Gas phase oxygen adsorption has been investigated with several experimental techniques including photoemission spectroscopy,[82,83] electron energy loss spectroscopy (EELS),[84] near-edge X-ray absorption fine structure spectroscopy (NEXAS),[85,86] and scanning tunneling microscopy.[87] The conclusion is that oxygen adsorption is a complex process as physisorption, molecular adsorption, dissociative chemisorption, and oxide formation are all possible. On the Pt(1 1 1) surface, molecular oxygen is predominantly adsorbed on Pt(1 1 1) at temperatures lower than 120 K via chemical adsorption. These chemisorption states may easily become unstable and dissociate at the temperatures ranging from 150 to 500 K and transfer to atomic adsorption states. On Pt(1 1 1) surface, NEXAFS revealed O_2^- ions.[85,86] EELS observed the existence of two different O–O stretching frequencies.[84] The EELS results were first interpreted in terms of non-magnetic O_2^{2-} ions. The actual state of the adsorbed oxygen is a very important issue of atomistic modeling of the OER. However, the experimental results indicate that the adsorption states vary with experimental conditions. This increases the complexity of the atomistic modeling of OER.

To date, very little experimental and modeling work has been done on the adsorption of oxygen in the presence of water and polymer electrolyte although it is strongly anticipated that water, polymer, and specifically adsorbed species have a great impact on adsorption of oxygen. Most experimental and simulation studies were carried out in so-called vapor phase. In these simulations, only a few gas molecules and a small-size lattice of the catalyst were considered.

3.2.1. Bond Strength of Adsorbed Oxygen Atom

Extensive computational studies of bond strength of atomic oxygen adsorption, especially by Hammer and Nørskov, have been conducted in order to understand the properties that affect adsorption energy.[98] They systematically investigated the property of transitional metal on oxygen adsorption. They used DFT method with GGA-PW91 functional and the slab model. For adsorption on a Pt(1 1 1) surface, their calculation reproduced the experimental equilibrium structure within an error of 1% and the calculated heat of adsorption was 2.68 eV which is in reasonable agreement with the experimental number of 2.4 eV/O_2 for a quarter of a monolayer of oxygen. They also investigated atomic oxygen adsorption on transition metals including Ru, Ni, Pd, Pt, Cu, Ag, and Au.[89–91]

3.2.2. Adsorption Process on Transition Metals

There are three adsorption models for molecular oxygen adsorption (Fig. 9):

- the Griffiths model – in which O_2 interacts with two bonds on a single substrate atom;
- the Pauling model – with end-on adsorption of the oxygen molecule through a single bond;
- the Yeager model – a bridge-like adsorption with two bonds interacting with two sites.

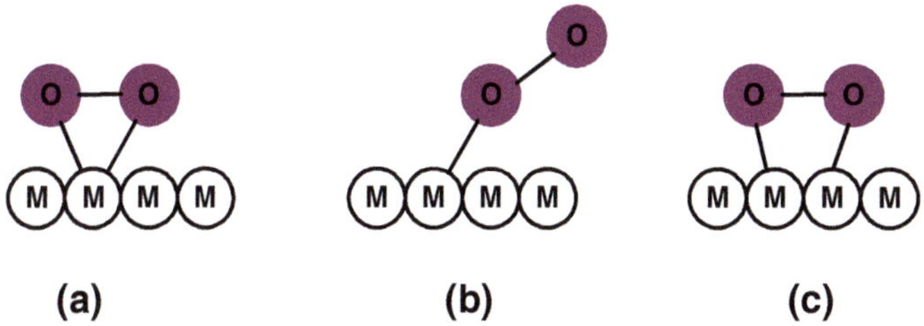

Figure 9. Scheme of oxygen adsorption models: (a) the Griffiths model, (b) the Pauling model, and (c) the Yeager model.

Computational studies of oxygen adsorption on Pt(1 1 1) was reported by Eichler and Hafner.[92] They used the DFT method with the GGA-PW91 functional and slab model. They identified two distinct but energetically degenerate chemisorbed molecular precursor states for O_2 on Pt(1 1 1) at distances of 1.8–1.9 Å. The first state was a superoxo-like paramagnetic precursor formed at the bridge site (t-b-t) with the molecule parallel to the surface (Fig. 9). The O–O bond length was 1.39 Å and the O–O stretching frequency was 850 cm^{-1}. The calculated adsorption energy was 0.72 eV. The second state was a peroxo-like non-magnetic precursor formed in the threefold hollow, with the atom slightly canted in a top-hollow-bridge geometry. The O–O bond length was 1.43 Å for t-f-b (1.42 Å for t-h-b), the O–O stretching frequency was 690 cm^{-1} (710 cm^{-1} for t-h-b) and the adsorption energy was 0.68 eV (0.58 eV for t-h-b).

3.2.3. *On Bimetallic Alloys*

Platinum-based bi-metallics (Pt M, M = Ti, Cr, V, Mn, Fe, Co, Ni, Cu, etc.) have been shown to exhibit enhanced activity toward the OER. Several rationales have been proposed including: (1) enhanced chemisorption of intermediates; (2) a lattice change of Pt that results in the shortening of Pt–Pt interatomic distances by alloying; (3) the formation of skin Pt which has increased d-electron vacancy of the thin Pt surface layer caused by the underlying alloy and the anchor effect of alloy metals on a carbon carrier.[93,94]

Xu et al. carried out self-consistent periodic density functional theory calculations (GGA-PW91) to study the adsorption of atomic oxygen and molecular oxygen, and the dissociation of O_2 on the

(1 1 1) facets of ordered Pt_3Co and Pt_3Fe alloys, and on monolayer Pt covering these two alloys.[95] They also investigated explicitly the strain effect by a 2–3% compression (corresponding to the lattice constant in Pt_3Co and Pt_3Fe) of the equilibrium lattice of Pt(1 1 1). They revealed that there is a linear relationship between the atomic oxygen binding energy and the oxygen dissociation barrier on the transition metals and alloys. The stronger the bond between a metallic catalyst and an atomic oxygen, the more effective the catalyst is in dissociating molecular oxygen. So instead of conducting a complicated and expensive transition-state study, evaluation of the atomic binding energy can be less resource demanding but a feasible approach for selecting and designing effective catalysts for OER. Through DFT calculations, they discovered that Co atoms on the Pt_3Co surface allowed O_2 dissociate more easily than on Pt(1 1 1). The lowest activation energy for O_2 adsorption on Pt_3Co was 0.24 eV whereas that for O_2 on pure Pt was 0.77 eV. The lowest activation energy for O adsorption on Pt_3Co was 4.29 eV whereas that for O adsorption on pure Pt was 3.88 eV.

Balbuena et al. conducted studies of alloys with clusters of Co, Ni, or Cr embedded in a Pt matrix.[96] They used the DFT cluster method with the B3PW91 functional combined with LANL2DZ pseudopotential and basis set. They postulated that Co and Cr could act as active sites for O_2 dissociation instead of being oxidized as "sacrificial sites." They identified XPt and XXPt (X = Co and Cr) being the best active sites to promote O_2 dissociation. While atomic models involving Ni atoms did show acceleration of the O_2 dissociation compared with that of pure Pt.

Wei et al. studied the effect of carbon support to the catalysts (Pt/C and Pt_3Fe/C) using a DFT method with the B3LYP functional.[97] A cluster model consisting of carbon atoms and Pt or Fe on the three major surfaces (1 0 0), (1 1 0), and (1 1 1), of Pt and Pt_3Fe was used to simulate the carbon-supported Pt catalysts. They found that the Pt_3Fe/C alloy catalyst had a lower total energy compared with Pt/C. The adhesion force between the Pt_3Fe alloy catalyst and the C was stronger than that seen between the Pt catalyst and C substrate.

Ruban et al. established an atomic d-band center model.[91] Based on this model, they examined the impact of surroundings to the activity and how the activity can be altered. They revealed that the shift of the d-band depended on the difference in the size of the two alloying metals. The important effect of depositing a layer of one kind of metal atoms on top of a substrate of another type of atoms was because this bilayer structure altered the electron density or the electronic orbitals. This in turn changed the center of the d-bands. In the case that a "small" metal atom is moved into the lattice of a "larger" one, d-band width at the atom becomes smaller than at the surface of the elemental metal. This results in an up-shift in the d-band center in order to maintain the same d-band filling locally. It is commonly accepted that a surface with a higher lying center of d-band (ε_d) can enable stronger bonds between adsorbates and catalysts whereas a surface with low lying ε_d will enable weaker bonds between adsorbates and catalysts. The most active platinum monolayer should have an ε_d with an intermediate value. How the specific adsorbates and polymer impact the state of d-band in a more realistic model is an interesting topic.

Jacob et al. systematically studied Pt/Ni alloy clusters with more than 600 atoms using a DFT method in order to explain why the OER activity of the Pt/Ni catalysts is enhanced in comparison to that of Pure Pt catalysts.[30] By examining the oxygen chemisorption at all surface sites, they found that the greatest atomic binding energy was 3.50 eV at a FCC site with one Ni and two Pt atoms. On pure Pt oxygen is most stable at the FCC site with three Pt atoms. Presence of the Ni atoms in the alloy system causes a drastic change of the electronic configuration of the cluster and therefore the binding energy of an oxygen atom at the FCC site. On the (1 1 1) face of Pt/Ni, oxygen is much more localized than on the (1 1 1) face of pure Pt. The modeling/simulation results seem to support the claim that the combination of a stronger bond and higher extent of localization correlates with the OER activity of the catalysts. However, in order to prove that this is true, more catalyst systems have to be examined.

3.3. Oxygen Electroreduction Reaction with an Emphasis on Charge Transfer at Metal/Water Interface

The research on charge transfer at metal/water interface is considered as a first stage towards a more realistic model including polymer and is considered as being relevant to the study of PEFC.

A generic quantum mechanics description of the oxygen electroreduction and charge transfer in EDL is as follows.

Striking characteristics of such an EDL at a metallic electrode/aqueous liquid electrolyte interface (Fig. 10) include: (1) a very strong electric field in the compact layer ($\sim 2.5\times 10^7$ V cm^{-1} for a potential drop of only 1 V across the interface); (2) a relatively smaller dielectric constant (~5–7 versus 78 in bulk of an aqueous solution); (3) a high concentration of charges in the forms of ions and dipoles (20–30 μC cm^{-2} or 0.11–0.17 e per surface atom) forming a barrier for Faradaic reaction, a property utilized in an electrochemical supercapacitors; (4) electrons of the metal spillover the metal surface into the compact layer giving rise to a possibility of electron tunneling or electron transfer at the interface. In a description of quantum mechanics, charge transfer occurs when

$$\Phi - eV = I - \Delta G_s \tag{32}$$

where Φ is the work function of the metal electrode, V is the potential difference between the electrode and electrolyte at OHP, I is

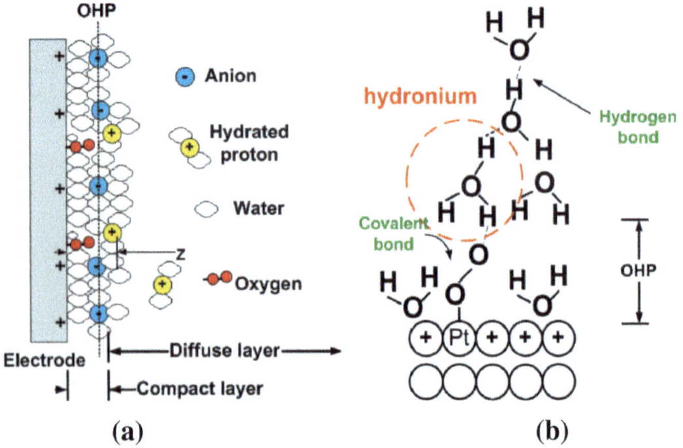

Figure 10. (a) Schematic illustration of EDL at solid/liquid interface. (b) Configuration of reactants in oxygen reduction reaction (ORR) in ELD at Pt/acidic solution interface.

the ionization potential of the reactant, ΔG_S is the change of its solvation (hydration) energy upon electron transfer or charge transfer.[55] Equation (32) expresses that at the equilibrium state of the charge transfer reaction at the interface the Fermi's level in the electrode should match the lowest unoccupied molecular orbital (LUMO) or highest occupied molecular orbitals (HOMO) of the reactants. In a general description of quantum statistical theory, the electrochemical current density, j, is given by literatures.[98,99]

$$j = \sum_{\Omega} [C] \left(\frac{\omega_{\text{eff}}}{2\pi} \right) \exp(\sigma_{\text{t}}) \int_{z_{\min}}^{\infty} k_{\text{e}}(z, \Omega) \exp\left(-\frac{\Delta E_{\text{a}}(z, \Omega)}{kT} \right) dz \qquad (33)$$

where Ω is the orientation degree of freedom, z is the reactant-metal surface distance, $z_{\min}$ is the minimum distance of the reactant to metal surface, $[C]$ is the bulk reactant concentration, ω_{eff} is the effective frequency of classic vibration modes of reacting cluster, k_{e} is the electron transfer coefficient, σ_{t} is the electron tunneling factor, and ΔE_{a} is the activation energy. Thus, the electron transfer at the interface is dictated by the EDL structure described with the HOMO or LUMO levels, the wave functions (Ψ_i^{tr} and Ψ_f^{tr}), vibration of reacting molecules, the electron tunneling factor, and $z_{\min}$. In the liquid electrolyte, $z_{\min}$ is similar to the OHP and only ~0.5 nm from the metal surface (Fig. 10b). The activation energy, ΔE_{a}, is a function of the potential drop across the EDL or overpotential and the detailed equilibrium and dynamic structures of EDL. Most of simulation studies in this topic are in fact purported to evaluate the activation energy but rarely to evaluate the electron transfer coefficient. Without a correct description of the electron transfer coefficient but only a good description of the activation energy, it would be difficult to predict the OER behavior of a given system without a fitting to experimental results. Thus, the ab initio models cannot be verified and the modeling methods are not reliable tools for designing and evaluating a new catalyst/polymer electrolyte system.

The OER is a multi-electron reaction which may include a number of elementary steps and involve different reaction intermediates. There are several pathways for O_2 electroreduction: (1) a "direct" four-electron reduction to H_2O (in low pH media) or to OH^- (in high pH media); (2) a two-electron pathway involving reduction to hydrogen peroxide; (3) a "series" pathway with two- and four-electron reduction; (4) a "parallel" pathway that is a combination of (1)–(3); (5) an "interactive" pathway in which the diffusion of species from (2) or (3) path into a "direct" path is possible. For Pt catalyst, four-electron reduction is believed to be the dominant one. However, the exact mechanisms are still unclear for all of these paths. Even for the first electron transfer step, there are two different views based on ab initio simulation results.[100,101] One of major problems for previous studies is that the models are too much simplified to reveal the effect of polymer. In fact, there has hardly been any simulation work on OER at the catalyst/polymer interface. Most of relevant studies use either vapor phase interface models or catalyst/water interface models. The electrode potential was usually not established explicitly in the way used by Taylor et al. and Janik et al. [102,103]

Anderson and his coworker carried out a series quantum chemistry studies of oxygen reduction reactions.[52–57] Anderson and Abu first studied reversible potential and activation energies for uncatalyzed oxygen reduction to water and the reverse oxidation reaction using the MP2/6-31G** method. The electrode was modeled by a non-interacting electron donor molecule with a chosen ionization potential (IP). The primary assumption is that when the reactant reaches a point on the reaction path where its electron affinity (EA) matched the donor IP, an electron transfer is initialized. The donor's IP or reactant's EA was related to the electrode potential by,

$$U = \mathrm{IP/eV} - 4.6\mathrm{eV} = \mathrm{EA/eV} - 4.6\mathrm{eV} \tag{35}$$

where 4.6 is the reference hydrogen electrode. They calculated the reaction energies of the following four-electron transfer reduction and two-electron transfer reduction:

$$O_2(\mathrm{g}) + 4\mathrm{H}^+(\mathrm{aq}) + 4\mathrm{e}^-(U) \leftrightarrow 2\mathrm{H_2O(aq)} \tag{36}$$

$$\mathrm{O_2(g)} + 2\mathrm{H}^+(\mathrm{aq}) + 2\mathrm{e}^-(U) \leftrightarrow \mathrm{H_2O_2(aq)} \tag{37}$$

With temperature, entropy, and enthalpy correction, they obtained free energy change ΔG^0. The reversible electrochemical potential was calculated by the equation

$$U^0 = -\frac{\Delta G^0}{nF} \tag{38}$$

where n is the number of electrons transferred in the reaction and F is the Faraday constant. The calculated reversible electrochemical potential was 1.18 V for four-electron transfer compared with 1.23 eV for the experimental number. For the two-electron transfer the electrochemical potential was 0.61 eV and the experimental number was 0.70 eV. They calculated the activation energy of the four one-electron transfer steps:

$$O_2(g) + H^+(aq) + e^-(U) \rightarrow HO_2\bullet(aq) \tag{39}$$

$$HO_2\bullet(aq) + H^+(aq) + e^-(U) \rightarrow H_2O_2(aq) \tag{40}$$

$$H_2O_2(aq) + H^+(aq) + e^-(U) \rightarrow HO\bullet(g) + H_2O(aq) \tag{41}$$

$$HO\bullet(g) + H^+(aq) + e^-(U) \rightarrow H_2O(aq) \tag{42}$$

The charge transfer reaction was modeled with a configuration shown in Fig. 11.

One of the important issues is how to establish a correct electrical field in the reaction center. Because in reality, the electrolyte

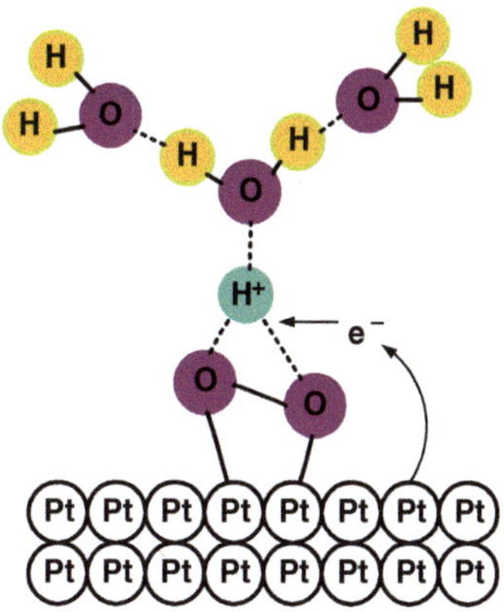

Figure 11. Structure model for charge transfer reaction at a Pt site

contains many ions; the contribution of all of them is estimated as a function of electrolyte ionic concentration by assuming that the hydronium ion at the reaction center experiences an average potential due to interactions with all the anions and cations in the electrolyte. Thus, Madelung sums of the potentials due to ions in a various regular arrays (crystal lattices) were calculated.[55] In these sums, the total infinite sum of interactions is equal to a single coulomb interaction. Anderson et al. assumed a rock-salt structure, for which the coulomb potential is given. The Madelung sum condenses all of the contributions to the potential at the reaction center to a single point charge as shown in Fig. 12.

Using this interface model and the method for establishing an electrode potential, Anderson et al. evaluated the potential energy surface of each of the elementary reactions and identified the transition state.[55] The lowest energy of the system had electron affinity equal to 4.6 eV + eU (U, the potential of the electrode). They studied the elementary steps in the electrode potential range of 0–2 V (standard hydrogen electrode) and noticed that H_2O_2 reduction (Eq. 41) had the highest activation energy. These results were

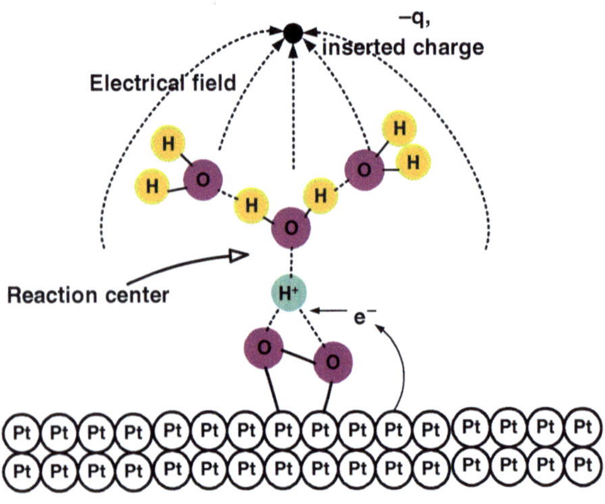

Figure 12. The Madelung sum of all the coulomb interactions of the electrolytes and their counterparts in base reduce to interactions between point charges as shown and the reaction center.

consistent with experimental observation of H_2O_2 generation over the weakly interacting electrodes like mercury and gold. The activation energies increased as the electrode potential increased. The activation energies for the four steps are in the following order: third step (Eq. 41) > first step (Eq. 39) > second step (Eq. 40) > fourth step (Eq. 42).

Anderson et al. reported the effect of platinum on oxygen reduction using a similar MP2 method.[104] A single platinum atom was used for coordinating with O_2, HO_2•, H_2O_2 and HO•:

$$\mathrm{Pt\,O_2 + H^+(aq) + e^-}(U) \rightarrow \mathrm{Pt\,OOH} \tag{43}$$

$$\mathrm{Pt\,OOH + H^+(aq) + e^-}(U) \rightarrow \mathrm{Pt\,(OHOH)} \tag{44}$$

$$\mathrm{Pt\,OHOH + H^+(aq) + e^-}(U) \rightarrow \mathrm{Pt\,OH + H_2O} \tag{45}$$

$$\mathrm{Pt\,OH + H^+(aq) + e^-}(U) \rightarrow \mathrm{Pt\,OH_2} \tag{46}$$

With this simple model, they found that binding the Pt atom to HO_2• and H_2O_2 stretched O–O bonds. Order of the effect was H_2O_2 >O_2 >HO_2•. The Pt atom had a significant effect on the most difficult reduction step, the reduction of HOOH to HO• +H_2O (Eq. 45).

Sidik and Anderson further studied the oxygen reduction when bonded to a Pt dual site using the B3LYP functional.[54] They observed that the O_2 adsorption energy for the Pauling type was 0.43 eV and for the Griffith type was 0.94 eV. The dissociation energy for Griffith type adsorbed O_2 was 0.74 eV. While the activation barrier for the first reduction step to OOH was less than 0.60 eV at 1.23 V electrode potential. In other words, the first electron transfer has a smaller barrier than that of O_2 dissociation. Furthermore, the dissociation barrier for the first electron transfer product OOH was much smaller, 0.06 eV. Based on the results, the authors concluded that O_2 did not dissociate before the first reduction step, and OOH easily dissociated once formed after the first electron transfer step which, they suggested, was the rate-determining step.

Li and Balbuena conducted a study of the first electron transfer step (Eq. 47) using the DFT method with a catalyst model of five Pt atoms.[105]

$$\mathrm{Pt_5O_2(ads) + H^+(aq) + e^-} \rightarrow \mathrm{Pt_5OOH} \tag{47}$$

The solvation effect was modeled through hydration of a proton by three water molecules (H_3O+$(H_2O)_2$). The effect of the electrode potential on the Pt/adsorbate/hydronium complex was considered by assuming that the reactant complex ($Pt_5O_2{\cdot}H_3O$) was electrically neutral. In the model, the potential energy surface of the reaction is a function of the bond length of the adsorbed O_2 molecule, the shortest distance between the adsorbed oxygen and the water oxygen, and the shortest distance between the proton and the water oxygen. They revealed that the heat of the reaction (Eq. 47) was 9.44 eV. They claimed that the proton transfer may not be involved in the rate-determining step and that a negative charge on the cluster/adsorbate complex results in a significant decrease in the activation energy.

Nørskov et al. studied the binding energy of the reactants and intermediates using a DFT method involving the following one-electron reaction in a dissociation mechanism as follows.

$$0.5O_2 + {}^* \rightarrow O^* \tag{48}$$

$$O^* + H^+ + e^- \rightarrow HO^* \tag{49}$$

$$HO^* + H^+ + e^- \rightarrow H_2O + {}^* \tag{50}$$

where "*" denotes a site on the surface.[106] They also investigated the following one-electron reactions in an association mechanism that involved the adsorbed molecular oxygen in the electron transfer as follows.

$$O_2 + {}^* \rightarrow O_2{}^* \tag{51}$$

$$O_2^* + (H^+ + e^-) \rightarrow HO_2^* \tag{52}$$

$$HO_2^* + (H^+ + e^-) \rightarrow H_2O + O^* \tag{53}$$

$$O^* + (H^+ + e^-) \rightarrow HO^* \tag{54}$$

$$HO^* + (H^+ + e^-) \rightarrow H_2O + {}^* \tag{55}$$

They calculated energies of different intermediate states. The results demonstrated that there was a correlation between oxygen reduction activity and the binding ability of O and OH. They further claimed that at high potential adsorbed oxygen was very stable and that proton and electron transfer was impossible. In addition, they proposed that the origin of the overpotential for Pt was the O and OH adsorption and that both dissociative and associative reaction paths may contribute to the OER.

Zhang et al. reported their study on the electrocatalytic activity for OER of platinum monolayer supported on Au(1 1 1), Rh(1 1 1), Pd(1 1 1), Ru(0 0 0 1), and Ir(1 1 1).[107] Using the DFT method, they calculated the activation energies for the O_2 dissociation and oxygen hydrogenation reaction. The activation energies for Pt(1 1 1) is close to the intercept of the two curves of oxygen dissociation and hydrogenation. They suggested that the ideal catalyst should have the activation energy at the intercept.

Jinnouchi and Okazaki performed AIMD studies of the first-electron transfer reaction with 1 hydronium ion, 9 water molecules, and 12 Pt atoms at 350 K as shown in Fig. 13.[108] The adsorbed water molecules and the hydronium ion hydrated the adsorbed oxygen atoms, and proton transfer through the constructed hydrogen bonds frequently occurred. When the conformation of these species satisfied certain conditions, the oxygen dissociation with the proton transfer reaction was induced and three OH were generated on the platinum surface (Fig. 14). The authors concluded that the oxygen dissociation tendency is one of the dominant factors for the reactivity of the cathode catalyst. This work demonstrates the power of AIMD that does not require specific assumption in order to describe charge transfer.

Wang and Balbuena performed an AIMD study of this one-electron reaction O_2+$H^+(H_2O)_3$/Pt(1 1 1) at 350 K. [109] They discovered that the proton transfer took place first and then chemisorption was observed. This first step induced the electron transfer from the slab. Finally, the HO–O–Pt dissociated into H–O and O

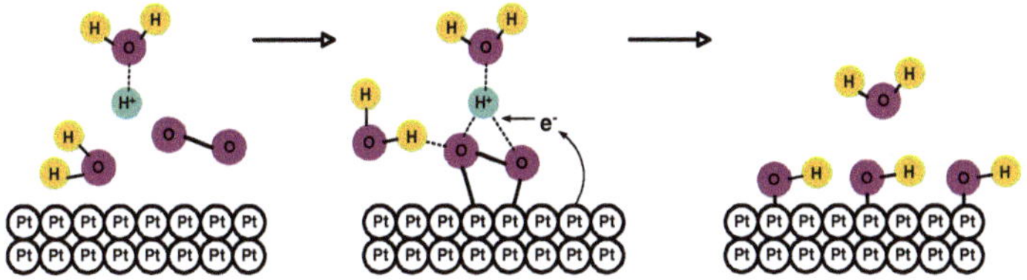

Figure 13. Mechanism of the first electron transfer in an oxygen reduction reaction.

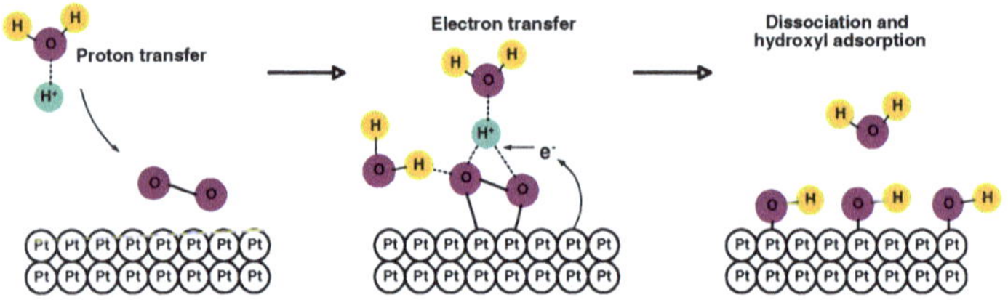

Figure 14. Proposed mechanism for the first electron transfer of OER.

without a clear barrier. To study the effect of an electrical field, they modeled the system of O_2 + $H^+(H_2O)_3$/Pt(1 1 1) with an extra electron. The proton transfer intermediate was formed in the earliest stage. This finding is similar to results reported by Jinnouchi.[108] Wang and Balbluena suggested that the mechanism for the first electron transfer involved the three major steps with the following order: (1) proton transfer, (2) electron transfer, and (3) dissociation and hydroxyl adsorption (Fig. 14). They claim that their simulation results validate that the O_2 reduction on a Pt surface may proceed via a parallel pathway. Perhaps, the direct path is the dominant path. Although their work are considered as being more realistic than earlier attempts while electron transfer is not explicitly described with self-consistent equations, there are four major unsolved problems. First, the models are so-called "vapor phase" models without an EDL. The very important dielectric properties of the EDL are not considered. The electrical field and its response to variation of the surface charge in the models must be very much different from the reality. Second, there are no polymer atoms involved in the model. The effect of interaction between the polymer atoms and the reactants is an important but an unresolved mystery. Third, after electron transfers from Pt slab to the reaction cluster, the total charge in the Pt slab increases. In reality, the loss of electron can be quickly compensated or averaged over the entire lattice of the catalyst. Therefore, the increase of the surface charge is negligible. Finally, although the activation energy and binding energy can be evaluated as functions of surface charge or potential in all these efforts, the electron transfer coefficient has not been evaluated. Thus, these models may be used to compare similar systems in terms of the reaction rate but not between systems involve very different materials.

Above simulations of the OER are based on models of a few atoms or a cluster of reaction center. Moreover, the electrode potential and the EDL are not explicitly established in the models or are established using ways of seemly oversimplification. Janik et al. conducted periodic density functional theoretical calculations with the double-reference method developed by Filhol and Neurock[62,110] to determine the potential dependence of the reaction energy and activation barrier for the reduction of $O_2{}^*$ to OOH^* on the fully hydrated Pt(1 1 1) surface. Their method allows for tuning the electrode potential with a slab representation of the electrode surface. They found that electron transfer precedes the protonation of the adsorbed O_2 molecule. This occurs when the proton completely joins a water molecule to form a hydronium that is connected to the adsorbed O_2 molecule by hydrogen bonding through two additional water molecules. The observation that the electron transfer precedes the proton transfer is different from that found by Wang and Balbuena.[109] The overall barrier to the first reduction step is the sum of the energies for the proton and for the electron to be incorporated within the reactive complex. They suggested that in order to simulate the OER process reliably it is necessary to include the extended electronic structure of the electrode, sufficient number of water molecules, and the approaching proton in the DFT model system. With regard to this requirement, this study is more realistic than the other studies. Yet, even in this work, polymer atoms were not included apparently due to lack of suitable computational hardware or software or both at that time.

Buehler et al. presented a preliminary study on formation of water from molecular oxygen and hydrogen using a series of atomistic simulations based on ReaxFF MD method.[111] They described the dynamics of water formation at a Pt catalyst. By performing this series of studies, we obtain statistically meaningful trajectories that permit to derive the reaction rate constants of water formation. However, the method requires calibrations with either ab initio simulation results in order to correctly evaluate the energetics of OER on Pt. Thus, this method is system specific and less reliable than the ab initio methods and will not replace ab initio methods. Nevertheless, this work demonstrates that atomistic simulation to "continuum description" can be linked with the ReaxFF MD in a hierarchical multiscale model.

4. MODELING OF OXIDATION OF CARBON MONOXIDE AND METHANOL

4.1. "Vapor Phase" Model

Two reasons for studying the chemical interaction between carbon monoxide on catalyst surface are (1) formation of CO on the catalysts in the oxidation of methanol to CO_2 in the direct methanol fuel cell result in inhibition of the catalysts and (2) carbon monoxide contaminants in the H_2 reformate fuel in proton exchange membrane fuel cells (PEMFCs) also inhibit the catalysts. Poisoning of the catalysts by strongly adsorbed CO results in a significant overpotential. Similar to the status of modeling of the OER at the cathode, modeling studies on oxidation of monoxide at the anode are mainly based on "gas phase" or "vapor phase" models that only involve CO and a catalyst cluster or slab hardly with the polymer electrolyte molecules and water molecules. However, ab initio simulations with the "vapor phase" models with a few atoms did qualitatively explain some experimental observations that bi- or trimetallic catalysts improved the CO tolerance of the hydrogen oxidation or enhanced methanol oxidation activity.[112] So far, the atomistic modeling or theoretical analyses provided two mechanisms for the observed enhancement on the catalytic activities for alloyed metallic catalysts: (1) the bifunctional effect, in which the unique catalytic properties of each of the elements in the alloy combine in a synergetic fashion to yield a more active surface and (2) the ligand or electronic effect, in which interaction between dissimilar atoms yield altered electronic states and hence a more active catalytic surface.

Anderson and Grantscharova conducted one of the first studies on the CO oxidation on Pt using a molecular orbital theory with a simple molecular model.[113] They used an atom superposition and electron delocalization molecular orbital (ASED-MO) method to investigate the electrochemical oxidation of the adsorbed CO on the Pt anodes. They found that the interaction of CO(ads) with the oxidant OH(ads) was effective only at high surface coverage.

Watanabe and Motoo, Janssen and Moolhuysen, Bockris and Wroblowa, and Shubina et al. and Shubina and Koper investigated the surfaces of catalysts as slabs with multiple layers and calculated the binding energies and geometrical configurations of CO and OH on Pt, Au, Pt/Ru, Pt/Sn, and Pt/Mo with a periodic DFT method. Based on the values of the binding energies, they concluded as follows.[6–8,114,115]

(1) Presence of Ru atoms in the Pt lattice results in weaker bonds of CO and OH to the Pt sites. However, presence of Pt atoms in Ru lattice causes a stronger bond of CO and OH to the Ru sites.
(2) For Pt/Sn alloys, CO adsorbed strongly only to the Pt sites but not to the Sn sites. OH tends to adsorb to the Sn sites. This implies that Pt/Sn should be a good CO-tolerant catalyst.
(3) Alloying Pt with Mo results in weakly adsorbed CO on both Pt and Mo sites. OH strongly adsorbed only on Mo site, which suggests that Pt/Mo could be a better bifunctional catalyst for CO oxidation than Pt/Ru.

Shubina and Koper also claimed that for the electrochemical oxidation of CO on bimetallic catalyst surfaces, the bifunctional effect is the most dominant mechanism. The more oxophylic element Ru or Sn seems to provide the oxygen donor, which is commonly believed to be adsorbed hydroxyl group, via activation of adsorbed water at a smaller electrode overpotential,

$$H_2O + {}^*Ru/Sn \rightarrow OH_{Ru/Sn} + H^+ + e^- \qquad (56)$$

Then the active OH radicals adsorbed on Ru or Sn react with the CO to produce CO_2:

$$CO_{Pt} + OH_{Ru/Sn} \rightarrow CO_2 + H^+ + e^- + {}^*Pt + {}^*Ru/Sn \qquad (57)$$

Liu et al. used a DFT method to describe the adsorption of H_2, CO, and H_2O on the (1 1 1) surfaces of Pt, Ru, Pt/Ru, and Pt/Sn.[116] Their calculations reconfirm the results by Shubina and Koper. The effect of the bond energy is attributed to the changes in the electronic structure of the surface atoms due to interactions of the catalyst surface with the surroundings. The reduction of the overpotentials for initializing the CO oxidation reaction for Ru, Pt/Ru, and Pt/Sn is due to the fact that the presence of Ru or Sn in the Pt lattice results in a steady dissociation of H_2O and formation of an active OH group. They also investigate the alloying effect on hydrogen oxidation at anode in the presence of CO, on the other hand. According to their reasoning, the enhancement of hydrogen oxidation rate is mainly due to decrease of the CO coverage.

Gao et al. used both experimental and DFT modeling methods to study the dual path mechanism for methanol decomposition on well-defined low Miller index platinum single crystal planes and to determine electrode potential range when the adsorbed intermediate, COad, is stable.[117] DFT calculations indicate that the path to form stable COad occurs via an initial exothermic C–H cleavage step to form adsorbed hydroxymethyl over a wide potential range, followed by exothermic dehydrogenation steps to form COad. Based on some preliminary results for the initial C–H and O–H cleavage steps over a stepped Pt(2 1 1) surface seems to support that the dominant parallel path toward positive potentials is initial O–H cleavage over defect sites.

Zhang et al. carried out DFT calculations for CO oxidation on Rh. In the calculations, they used a generalized-gradient approximation and the electronic wave functions were expanded in a plane wave basis set whereas the ionic cores were described with ultrasoft pseudopotentials.[120] Their study provides a detailed information regarding the electronic states. They attempted to explain why the activation of O atom is very important in CO oxidation. In a normal state, the saturation of O 2p orbitals in the threefold hollow site gives inactive O atom. They show that the chemistry of CO oxidation is determined by the 2p orbitals of oxygen that play an important role in determining the reactivities and reaction pathways of other molecules and fragments. They also propose that the activation of oxygen as a key event for other oxidation reactions must go through a transition state (TS). Some TSs for several other catalytic reactions such as NH + H, CH_2 + H, C + O, and N + O were identified. It was found that the TS structures are similar to that in the CO oxidation.

Gong et al. used a DFT-slab model with GGA-PBE functional, plane wave basis set, and ultrasoft pseudopotentials to investigate CO oxidation on pure metals including Ru, Os, Ir, Rh, Pd, and Pt, and their corresponding metal oxides: RuO_2, OsO_2, IrO_2, RhO_2, PdO_2, and PtO_2.[118] Their results indicate that the barriers on metal oxides are generally lower than those of their corresponding metals for CO oxidation and the higher activity of metal oxides is mainly due to a surface geometric effect. They also found a general rule that the shorter the OC–O bond distance is at the transition state (TS), the higher the barrier. They claim that the results provide general rules not only for CO oxidation but also for other reactions.

In another systematic work aiming to address the water catalytic role in general, Gong et al. used DFT to investigate the CO oxidation in the presence of water on Pt(1 1 1) via evaluating energies of initial, transition, and final states for many possible elementary steps. Their findings suggest that in the presence of water the CO oxidation barrier can be significantly reduced.[119]

4.2. Realistic "Liquid Phase" Model

Desai and Neurock employed a non-local gradient corrected DFT-slab model including 23 water molecules or a so-called "liquid phase" model to evaluate the overall reaction energies along with the barriers for the activation of water and the oxidation of CO over Pt(111) and Pt/Ru(1 1 1) surfaces.[121] In this study, they conducted energy minimization, TS determination, and AIMD simulation to probe the mechanisms that control the bifunctional effect. Their findings include (1) the activation energy of water over Pt(1 1 1) in the vapor phase is energetically unfavorable with an high activation barrier of 142 kJ mol^{-1}, (2) the presence of water molecules reduces the barrier to about 75 kJ mol^{-1}, (3) the addition of Ru lowers the barrier for the activation of water in the vapor phase and in the liquid phase, and (4) the presence of water, however, markedly promotes the heterolytic activation that leads to the formation of an adsorbed OH^* surface intermediate and a proton that migrates into solution.

Taylor et al. conducted DFT simulations using a periodic model of the interface between water and various metal surfaces with an index of (1 1 1).[102] The chemistry of water at these charged interfaces was investigated and the parameters relevant to the macroscopic behavior of the interface, such as the capacitance and the potential of zero charge (PZC), were evaluated. They also examined the influence of co-adsorbed CO upon the equilibrium potential for the activation of water on Pt(1 1 1). They found that for copper and platinum there was a potential window over which water is inert. However, on Ni(1 1 1) surface water was always found in some dissociated form (i.e., adsorbed OH^* or H^*). The relaxation of water molecules via the flip/flop rotation is an important contribution to the interfacial capacitance. Their calculations for the co-adsorbed H_2O/CO system indicate that the adsorption of CO affected the binding energy of OH. This causes that water is activated at a higher equilibrium potential.

Janik and Neurock investigated the adsorption and electro-oxidation of CO on the surface of Pt using the double-reference potential method and a model with 20–30 water molecules.[103] They used a DFT method at a constant potential half-cell formalism to model the electro-oxidation of CO over Pt(1 1 1). The potential was tuned by adding or removing electrons from the system. The equilibrium potential for oxidation of adsorbed CO was calculated to be 0.62 V (NHE), which is in good agreement with the onset potential of CO oxidation in stripping voltammetry experiments. The activation barrier for CO^*+OH^* coupling is substantially less than that of CO^* +O^* coupling, indicating that the majority of CO oxidation occurs through a surface hydroxyl intermediate. The coupling of carbon monoxide and hydroxyl intermediates on the surface proceeds with the formation of a metastable hydrocarboxyl species ($COOH^*$). The barrier for CO^* +OH^* coupling decreases from 0.55 to 0.3 eV with increasing electrode potentials. The results indicate that the coupling of co-adsorbed CO and OH controls the reaction rate.

Harting et al. reported the results of DFT and AIMD studies for the oxidation of methanol on the Pt(1 1 1) face in aqueous solution.[122] Their work reveals that the oxidation of methanol is initiated at the moment when a hydrogen bond forms between the OH group of the methanol and a water molecule. The initial step of the reaction is the cleavage of a CH bond with the bond direction points towards the platinum surface. This is followed by a rapid dissociation of the methanol OH bond, which leads to formation of a formaldehyde as a stable intermediate within the timescale of the simulation. Charge delocalization is achieved by the formation of a Zundel ion $H_5O_2^+$ in the aqueous phase.

Harting and Spohr investigated oxidation of methanol on the (2 1 1) face of a platinum single crystal.[123] Similar to the reaction pathway on the (1 1 1) crystal face water plays an important role. That is the adsorption of methanol on charged and uncharged surfaces is strongly enhanced by the formation of a hydrogen bond to a co-adsorbed water molecule. A methyl hydrogen atom of a methanol is first adsorbed. Scission of one of the CH-bonds is the first reaction step. In the presence of water, the onset of the oxidation reaction is favored by a co-adsorbed neighboring water molecule which forms a hydrogen bond with the methanol OH-group. At a minimum number of adjacent water molecules, the CH- and OH-bonds are cleaved on the charged surfaces. The protonic charge given by the dissociation of the methanol hydroxyl group is delocalized inside the aqueous cluster and formaldehyde is formed as a stable intermediate product.

Ludwig et al. investigated the dissociation of hydrogen on eight platinum surfaces, Pt(1 1 1), Pt(1 0 0), Pt(1 1 0), Pt(2 1 1), Pt(3 1 1), Pt(3 3 1), Pt(3 3 2), and Pt(5 3 3) using a MD method with a ReaxFF. The ReaxFF was derived from a training calculation using surfaces not included in this work.[124] The energetics of the eight surfaces in the absence of hydrogen at 0 K were first compared to previous DFT calculations and found to underestimate surface energy. However, if the Pt(111) is used as a reference state, the trend observed between surfaces in parallel to that obtained with the DFT methods. They claimed that these results justified a relative comparison between the reaction rates for the various Pt surfaces. However, the absolute magnitude of the reaction rate predicted using this ReaxFF method was too low. When normalized with the Pt reference, the simulations show the correct linear dependence. Although the method cannot correctly predict the absolute magnitude of the surface energy, it can simulate a large-scale system that is impossible to be analyzed using the ab initio methods.

Hartnig, Kulikovsky, and Spohr investigated the catalytic oxidation of methanol in realistic, i.e., fuel cell relevant environments, using DFT methods to describe the coupling between the electronic structures and dynamics of atoms.[123,125] They first investigated the oxidation of adsorbed methanol at a platinum/water interface with charged or uncharged Pt surface using an AIMD method. They found that the presence of water not only altered the reaction pathway for a corresponding reaction in gas phase, but also changed the sequence of reactive events by directly incorporating hydrogen ions into its hydrogen bond network, as illustrated in Fig. 15. These calculations are currently extended to investigate the influence of fuel cell environment on the later oxidation steps.

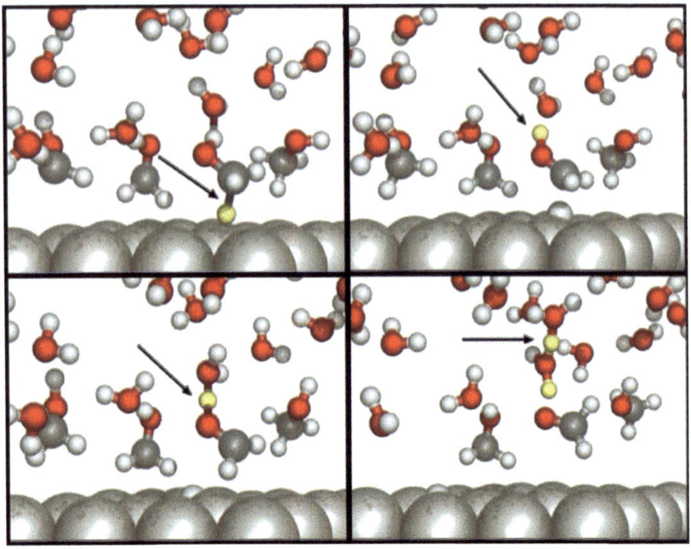

Figure 15. Chemical reaction dynamics of methanol oxidation to formaldehyde: one of the C–H bonds of the methyl group becomes elongated (top left) and eventually breaks (top right). The adsorbed hydroxymethyl group stabilized by forming a hydrogen bonded complex to a water molecule (bottom left) and dissociates rapidly into adsorbed formaldehyde and a hydronium ion (bottom right) which further stabilized by undergoing structural diffusion steps to form Zundel ions H_5O^+.[123,125] Reprinted from Chemical Physics, Vol. 319, C. Hartnig and E. Spohr, The role of water in the initial steps of methanol oxidation on Pt(1 1 1), p. 108, Copyright (2005), with permission from Elsevier.

5. MODELING OF TRANSPORT PROCESSES IN NAFION POLYMER ELECTROLYTES

5.1. Theoretical Views of Proton Transport in Aqueous Systems and in Hydrated Nafion Membranes

5.1.1. In Aqueous Solution

The mobility of protons in water is 5–8 times greater than that of other cations.[125] This phenomenon is thought to be very important for proton permeation through bio-membranes in biological organisms.[126] It is also assumed that this warrants a high conductivity of Nafion® series membranes that consist of nanoscale hydrophobic domains and water containing hydrophilic domains or channels.[127,128] This phenomenon has been studied since C.J.D. von Grotthuss firstly proposed a mechanism in 1806.[129] The transport of protons is determined by the rate at which the hydrogen bond between a hydronium ion and a water molecule forms (structural diffusion), rather than by the slower rate at which hydronium ions may migrate en masse (physical diffusion). Many detailed descriptions of the Grotthuss mechanism have been proposed as follows.

First of all, what was considered were bare hydronium $H_3{}^+O$ ions with three equivalent protons, a hydrated hydronium ion with three strongly bound water molecules (i.e., Eigen cluster $H_9O_4{}^+$), and the symmetric $H_5O_2{}^+$ complex in which a proton is shared between two water molecules (i.e., the Zundel ion). Many intermediates or more-complex states of the "hydrated proton," $H^+(H_2O)_n$, may also exist. All clusters have a finite lifetime and transform between each other during charge transport. Due to the variation of the relative abundance of these three basic states, proton transfer may occur via different pathways.

Firstly, the most straightforward transport mechanism for any proton is that it hops from one water molecule to a neighboring water molecule. It was initially speculated that when this is happening, a neighboring water molecule must approach one of the protons in the H_3O^+ ion and form an intermediate similar to the Zundel, where this proton is at the bridge position. From this state, there are two possible pathways in which $H_5O_2{}^+$ dissociates into a hydronium ion and a water molecule: (1) the proton moves out of the reach of the former host water molecule to join the other water molecule in the intermediate state resulting in a proton transfer and (2) the proton moves back to the former host water molecule without making a proton transfer. This hopping mechanism can also be envisaged in a $H_9O_4{}^+$ cluster.

Secondly, proton transfer can be realized via the structural diffusion of $H_5O_2{}^+$. A water molecule next to such a cluster approaches and interacts with the cluster in which the proton is at the center and shared by the two molecules. The interaction may drag a proton from one of the water molecules in the previous $H_5O_2{}^+$ to the bridge position and thus form a new $H_5O_2{}^+$ while the proton that used to be shared by the two molecules in the previous $H_5O_2{}^+$ joins into the new $H_5O_2{}^+$. Thus, the proton transfer is not completed with the hopping of the charge carrier from one location to another but by the motion of charge center via change of electronic states. This mechanism can also be envisaged for larger protonated water clusters. When both Zundel and Eigen ions account for a significant fraction of proton states, then these two proton-transport pathways become equivalent in the sense that proton transport arises from a chain of consecutive Eigen-to-Zundel and Zundel-to-Eigen transitions. In all these processes, the key process is the constant forming and breaking of hydrogen bonds, which allows proton to wonder among a network of water molecules.

Finally, the H_3O^+ ion can drift classically as a whole. However, this, also called physical diffusion, is only one of the contributions to the proton transport and is not the dominating contribution in bulk water.

5.1.2. In Hydrated Membrane (Nafion)

There must be modifications of the proton-transport mechanism in the membrane environment. First, the backbones and side-chains in a Nafion membrane form confinements for the motion of water molecules. In the water-filled pores, in particular, near the pore surface, the structure of water may be significantly different than in the bulk water. Water molecules are pretty much "frozen" in comparison to the bulk aqueous phase. As a consequence, higher barriers for molecular preorganization are expected and a higher effective activation energy for proton transfer. The new structures of water clusters could significantly modify the pathways and rates of proton transfer in the clusters. Second, it was thought that proton transfer following the paths on the surface of the pore with an array of SO_3^- ionic groups was very difficult because the distance between the neighboring SO_3^- groups was 0.7–1 nm too large for a proton hopping between the two neighboring equilibrium positions. This is called a Coulomb barrier that is smaller the denser the array of SO_3^- groups. However, the side chains carrying the SO_3^- groups can move ("bow") toward each other to facilitate symmetric proton transfer with a zero reaction free energy. The energetic expense for this process will be the work to bring two charged groups together up to the distances at which a proton transfer is possible. The overall non-concerted fluctuations of the side chains will lower the Coulomb barriers, thereby reducing the effective activation energy for the surface conductivity process.

5.2. Ab Initio Models

Urata et al. analyzed the intermolecular interaction energies between a water molecule and four models of pendant chains for perfluorinated ionomers, $CF_3OCF_2CF_2SO_3H$, $CF_3OCF_2CF_2SO_3^-$, $CF_3CF_2CF_2COOH$, and $CF_3CF_2CF_2COO^-$ using ab initio molecular orbital computational methods.[130] They optimized the geometries of isolated monomers and then employed the optimized geometries for the calculations of the energies of the complexes. The intermolecular interaction energies for 100 configurations were calculated at the MP2 level, and the others (400) were calculated at the HF level. From these analyses, they obtained the following conclusions. (1) Because the ether oxygen in pendant chain cannot bind strongly with water due to fluorination, the oxygen in the pendant chain makes little contribution to absorption of water in perfluorinated ionomer membranes. (2) Ionization of sulfonic and carboxylic acids is the primary reason for an extended region with water molecules attractively bind with pendant chains.

They also conducted analyses to reveal the intermolecular interaction energies between methanol and the perfluorinated ionomers using similar approaches.[131] Intermolecular interaction energy of the most stable complex for $CF_3OCF_2CF_2SO_3$ + CH_3OH is –10.38 kcal mol^{-1} and almost the same as that of $CF_3OCF_2CF_2SO_3$ + H_2O complex (–10.58 kcal mol^{-1}). Based on this result, they concluded that, similarly to water, methanol would likely populate around the acidic site. However, according to a systematic analysis for 500 random configurations, more dissimilar intermolecular interaction energy was observed for methanol than for water. The authors also concluded that the tendency for methanol approaching to at the hydrophobic sites is greater than that for water approaching to the hydrophobic sites although methanol molecules concentrate around the hydrophilic sites.

Paddison and Elliott investigated the flexibility of the side chain and effects of conformational changes in the backbone on hydration and proton transfer in a short-side-chain of perfluorosulfonic acid fuel cell membrane or $CF_3CF(\text{-}O(CF_2)_2SO_3H)\text{-}(CF_2)_7\text{-}CF(\text{-}O(CF_2)_2SO_3H)CF_3$ via ab initio methods.[132] Computational results of energies of the same polymeric fragment with 4–7 water molecules indicate that the perfluorocarbon backbone may adopt either an elongated geometry, with all carbons in a trans configuration, or a folded conformation as a result of the hydrogen bonding of the terminal sulfonic acids with the water. These electronic structure calculations reveal that the fragments having the "kinked" backbone possessed a higher binding energy with water through the sulfonic acid groups. This suggests that the side chains may interact differently with surrounding water molecules and proton dissociation and proton conductivity may be promoted by altering the geometrical configurations.

They also revealed that the number of water molecules needed to form water network between the sulfonic acid groups is a function of the number of fluoromethylene groups in the backbone. Generally speaking, one, two, and three water molecules are required to connect the sulfonic acid groups with number of fluoromethylene groups of 5, 7, and 9 in the backbone respectively.[133] In addition, proton dissociation is independent of the distance between the side chains.

Because of the complexity of hydrated PEMs, a full atomistic modeling of proton transport is impractical. The generic problem is a disparity of time and space scales. While elementary molecular dynamics events occur on a femtosecond time scale, the time interval between consecutive transfer events is usually 3 orders of magnitude greater. The smallest pore may be a few tenth of nanometer while the largest may be a few tens of nanometers. The molecular dynamics events that protons transfer between the water filled pores may have a timescale of 100–1000 ns. This combination of time and spatial scales are far out of the domain for AIMD but in the domain of MD and KMC as shown in Fig. 2. Because of this difficulty, in the models the complexity of the systems is restricted. In fact in many models the dynamics of excess protons in liquid water is considered as an approximation for proton conduction in a hydrated Nafion membrane. The conformations and energetics of proton dissociation in acid/water clusters were also evaluated as approximations for those in a Nafion membrane.[16,19,20,22,24,25]

Eikerling et al. investigated a model trifluoromethane sulfonic acid for proton transfer in PEMs, relevant to a regime of minimal water content and high concentration of anionic groups in a PEM, using the VASP (Vienna Ab-initio Simulation Package) DFT based total-energy code, under a Generalized-Gradient Approximation (GGA) with ultra-soft pseudopotentials for all atoms.[28–33,134] In these simulations a defect involving the formation of an $H_5O_2^+$

ion and the reorganization of neighboring sulfonic acid groups was identified. In this defect structure a proton is shared between the oxygen atoms of the anionic sites (Fig. 16). The energy of formation of this defect is 0.3 eV in agreement with the experimentally determined activation energy for proton transport in minimally hydrated Nafion of 0.36 eV. They claimed that proton transport in Nafion was related to the formation of this type of defects.

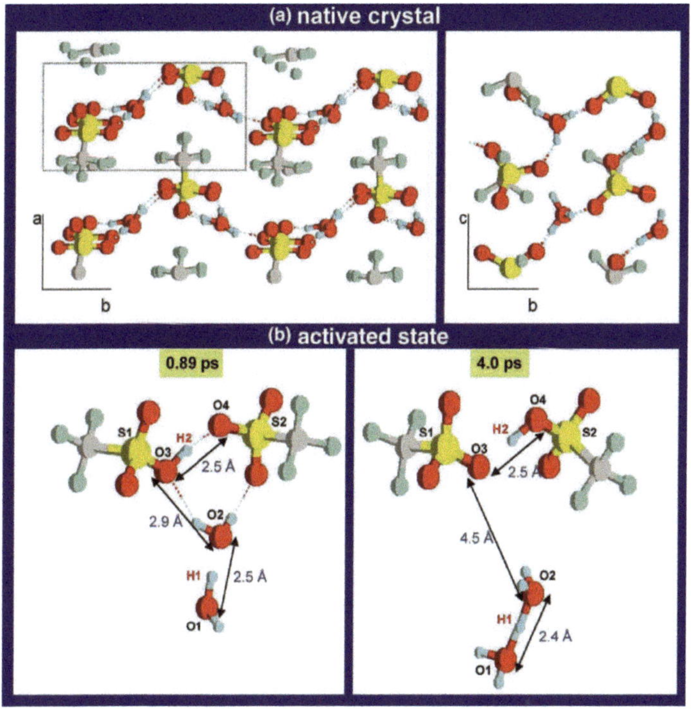

Figure 16. (a) The regular structure of the crystal is depictured, as observed during microcanonical sampling. (b) Snapshots (quenched configurations) along the trajectory, taken at 0.89 and 4.0 ps, give a pictorial impression of this defect state.[134] Reprinted from Chemical Physics Letters, Vol. 368, Michael Eikerling, Stephen J. Paddison, Lawrence R. Pratt and Thomas A. Zawodzinski, Defect structure for proton transport in a triflic acid monohydrate solid, p. 185, Copyright (2005), with permission from Elsevier.

Paddison and Zawodzinski presented results of theoretical calculations on the side chain of Nafion in order to establish a microscopic picture for water and proton transport in the membrane.[135] They optimized geometries for the trifluoromethane sulfonic acid fragment (CF_3-SO_3-H), the di-trifluoromethane ether fragment (CF_3-O-CF_3), and the side chain (CF_3-OCF_2-CF_2CF_2-OCF_2-CF_2-SO_3H) by means of both an ab initio method with second order Møller–Plesset electron correlation corrections, and DFT with Becke's three-parameter hybrid method. Several rotational potential energy surfaces were calculated to assess chain flexibility and proton accessibility. They added a water molecule as a probe to identify and characterize hydrophilic sites. These calculations confirmed that the sulfonic acid group is hydrophilic and the ethers are hydrophobic.

Paddison summarized studies of sulfonic acid-based polymer electrolytes or model electrolytes under low hydration levels.[135] He found that no dissociation of the proton was observed with either CF_3SO_3H or $CH_3C_6H_4SO_3H$ until three water molecules were added. The results reveal that the formation of a hydronium ion is initiated by formation of hydrogen bonds with at least two water molecules and one of the oxygens of the triflate anion. The analysis also reveals that the dissociation is a result that the excess positive charge is stabilized in the hydrogen bonding network and the excess electron density due to the breaking of the $-SO_3$–H bond is delocalized by the neighboring groups such as $-CF_3$ or $-C_6H_4CH_3$. Another interesting finding is that the separation of the proton from the anion is about the mean of that observed in the Eigen (H_9OC_4, 2.60 Å) and Zundel (H_5OC_2, 2.50 Å) ions in bulk water for the perfluoro system.

Although ab initio methods provide elegant solutions to problem of simulating proton transfer with both vehicular and Grotthuss mechanism, the scheme is computationally too expensive to be used. In other words, the size of the system is too small and simulation time scale is too short to reflect a macroscopic transport process. The primary problem for these ab initio models is that the computational results cannot be confirmed with a huge library of experimental diffusivity and conductivity data or existing polymer electrolytes. This is one of major reasons that heavily impede the progress of the atomistic modeling research of polymer electrolytes. If the simulation results cannot be compared with the experimental results, atomistic modeling is mostly a philosophy but not a science.

5.3. Classic MD Models

A series of articles were published by Ennari et al. on MD simulation of transport processes in Poly(Ethylene Oxide) and sulfonic acid-based polymer electrolyte.[136,137] The work was started by the determination of the parameters for the ions missing from the PCFF forcefield made by MSI (Molecular Simulations Inc.), to create a new forcefield, NJPCFF. In the models, the proton is represented as a hard ball with a positive charge. Zhou et al. used the similar approach to model Nafion.[138] The repeating unit of Nafion (Fig. 17) was optimized using ab initio VAMP scheme. The protons were modeled with hydronium ions. Three unit cell or molecular models were used for the MD simulation. The unit cell contains 5000 atoms 20 pendent side chains, and branched Nafion backbone created with the repeating unit. Their water uptakes or water contents were 3, 13, or 22 H_2O/SO_3^-, which correspond to the room temperature water uptakes at 50% relative humidity (RH), at 100% RH, and in liquid water respectively.[18] The temperature was initially set at a value between 298.15 and 423.15 K under NVE ensemble with constant particle number, constant volume (1 bar), and constant energy.

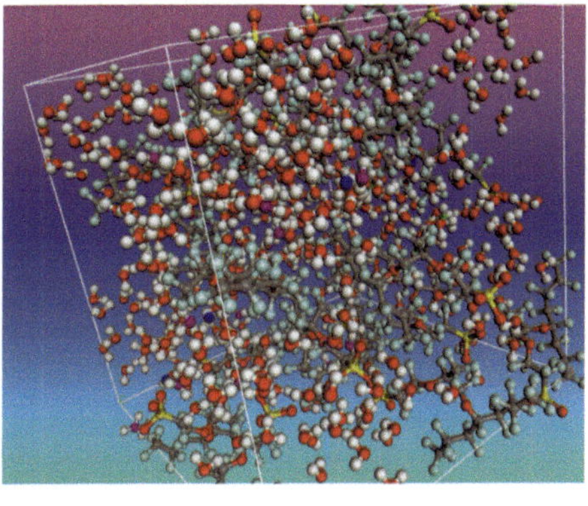

Figure 17. An atomistic model for molecular dynamic simulation of physical diffusion of ions and mediator in Nafion clusters.[138] Reprinted from Journal of Electrochemical Society, X. Zhou, Z. Chen, F. Delgado, D. Brenner, R. Srivastava, J. Electrochem. Soc. **154**, B82 (2007)– reproduced by permission of the Electrochemical Society.

However, the final temperature was usually different from the initial temperature. The initial velocity was randomly determined according to Boltzmann statistics. The time-step was set at 1.0 fs and duration was 100 ps. Each run of the molecular dynamic (MD) computation lasts about 72 h using a Dell Pentium IV personal computer. The position vector of each particle, $R_i(t)$, is saved as a function of time. The diffusivity of a specific molecule or ion was evaluated using[136,137]:

$$D_a = \frac{1}{6N_a} \lim_{t\to\infty} \frac{d}{dt} \sum_{i=1}^{Na} \left\langle [R_i(t) - R_i(t)]^2 \right\rangle \tag{58}$$

where N_a is the number of diffusing molecules. The sum term on the right side divided by N_a is the mean square displacement (MSD). The Einstein equation was used to calculate the ionic conductivity,

$$\sigma = \frac{e^2}{6tVkT} \left(\sum_i z_i^2 \left\langle [R_i(t) - R_i(0)]^2 \right\rangle + 2 \sum_{j>i} z_i z_j \left\langle [R_i(t) - R_i(0)][R_j(t) - R_j(0)] \right\rangle \right) \tag{59}$$

where t is the time, V the volume of the unit cell, e the electronic charge, k the Boltzmann's constant, T the temperature, and z is the charge of the ions. The movement of protons was investigated by tracking the trajectories of the protons in the unit cells. The proton is entangled with a water molecule in its path moving to and away from a sulfonic acid group. This might explain why the diffusion coefficient of proton is not very different from that of hydronium.

As shown in Fig. 18, the simulation conductivity data were generally consistent with the experimental results. However, there are appreciable differences between the simulation and experimental results. At some points, the differences can be 100%. The other observation is that the simulation must underestimate the activation energies of the conduction. The primary reason for this discrepancy is that these simulation models do not take into account interaction between the membrane itself and its environment. In reality, the water uptake at elevated temperatures may be greater than that at room temperature. In the simulations, it was assumed that both

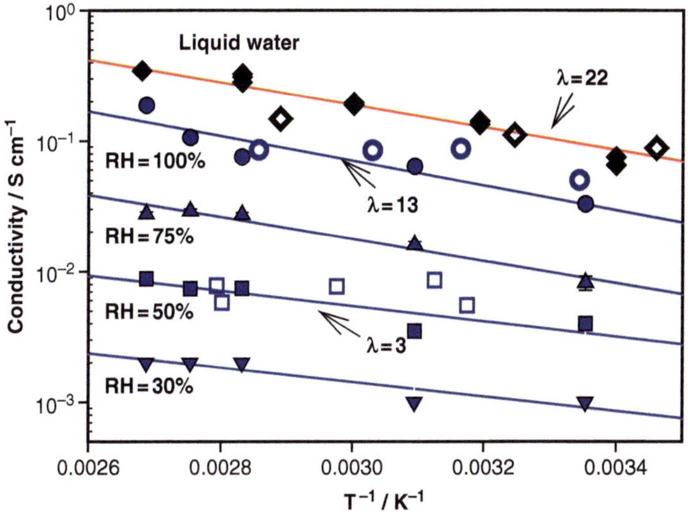

Figure 18. Conductivity as a function of temperature and relative humidity and results of atomistic simulation. The closed symbols are experimental data points.[138] The open symbols are simulation data points.[138] Reprinted from Journal of Electrochemical Society, X. Zhou, Z. Chen, F. Delgado, D. Brenner, R. Srivastava, J. Electrochem. Soc. **154**, B82 (2007)– reproduced by permission of the Electrochemical Society.

water content and density were constants for varying temperature. The second reason may be that the classical MD simulations do not involve the thermally activated chemical reactions that may contribute to the conduction of protons, structural diffusion, and dissociation of protons from the sulfonic acid groups.

The simulation models also correctly predicted the diffusivities of hydronium and methanol in a wide range of temperature (Fig. 19). Methanol is a neutral species and weakly interacts with Nafion backbone. It is not surprising that the present MD models that do not consider chemical interaction between the molecules can still correctly evaluate the diffusivity of methanol. Because the present experimental setup is limited for liquid samples, whether or not the permeability of diffusivity is strongly depends on water content has not been examined. In summary, this work provided benchmark for the atomistic simulation of the transport processes in Nafion at water content above 3 although at some points, the errors can be 100%.

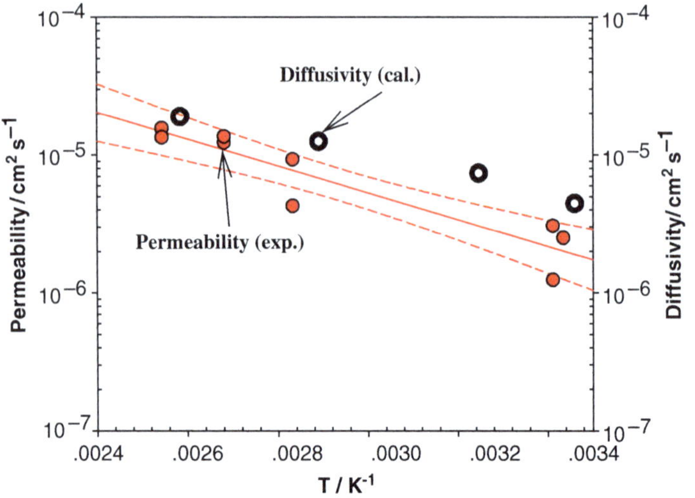

Figure 19. Methanol permeability as a function of temperature and diffusivity data evaluated using the atomistic simulation method.[138] Reprinted from Journal of Electrochemical Society, X. Zhou, Z. Chen, F. Delgado, D. Brenner, R. Srivastava, J. Electrochem. Soc. **154**, B82 (2007)– Reproduced by permission of the Electrochemical Society.

Jang et al. investigated the effect of monomeric sequence of Nafion chain on the diffusivity of water at a water content of 15 using a classical MD method.[139] They examined two extreme cases: blocky case where the distance between adjacent sulfonic acid groups would be 6 Å and dispersed case where the ionizable monomeric unit is uniformly distributed along the chain and the distance between adjacent sulfonic acid groups would be 22 Å. The distribution of the hydrophilic and the hydrophobic patches at the interface is much more segregated for blocky Nafion. This leads to a water diffusion coefficient for the dispersed case that is 25% smaller than for the blocky case (0.46×10^{-5} versus 0.59×10^{-5} cm^2 s^{-1} at 300 K). Both values are within the error range of the experimental value (0.50×10^{-5} cm^2 s^{-1}). This outcome again demonstrates that the classic MD methods can provide simulation results that can be compared with the experimental results. MD modeling can generate useful guidance for optimizing the properties of Nafion and may point out the route for developing new membranes to replace Nafion in PEMFC.

5.4. Empirical Valence Bond and ReaXFF Models

Dokmaisrijan and Spohr investigated the proton distributions in slab simulation models of aqueous pores in polymer electrolyte membranes.[140] They considered three different models. Model A includes walls of sulfonic acid groups placed on the position of the divergence of the Lennard–Jones wall. In Model B, the structure of pores is modified in such a way that the sulfonic acid group protrudes by 2 Å into the liquid phase. In addition the sulfonic acid groups can rotate around their equilibrium position defined by a tether force field. This is a more realistic model than Model A because the sulfonic acid groups are solvated. Model C include rigid water molecules and hydronium ions described by a rigid version of the hydronium potential function while Models A and B are based on an EVB bond description. They used these models in order to identify the common characteristics independent of force field and geometrical structures. They observed that the activation energy for proton diffusion from one sulfonic acid group to next was only 1.5 kT and similar to that in bulk acids. They envisage that for more disordered real pores proton diffusion is faster. The energy barriers are much smaller due to the combined electrostatic effect of sulfonic acid groups. Thus, above analyses show that within a pore there is a continuous low-energy pathway for proton motion along the surface of large and small pores.

Seeliger et al. investigated the transport of proton in the pores of Nafion using an EVB method that allows continuous breaking and forming of O–H bonds.[141] Thus three possible mechanisms: Grotthuss mechanism, physical diffusion, and surface diffusion can be examined. Several sets of simulations were performed. Simulations with the EVB model and the flexible all-atom representation of Nafion have been carried out for two different water contents $\lambda = 5$ (E5) and $\lambda = 10$ (E10) at three different temperatures (298, 348 and 398 K). The pore models are created by equilibrating a Nafion phase without water and later added aqueous phase for a nanosecond. This Nafion phase is more realistic than those in Dokmaisrijan and Spohr's simulation work. They conclude that for λ between 5 and 10 the proton conduction mechanism is dominated by the structural diffusion mechanism (Grotthus mechanism). In addition, the results seem to support the conjecture that the observed increase in activation energy at low humidity λ is not associated with a specific surface conduction but the proposed bridging mechanism that Nafion clusters temporarily open and close to aqueous phase. However, although the simulations ran up to 30 ns, this bridging effect could not be confirmed.

Solvation and solvation properties of hydrated proton are issues that cannot be treated with classic MD method.[142] However, solution properties can be analyzed using a multistate empirical valence bond (MS-EVB) methodology, which enables proton transport through the Grotthuss hopping mechanism. Peterson et al. employed MS-EVB to investigate two degrees of hydration of Nafion polymer.[142] This work highlights great differences between the results by classic MD method in which the hydroniums are non-dissociable and those by MS-EVB methods in which the bonds of hydrated proton clusters (e.g., $H_5O_2^+$ and $H_9O_4^+$ ions) can break and form. For the MD simulation, there appears an artificial peak in the classical hydronium/sulfonic oxygen radial distribution. The peak is thought to be the consequence of inability of the classical model to transfer protons between the two moieties of the Zundel cation via the Grotthuss mechanism.

Peterson and Voth studied solvation and transport properties of the sulfonic acid–hydronium ion pair in hydrated Nafion using an improved self-consistent multistate empirical valence bond (SCI-MS-EVB) molecular dynamics simulation.[143] The nature of the sulfonic acid–hydronium ion pair was characterized through analysis of free-energy profiles. They found that the diffusion of proton is a result of highly correlated vehicular motion of the transient state and fluctuation of bonds. Similar to that was found in ab initio modeling sulfonic acid ions seem to be a proton "trap" or barriers for diffusion of hydronium in hydrophilic domains or water filled pores of Nafion. The authors suggest that short pendant chain may restrain the sulfonic acid groups from trapping the excess protons in the bulk water region and may result in faster diffusion of proton.

Goddard et al.[18] are currently carrying out exploratory calculations on both OER and proton conduction processes using the ReaXFF MD method. They anticipate that this will establish a very realistic MD model of a tiny fuel cell and will allow determining how the performance changes as the composition, configuration, and other conditions are altered.

6. SUMMARIZING REMARKS

(1) Modeling or simulation of PEFC is a multiscale problem and should invoke both atomistic and continuum approaches.

(2) Two primary goals of atomistic modeling of PEFC are, first, to supplement the experiments performed in laboratory to study what has not been or cannot be experimentally studied; and second, to conduct a computer design followed by virtual tests that the experiment in laboratory is difficult or impossible to be performed under current status of technology. This brings about a dilemma: on the one hand, atomistic models can describe a phenomenon as microscopic as possible at the atomic level and on the other hand cannot describe the phenomenon in a system with a size as macroscopic as possible and in a timescale as long as possible.

(3) Although a first principle or ab initio atomistic simulation of a one-million atom system is being attempted with one of the most powerful computers, ab initio atomistic modeling of a macroscopic (10^{24} atoms) in a long time scale (10^3 s) is seemly not possible in a foreseeable future. The hierarchical multiscale simulation methods are implementable options for the time being.

(4) The atomistic modeling methods can be divided into three major categories. Firstly, ab initio or first principle methods are based on Schrödinger equation and hence are "exact" in the sense that they can be as accurate as possible, only subject to computational time or computational resource limitations. Ab initio molecular dynamics (AIMD) methods can evaluate the forces acting on the nuclei from the electronic structure calculations that are performed as the trajectory of the molecules is generated. The ab initio methods are considered as the foundation of atomistic modeling. Secondly, Classical dynamic simulation methods or classical molecular dynamic (MD) methods are based on the Newton's second law and inter-atom force fields (FFs) that are evaluated using quantum mechanics. This method is a powerful tool for analyzing transport processes requiring a moderate computational resource but usually do not allow analysis of the chemical reactions that involve chemical bond formation/breaking. The empirical valence bond (EVB) and ReaXFF methods are essentially based on Newton's second law and the interaction

potential energy but are able to describe bond formation/breaking via explicitly expressing the interaction potential energy as a function of the bond order. They also require a moderate computational resource. Finally, Monte Carlo (MC) methods based on the deterministic equations of the MD methods explore the system phase space with stochastic algorithms. Both MC and KMC methods can address the “time gap” problem of MD methods but are unable to generate the trajectories of the atomic system.

(5) A majority of literatures on atomistic modeling of PEFC are about Nafion polymer electrolyte based systems. The predominant issues are (1) OER at cathode, (2) oxidation of CO and methanol, and (3) transport processes in Nafion polymer electrolyte.

(6) So far, the existing ab initio interface structure models can include a catalyst slab and a layer of water molecules and adsorbates with a total 100 atoms or so. Neither polymer atoms nor specific adsorbates can be involved. The electrical field was initially simulated using either a uniformed distributed charges or inserted single charge calculated via Madelung sums of the potentials. Notable achievement is the establishment of the double-reference method for defining electrode potential with respect to the standard hydrogen potential.

(7) The existing MD interface structure models can involve polymer clusters and a catalyst slab with thousands of atoms. These models, however, do not establish a correct EDL structure and electrode potential. Notable finding is that the presence of polymer alters the transport and adsorption oxygen at the interface and water distribution on the catalyst surface.

(8) Early studies on adsorption of oxygen on catalysts are based on the so-called “vapor phase” models that only include a few molecules on top of a catalyst slab without participation of a water layer and any polymer atom. Correlation of the O–O and O–M bond strength to the experimental observations is revealed in these studies.

(9) Ab initio modeling of charge transfer process in the oxygen electroreduction reaction (OER) has achieved some notable stepwise progresses in complexity and in proximity to realistic systems. Initial models only involved a few Pt and reactant atoms. Later, a catalyst slab replaced the Pt cluster. The

electron transfer is initially imposed by an artificial operation. Later, AIMD allow natural transfer of electron from the catalyst to the reaction center. Initially, the reaction is simulated without a well-defined EDL structure and electrode potential. Later, the EDL was defined with layers of water molecules and the electrode potential is established using a two-reference method. A notable observation is that the "vapor phase" AIMD models point out that proton transfer is prior to electron transfer while the "liquid phase" models point out, vice versa. To date, there has been no ab initio model involving polymer clusters. The simulation results still cannot be compared with experimental observations because the simulations are conducted with very small systems in very short timescale thereby only representing a few possibilities from a huge pool of possibilities for a real macroscopic system. It has been attempted to use a ReaXFF MD method to simulate OER in a much larger system. However, one of the understandings of the ReaXFF method is that it requires calibrations by using an ab initio method as a reference.

(10) So far, the atomistic modeling on oxidation of CO and methanol has been aimed to elucidate mechanisms for (1) the bifunctional effect, in which the unique catalytic properties of each of the elements in the alloy combine in a synergetic fashion to yield a more active surface and (2) the ligand or electronic effect, in which the interaction between dissimilar atoms yield alters electronic states and hence results in a more active catalytic surface. In parallel to the study on the OER, study of oxidation of CO and methanol has seen a progress from "vapor phase" models to "liquid phase" models. However, polymer cluster has not been involved in the ab initio models.

(11) Ab initio methods provide elegant solutions to the problem of simulating proton diffusion and conduction with the vehicular and Grotthuss mechanism. Modeling of water and representative Nafion clusters has been readily performed. Notable findings include the formation of a defect structure in the ordered liquid water cluster. The activation energy for the defect formation is similar to that for conduction of proton in Nafion membrane. Classical MD methods can only account for physical diffusion of proton but can create very realistic model

for Nafion membrane. The simulation results can be compared with the experimental observations. Good agreement has been found in a large range of water content ($\lambda = 3.0$–15.0). From the point that the simulation results can be compared directly with the experimental results when structural diffusion is negligible, the classical MD modeling can serve as a reference for further development of the atomistic modeling of Nafion. EVB and ReaXFF MD methods have also been used to simulate the transport processes in Nafion. These methods require less computational resources than the ab initio methods but also require a calibration by the ab initio methods. Both structural and physical diffusions of proton can be described with EVB and ReaXFF MD methods.

(12) The reason why atomistic modeling study of PEFCs is considered as being in its initial stage of progress is that the system size and timescale in the simulations are often very limited. With the AIMD approach that is considered to be "exact" can reveal some very important mechanisms for OER, oxidation of CO and methanol, and proton transfer in Nafion. However, because the simulations are conducted with very small system (<200 atoms) and with a timescale of several picoseconds, it may be an overstatement when one claims that a phenomenon observed experimentally can be explained with mechanisms found in the simulations.

ACKNOWLEDGMENT

The authors would like to thank the National Science Foundation for financial support for this work via Grant # CBET-0933393.

REFERENCES

[1] K.D. Kreuer, S.J. Paddison, E. Spohr, M. Schuster, Chem. Rev. **104**, 4637 (2004)
[2] K.D. Kreuer, Solid State Ionics **94**, 5562 (1997)
[3] E. Spohr, P. Commer, A.A. Kornyshev, J. Phys. Chem. B **106**, 10560 (2002)
[4] D. Seeliger, C. Hartnig, E. Spohr, Electrochim. Acta **50**, 4234 (2005)
[5] S. Walbran, A.A. Kornysheva, J. Chem. Phys. **114**, 10039 (2001)
[6] M. Tuckerman, K. Laasonen, M. Sprik, and M. Parrinello, J. Phys. Chem. **99**, 5749 (1995)
[7] J.A. Morrone, M.E. Tuckermana, J. Chem. Phys. **117**, 4403 (2002)

[8] A. Zunger, A. Franceschetti, G. Bester, W.B. Jones, K. Kim, P.A. Graf, L-W. Wang, A. Canning, O. Marques, C. Voemel, J. Dongarra, J. Langou, S. Tomov, J. Phys. B: Conference Series **46**, 292 (2006)
[9] S.E. Thompson and S. Parthasarathy, Materials Today **9**, 20 (2006)
[10] W. Goddard III, B. Merinov, A.V. Duin, T. Jacob, M. Blanco, V. Molinero, S.S. Jang, Y.H. Jang, Mol. Simul. **32**, 251 (2006)
[11] J. Hafner, Acta Mater. **48**, 71 (2000)
[12] P. Hohenberg, W. Kohn, Phys. Rev. B **136**, 864 (1964)
[13] W. Kohn, L.J. Sham, Phys. Rev. A **140**, 1133 (1965)
[14] U.V. Barth, L. Hedin, J. Phys. C: Solid St. Phys. **5**, 1629 (1972)
[15] J.P. Perdew, A. Zunger, Phys. Rev. B **23**, 5075 (1981)
[16] J.P. Perdew, Phys. Rev. B **33**, 8822 (1986)
[17] J.P. Perdew, K. Burke, M. Ernzerhof, Phys. Rev. Lett. **77**, 3865 (1996)
[18] W.A. Goddard III, T. Cagin, M. Blanco, N. Vaidehi, S. Dasgupta, W. Floriano, M. Belmares, J. Kua, G. Zamanakos, S. Kashihara, M. Iotov, G. Gao, Comput. Theor. Polym. Sci. **11**, 329 (2001)
[19] R. Car, M. Parrinello, Phys. Rev. Lett. **55**, 2471 (1985)
[20] M.P. Allen, Computational Soft Matter: From Synthetic Polymers to Proteins, **23**, 1 (2004)
[21] L. Mayo, B.D. Olafson, W.A. Goddard III, J. Phys. Chem. **94**, 8897 (1990)
[22] P. Jannasch, Curr. Opin. Colloid Interface Sci. **8**, 96 (2003)
[23] E. Chalkova, X. Zhou, C. Ambler, M.A. Hofmann, J.A. Weston, H.R. Allcock, S.N. Lvov, Electrochem. Solid-State Lett. **10**, 221 (2002)
[24] P. Spinelli, C. Francia, E.P. Ambrosio, M. Lucariello, J. Power Sources **178**, 517 (2008)
[25] M.D. Macia, J.M. Campina, E. Herrero, J.M. Feliu, J. Electroanal. Chem **564**, 141 (2004)
[26] M. Watanabe, S. Motto, J. Electroanal. Chem. **60**, 275 (1975)
[27] J. Kua, W.A. Goddard III, J. Am. Chem. Soc. **121**, 10928 (1999)
[28] A.B. Anderson, E. Grantscharova, S. Seong, J. Electrochem. Soc. **148**, 2075 (1996)
[29] A.B. Anderson, Int. J. Quantum Chem., **49**, 581 (1994)
[30] T. Jacob, B.V. Merinov, W.A. Goodard III, Chem. Phys. Lett. **385**, 374 (2004)
[31] G. Barabino, C. Gavotti, M. Marchesi, Chem. Phys. Lett. **104**, 478 (1984)
[32] C.Y. Lee, J.A. McCammon, P.J. Rossky, J. Chem. Phys. **80**, 4448 (1984)
[33] E. Spohr, K. Heinzinger, Electrochim. Acta **33**, 1211 (1988)
[34] P.S. Crozier, R.L. Rowley, D. Henderson, J. Chem. Phys. **113**, 9202 (2000)
[35] C. Hartnig, M.T.M. Koper, J. Phys. Chem. B **108**, 3824 (2004)
[36] E. Spohr, J. Phys. Chem. **93**, 6171 (1989)
[37] J.I. Siepmann, M. Sprik, J. Chem. Phys. **102**, 511 (1995)
[38] I.-C. Yeh, M.L. Berkowitz, Chem. Phys. Lett. **301**, 81 (1999)
[39] E. Spohr, J. Chem. Phys. **107**, 6342 (1997)
[40] E. Spohr, Electrochim. Acta **49**, 23 (2003)
[41] J.P. Badiali, M.L. Rosinberg, J. Goodisman, J. Electroanal. Chem. Interfacial Electrochem. **130**, 31 (1981)
[42] W. Schmickler, J. Electroanal. Chem. Interfacial Electrochem. **150**, 19 (1983)
[43] J.W. Halley, D. Price, Phys. Rev. B **35**, 9095 (1987); K. Rice, Phys. Rev. **31**, 1051 (1928)
[44] D.L. Price, J.W. Halley, Phys. Rev. B **38**, 9357 (1988)
[45] J. Goodisman, J. Chem. Phys. **90**, 5756 (1989)

[46] D.L. Price, J.W. Halley, J. Chem. Phys. **102**, 6603 (1995)
[47] J.W. Halley, A. Mazzolo, Y. Zhou, and D. Price, J. Electroanal. Chem. **450**, 273 (1998)
[48] S. Izvekov, A. Mazzolo, K. VanOpdorp, G.A. Voth, J. Chem. Phys. **114**, 3248 (2001)
[49] P. Vassilev, R.A. van Santen, M.T.M. Koper, J. Chem. Phys. **122**, 54701 (2005)
[50] A.B. Anderson and N.C. Debnath, J. Am. Chem. Soc. **105**, 18 (1983)
[51] S. Seong, A.B. Anderson, J. Phys. Chem. **100**, 11744 (1996)
[52] A.B. Anderson, T.V. Albu, J. Am. Chem. Soc. **121**, 11855 (1999)
[53] T.V. Albu, A.B. Anderson, Electrochim. Acta **46**, 3001 (2001)
[54] R.A. Sidik, A.B. Anderson, J. Electroanal. Chem. **528**, 69 (2002)
[55] A.B. Anderson, N.M. Neshev, R.A. Sidik, P. Shiller, Electrochim. Acta **47**, 2999 (2002)
[56] A.B. Anderson, Electrochim. Acta **48**, 3743 (2003)
[57] J. Narayanasamy, A.B. Anderson, J. Phys. Chem. B **107**, 6898 (2003)
[58] M.F. Toney, J.N. Howard, J. Richer, G.L. Borges, J.G. Gordon, O.R. Melroy, D.G. Wiesler, D. Yee, and L.B. Sorensen, Nature **368**, 444 (1994)
[59] Y. Lozovoi, A. Alavi, J. Kohanoff, R.M. Lynden-Bell, J. Chem. Phys. **115**, 1661 (2001)
[60] J.-S. Filhol, M. Neurock, Angew. Chem., Int. Ed. **45**, 402 (2006)
[61] C.D. Taylor, S.A. Wasileski, J.-S. Filhol, M. Neurock, Phys. Rev. B **73**, 165402 (2006)
[62] M.J. Janik, C.D. Taylor, M Neurock, J. Electrochem. Soc. **156**, B126 (2009)
[63] K. Raghavan, K. Foster, K. Motakabbir, M. Berkowitz, J. Chem. Phys. **94**, 2110 (1991)
[64] E. Spohr, J. Chem. Phys. **93**, 6171 (1989)
[65] S. Senapati, A. Chandra, J. Chem. Phys. **113**, 8817 (2000)
[66] X. Xia, M.L. Berkowitz, Phys. Rev. Lett. **74**, 3193 (1995)
[67] I.-C. Yeh, M.L. Berkowitz, J. Chem. Phys. **112**, 10491 (2000)
[68] P.G. Kusalik, I.M. Svishchev, Science **265**, 1219 (1994)
[69] I.M. Svishchev, P.G. Kusalik, J. Wang, R.J. Boyd, J. Chem. Phys. **105**, 4742 (1996)
[70] M.F. Toney, J.N. Howard, J. Richer, G.L. Borges, J.G. Gordon, O.R. Melroy, D.G. Wiesler, D. Yee, L.D. Sorensen, Surf. Sci. **335**, 326 (1995)
[71] J.I. Siepmann, M. Sprik, J. Chem. Phys. **102**, 511 (1995)
[72] J.W. Halley, A. Mazzolo, Y. Zhou, D. Proce, J. Electroanal. Soc. **450**, 273 (1998)
[73] P.B. Balbuena, E.J. Lamas, Y. Wang, Electrochim. Acta **50**, 3788 (2005)
[74] E.J. Lamas, P.B. Balbuena, Electrochim. Acta **51**, 5904 (2006)
[75] A.P. Sutton, J. Chen, Philos. Mag. Lett. **61**, 139 (1990)
[76] H.J.C. Berendsen, J.R. Grigera, T.P. Straatsma, J. Phys. Chem. **91**, 6269 (1987)
[77] I. Kusaka, Z.G. Wang, J.H. Seinfeld, J. Phys. Chem. **108**, 6829 (1998)
[78] A.K. Rappe, W.A. Goddard, J. Phys. Chem. **95**, 3358 (1991)
[79] MSI, Force-field based simulations, San Diego, 1997
[80] S.J. Paddison, T.A. Zawodzinski, Solid State Ionics **115**, 333 (1998)
[81] A.C.T. van Duin, S. Dasgupta, F. Lorant, and W.A. Goddard, J. Phys. Chem. A, **105**(41), 9396 (2001)
[82] H. Steininger, S. Lehwald, H. Ibach, Surf. Sci. **123**, 1 (1982)
[83] C. Puglia, A. Nilsson, B. Hernnaes, O. Karis, P. Bennich, N. Martensson, Surf. Sci. **342**, 119 (1995)
[84] J.L. Gland, B.A. Sexton, G.B. Fisher, Surf. Sci. **95**, 587 (1980)

[85] D.A. Outka, J. Stoehr, W. Jark, P. Stevens, J. Solomon, R.J. Madix, Phys. Rev. B **35**, 4119 (1987)
[96] W. Wurth, J. Stoehr, P. Feulner, X. Pan, K.R. Bauchspiess, Y. Baba, E. Hudel, G. Rocker, D. Menzel, Phys. Rev. Lett. **65**, 2426 (1990)
[97] B.C. Stipe, M.A. Rezaei, W. Ho, Science **279**, 1907 (1998)
[98] B. Hammer, J.K. Nørskov, Adv. Catal. **45**, 71 (2000)
[99] B. Hammer, L.B. Hansen, J.K. Nørskov, Phys. Rev. B **59**, 7413 (1999)
[90] J. Greeley, J.K. Nørskov, M. Mavrikakis, Annu. Rev. Phys. Chem. **53**, 319 (2002)
[91] A. Ruban, B. Hammer, P. Stoltze, H.L. Skriver, J.K. Nørskov, J. Mol. Catal. A **115**, 421 (1997)
[92] A. Eichler, J. Hafner. Phys. Rev. Lett. **79**, 4481 (1997)
[93] T. Toda, H. Igarashi, H. Uchida, M. Watanabe, J. Electroanal. Chem. **146**, 3750 (1999)
[94] A. Ruban, H.L. Skriver, J.K. Nørskov, Phys. Rev. B **59**, 15990 (1999)
[95] Y. Xu, A.V. Ruban, M. Mavrikakis, J. Am. Chem. Soc. **126**, 4717 (2004)
[96] P.B. Balbuena, D. Altomare, L. Agapito, J.M. Seminario, J. Phys. Chem. B **107**, 13671 (2003)
[97] Z.D. Wei, F. Yin, L.L. Li, X.W. Wei, X.A. Liu, J. Electroanal. Chem. **541**, 185 (2003)
[98] R.R. Nazmutdinov, G.A. Tsirlina, O.A. Petrii, Y.I. Kharkats, A.M. Kuznetsov, Electrochim. Acta **45**, 3521 (2000)
[99] O.A. Petrii, R.R. Nazmutdinov, M.D. Bronshtein, G.A. Tsirlina, Electrochim. Acta **52**, 3493 (2007)
[100] A. Damjanovic, V. Brusic, Electrochim. Acta **12**, 615 (1967)
[101] E. Yeager, M. Razaq, D. Gervasio, A. Razaq, Electrocatalysis and Oxygen Electrochemistry Proc. **92**, 440 (1992)
[102] C.D. Taylor, M.J. Janik, M. Neurock, R.G. Kelly, Mol. Simul. **33**, 429 (2007)
[103] M.J. Janik, M. Neurock, Electrochim. Acta **52**, 5517 (2007)
[104] A.B. Anderson, T.V. Albu, J. Electrochem. Soc. **147**, 4229 (2000)
[105] T. Li, P.B. Balbuena, Chem. Phys. Lett. **367**, 439 (2003)
[106] J.K. Nørskov, J. Rossmeisl, A. Logadottir, L. Lindqvist, J.R. Kitchin, T. Bligaard, H. Jónsson, J. Phys. Chem. B **108**, 17886 (2004)
[107] J. Zhang, M.B. Vukmirovic, Y. Xu, M. Mavrikakis, R.R. Adzic, Ang. Chem. Int. Ed. **44**, 2132 (2005)
[108] R. Jinnouchi, K. Okazaki, Microscale Thermophys. Eng. **7**, 15 (2003)
[109] Y. Wang, P.B. Balbuena J. Phys. Chem. B **109**, 14896 (2005)
[110] J.-S. Filhol, M. Neurock, Angew. Chem. Int. Ed. **45**, 402 (2006)
[111] M.J. Buehler, A.C.T. van Duin, T. Jacob, Y. Jang, B. Merinov, W.A. Goddard III, MRS Proceedings Fall 2005
[112] M.T.M. Koper, Surf. Sci. **548**, 1 (2004)
[113] A.B. Anderson, E. Grantscharova, J. Phys. Chem. **99**, 9143 (1995)
[114] T.E. Shubina, C. Hartnigw, M.T.M. Koper, Phys. Chem. Chem. Phys. **6**, 4215 (2004)
[115] T.E. Shubina, M.T.M. Koper, Electrochim. Acta **47**, 3621 (2002)
[116] P. Liu, A. Logadottir, J.K. Nørskov, Electrochim. Acta **48**, 3731 (2003)
[117] D. Cao, G.-Q. Lu, A. Wieckowski, S.A. Wasileski, M. Neurock, J. Phys. Chem. B **109**, 11622 (2005)
[118] X. Gong, Z. Liu, R. Raval, P. Hu, J. Am. Chem. Soc. **126**, 8 (2004)
[119] X. Gong, P. Hu, R. Raval, J. Chem. Phys. **119**, 6324 (2003)
[120] C.J. Zhang, P. Hu, J. Am. Chem. Soc. **122**, 2134 (2000)

[121] S. Desai, M. Neurock, Electrochim. Acta **48**, 3759 (2003)
[122] C. Hartnig, J. Grimminger, E. Spohr, Electrochim. Acta **52**, 2236 (2007)
[123] C. Hartnig, E. Spohr, Chem. Phys. **319**, 185 (2005)
[124] J. Ludwig, D.G. Vlachos, A.C.T. van Duin W.A. Goddard, J. Phys. Chem. B **110**, 4274 (2006)
[125] A.A. Kulikovsky, E. Spohr, NIC Symposium **32**, 269 (2006)
[126] J. Choi, H. Lee, S. Moon, J. Colloid Interface Sci. **238**, 188 (2001)
[127] R. Pomes, B. Roux, Biophys. J. **71**, 19 (1996)
[128] Y.S. Park, T. Hatae, H. Itoh, M.Y. Jang, Y. Yamazaki, Electrochim. Acta **50**, 592 (2004)
[129] S. Banerjee, D.E. Curtin, J. Fluorine Chem. **125**, 1211 (2004)
[130] C.J.D. von Grotthuss. Ann. Chim. **8**, 54 (1806)
[131] S. Urata, J. Irisawa, A. Takada, S. Tsuzuki, W. Shinodab, M. Mikami, Phys. Chem. Chem. Phys. **6**, 3325 (2004)
[132] S.J. Paddison, J.A. Elliottb, Phys. Chem. Chem. Phys. **8**, 2193 (2006)
[133] S.J. Paddison, J.A. Elliottb, J. Phys. Chem. A **109**, 7583 (2005)
[134] M. Eikerling, S.J. Paddison, L.R. Pratt, T.A. Zawodzinski Jr, Chem. Phys. Lett. **368**, 108 (2003)
[135] S.J. Paddison, T.A. Zawodzinski Jr, Solid State Ionics **113–115**, 333 (1998)
[136] J. Ennari, M. Elomaa, F. Sundholm, Polymer **40**, 5035 (1999)
[137] J. Ennari, I. Neelov, F. Sundholm, Polymer **41**, 2149 (2000)
[138] X. Zhou, Z. Chen, F. Delgado, D. Brenner, R. Srivastava, J. Electrochem. Soc. **154**, B82 (2007)
[139] S.S. Jang, V. Molinero, T. Cüagın, W.A. Goddard III, J. Phys. Chem. B **108**, 3149 (2004)
[140] S. Dokmaisrijan, E. Spohr, Journal of Molecular Liquids **129**, 92 (2006)
[141] D. Seeliger, C. Hartnig, E. Spohr, Electrochim. Acta **50**, 4234 (2005)
[142] M.K. Petersen, F. Wang, N.P. Blake, H. Metiu, G.A. Voth, J. Phys. Chem. B **109**, 3727 (2005)
[143] M.K. Petersen, G.A. Voth, J. Phys. Chem. B **110**, 18594 (2006)

Index

U. Pasaogullari, C.-Y. Wang (eds.), *Modeling and Diagnostics of Polymer Electrolyte Fuel Cells*, Modern Aspects of Electrochemistry 49, DOI 10.1007/978-0-387-98068-3, © Springer Science+Business Media, LLC 2010

MIX
Papier aus verantwortungsvollen Quellen
Paper from responsible sources
FSC® C105338

If you have any concerns about our products,
you can contact us on
ProductSafety@springernature.com

In case Publisher is established outside the EU,
the EU authorized representative is:
Springer Nature Customer Service Center GmbH
Europaplatz 3, 69115 Heidelberg, Germany

Printed by Libri Plureos GmbH
in Hamburg, Germany